International Review of
Cytology
A Survey of
Cell Biology

VOLUME 187

International Review of Cytology

A Survey of Cell Biology

Edited by

Kwang W. Jeon
Department of Biochemistry
University of Tennessee
Knoxville, Tennessee

VOLUME 187

ACADEMIC PRESS
San Diego London Boston New York Sydney Tokyo Toronto

Cover photograph: Mature preprophase bands (PPBs) in late prophase cells of onion root tips. (For more details, see Chapter 1, Figure 1.)

This book is printed on acid-free paper. 

Academic Press
a division of Harcourt Brace & Company
525 B Street, Suite 1900, San Diego, California 92101-4495, USA
http://www.apnet.com

Academic Press
24-28 Oval Road, London NW1 7DX, UK
http://www.hbuk.co.uk/ap/

International Standard Book Number: 0-12-364591-3

PRINTED IN THE UNITED STATES OF AMERICA
99 00 01 02 03 04 EB 9 8 7 6 5 4 3 2 1

CONTENTS

The Pecten Oculi of the Chicken: A Model System for Vascular Differentiation and Barrier Maturation

Hartwig Wolburg, Stefan Liebner, Andreas Reichenbach, and Holger Gerhardt

Bacteriorhodopsin

Janos K. Lanyi

The Regulation of Apoptosis by Microbial Pathogens

Jeremy E. Moss, Antonios O. Aliprantis, and Arturo Zychlinsky

Invertebrate Opioid Precursors: Evolutionary Conservation and the Significance of Enzymatic Processing

George B. Stefano and Michel Salzet

CONTRIBUTORS

Numbers in parentheses indicate the pages on which the authors' contributions begin.

Antonios O. Aliprantis (203), *The Skirball Institute of Biomolecular Medicine, New York University Medical Center, New York City, New York 10016*

Søren S. L. Andersen (51), *Department of Molecular Biology, Princeton University, Princeton, New Jersey 08540-1014*

Holger Gerhardt (111), *Institute of Pathology, University of Tübingen, D-72076 Tübingen, Germany*

Janos K. Lanyi (161), *Department of Physiology and Biophysics, University of California, Irvine, California 92697-4560*

Stefan Liebner (111), *Institute of Pathology, University of Tübingen, D-72076 Tübingen, Germany*

Yoshinobu Mineyuki (1), *Department of Biological Science, Hiroshima University, Higashi-Hiroshima 739-8526, Japan*

Jeremy E. Moss (203), *The Skirball Institute of Biomolecular Medicine, New York University Medical Center, New York City, New York 10016*

Andreas Reichenbach (111), *Paul Flechsig Institute for Brain Research, University of Leipzig, D-04109 Leipzig, Germany*

Michel Salzet (261), *Laboratoire de Biologie Animale, Université des Sciences et Techniques de Lille, 59655 Villeneuve d'Ascq Cedex, France*

George B. Stefano (261), *Neuroscience Institute, State University of New York, College at Old Westbury, Old Westbury, New York 11568-0210*

Hartwig Wolburg (111), *Institute of Pathology, University of Tübingen, D-72076 Tübingen, Germany*

Arturo Zychlinsky (203), *The Skirball Institute of Biomolecular Medicine, New York University Medical Center, New York City, New York 10016*

The Preprophase Band of Microtubules: Its Function as a Cytokinetic Apparatus in Higher Plants

Yoshinobu Mineyuki
Department of Biological Science, Faculty of Science, Hiroshima University, Kagamiyama 1-3-1, Higashi-Hiroshima 739-8526, Japan

Features, development, and functions of preprophase bands (PPBs) of microtubules (MTs) are reviewed. The PPB is an array of cortical MTs in higher plants that appears in G_2 and prophase and predicts where the cell plate will be inserted (the division site). Experimental obliteration of the PPB causes misplacement of cell plate insertion, suggesting that the PPB is a determinant of the ultimate division site. Its development contains two elementary processes: Broad PPB formation first fixes the axis of division polarity in the cell, and PPB narrowing then defines the precise division site. The PPB disappears at the prophase/prometaphase transition stage, but it leaves information in some yet unidentified form at the division site. This information assists correct insertion of cell plates and maturation of new cell walls after cytokinesis. Several kinds of molecules are reported to occur in PPBs, but their roles are not yet understood. Actin and cyclin-dependent kinase homologs are suggested to be involved in the band narrowing MT, which is essential for PPBs to mature at the division site. Other possible functions of the PPB, such as premitotic nuclear positioning and prophase spindle orientation, are also reviewed.

KEY WORDS: Actin, Cyclin-dependent kinase, Cytokinesis, Division site, Microtubule, Plant cell, Preprophase band.

I. Introduction

The preprophase band (PPB) of microtubules (MTs) is a unique band of cortical MTs encircling the nucleus in premitotic cells of higher plants

0074-7696/99 $30.00

(Fig. 1). It is located in the cell cortex where fusion of the future cell plate occurs at the end of mitosis. Using the electron microscope (EM), many examples of PPBs were observed and the following possible functions of PPBs have been proposed (Newcomb, 1969; Pickett-Heaps, 1974; Hepler and Palevitz, 1974; Gunning and Hardham, 1982; Gunning, 1982): (i) a source of tubulins/MTs for mitotic spindles (Pickett-Heaps and Northcote, 1966a); (ii) establishment of the division site (Pickett-Heaps and Northcote, 1966b); (iii) response of the premitotic cell to factors inducing polarization

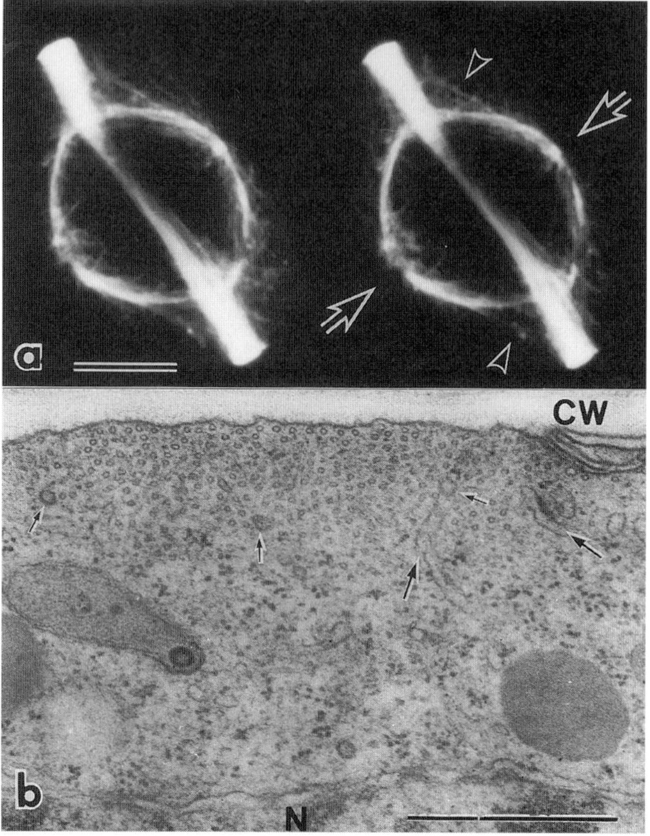

FIG. 1 Mature PPBs in late prophase cells of onion root tips. (a) Stereo pair images of tubulin immunofluorescence taken by a confocal laser scanning microscope. Note that the mature PPB positions are at right angles to the prophase spindle axis (arrows). Arrowheads show MTs connecting the PPB to the spindle pole region. Scale bar = 10 μm. (b) An electron micrograph of a cross section of a mature PPB. CW, cell wall; N, nucleus. Large arrows indicated smooth ER, and small arrows indicate small vesicles. Scale bar = 0.5 μm (photographs by A. Nogami).

(Pickett-Heaps, 1969a,b,c, 1974); (iv) premitotic nuclear migration and/or nuclear orientation (Burgess and Northcote, 1967; Jarosch, 1989, 1990); (v) premitotic nuclear anchoring (Mineyuki and Furuya, 1986); (vi) mitotic spindle orientation (Burgess and Northcote, 1967); (vii) localized cell wall deposition (Packard and Stack, 1976; Galatis and Mitrakos, 1979); (viii) guidance of edges of the growing cell plate (Gunning *et al.,* 1978b); (ix) prevention of the initiation of ingrowing cleavage furrow (O'Brien, 1983); and (x) deposition of the MT-organizing center (MTOC) after cell division (Hepler and Palevitz, 1974; Gunning *et al.,* 1978b; Gunning, 1980).

PPB studies were systematically reviewed by Gunning (1982). Since then, two reviews have appeared (Gunning and Wick, 1985; Wick, 1991). Advances in immunofluorescence microscopy (IFM), video and confocal microscopy, and microinjection techniques enable us to examine three-dimensional localization of components of PPBs in cells (Lloyd *et al.,* 1992; Gunning, 1992; Mineyuki, 1993) and to observe the behavior of associated molecules directly in a living cell (Zhang *et al.,* 1990; Hepler *et al.,* 1993). The aim of this review is to present recent views of PPBs and their roles, summarizing about 30 years of research on the topic.

II. The Division Site

A. Terminology

1. Division Site

Sometimes the term "division site" is used ambiguously as a place where cell division takes place. In this review, the division site in higher plants is defined as the cortical region where fusion of cell plate and parental cell walls occurs at the end of cytokinesis (Gunning, 1982). In animals the corresponding division site is the site where the cleavage furrow begins to form at the cell surface.

2. Preprophase

The term "preprophase" was first introduced by Pickett-Heaps and Northcote (1966a) to describe the stage of cell division cycle in which there is visible activity or organization in preparation for cell division, i.e., formation of the PPB. Pickett-Heaps (1969a) tried to define the stage of preprophase in terms of the stage of chromosome condensation and characteristic appearance of the nucleolus and of a PPB. Although the idea that preprophase equates with early prophase was implicit in some early EM works (Cronshaw and Esau, 1968; Esau and Gill, 1969; Roberts *et al.,* 1985; Apostolakos and Galatis, 1985a; Bakhuizen *et al.,* 1985; Simmonds, 1986), many others have described PPBs in interphase cells using a wide range of materials. The early appearance of the PPB was confirmed by IFM (Wick and Duniec, 1983, 1984; Mineyuki *et al.,* 1988b). Now many workers agree that PPBs are seen in most prophase cells and some interphase cells (see Section V). It is therefore not appropriate to use the term preprophase to refer to a specific stage of the cell cycle, such as the stage between G_2 and prophase. The term preprophase must be defined as a stage of a cell in which there is a visible activity or organization related to the division site establishment, i.e., formation of the PPB in higher plants and other kinds of apparatus comparable to a PPB in lower plants.

3. Position versus Orientation

In this review, the terms "position" and "orientation" are used in the narrow sense. For example, while the orientation of the cell plate is the same in symmetrical transverse division (Fig. 5a) and asymmetrical transverse division (Fig. 5b), the position of the cell plate is different in the two situations.

B. Division Site Regulation and Plant Morphogenesis

Many descriptive studies (Sinnott, 1960) documented the importance of division site regulation in morphogenesis. This view has been questioned since 1962 because some organogenesis can occur in the absence of cell divisions (Haber, 1962; Lyndon, 1990) or in mutants with altered division planes (Trass *et al.,* 1995; Smith *et al.,* 1996). However, the early view seems to be true in the case of terminal cell differentiation (e.g., in stomatal differentiation). Thus, the earliest sign of stomatal differentiation in onion is the appearance of asymmetrical distribution of nucleus and cytoplasm in the parental epidermal cell (Figs. 2b and 3). This cell divides asymmetrically to produce a small apical cell, called the guard mother cell (GMC) (Fig. 2c). Then, the GMC divides longitudinally to form two guard cells (GCs) (Fig. 2e). When the nucleus in the epidermal cell is displaced by basipetal centrifugation, the cell divides to form a small basal cell (Figs. 2g and 2h). However, neither the large apical cell nor the small basal cell starts to differentiate (Fig. 2i). In order to differentiate a stoma, the apical cell must divide again (Bünning and Biegert, 1953). A similar requirement for asymmetrical division and subsequent cell differentiation is known in pollen grain development (Bünning, 1952).

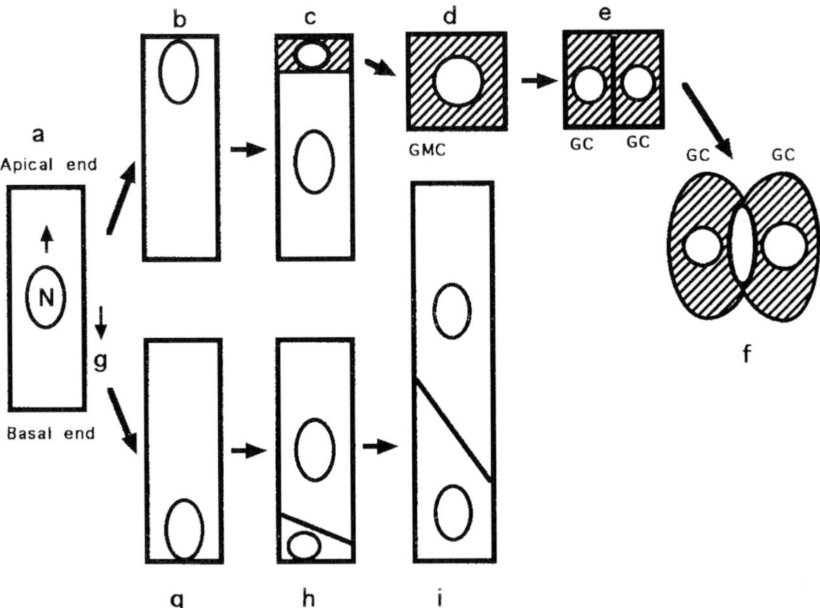

FIG. 2 Diagram illustrating Bünning and Biegert's (1953) observation on the stomatogenesis of onion seedlings (a–f) and the experimental manipulation of stomatal differentiation using basipetal centrifugation (a–g–h–i–j). (a) Elongated epidermal cell of cotyledon. The nucleus (N) is in the center of the cell. (b) The nucleus migrates to the apical end. (c) The cell divides asymmetrically to give rise to a small apical cell and a large basal cell. (d) The apical cell (GMC; guard mother cell) grows. (e) The GMC divides longitudinally to give rise to two guard cells (GCs). (f) The GCs grow and the intercellular space between these cells develops to become a stomatal pore. (g) The elongated epidermal cell is centrifuged basipetally to displace the nucleus to the basal part of the cell. (h) The nucleus divides in the basal part of the cell to form a small basal cell and a large apical cell. (i) Neither the small basal cell nor the large apical cell differentiates; they remain as epidermal cells. GMC and GC are shown hatched. An arrow in front of the nucleus in a shows the direction of the nuclear migration. An arrow with a small letter g shows the direction of the centrifugation (330–600 *g* for 30 min).

C. The Division Site Is Determined before Karyokinesis

The cytokinetic apparatus that establishes the division plane is quite different in the plant and animal kingdoms. Insertion of the final division plane in animal cells (cleavage furrow) is carried out by a contractile ring, an array of actomyosin encircling the cell cortex (Mabuchi, 1986), whereas in higher plants cell plate formation is achieved by a phragmoplast in which MTs play an essential role (Gunning, 1982). While cleavage furrows start on the cell surface and grow centripetally, cell plates usually originate in the middle of the cell, between daughter nuclei, and grow centrifugally.

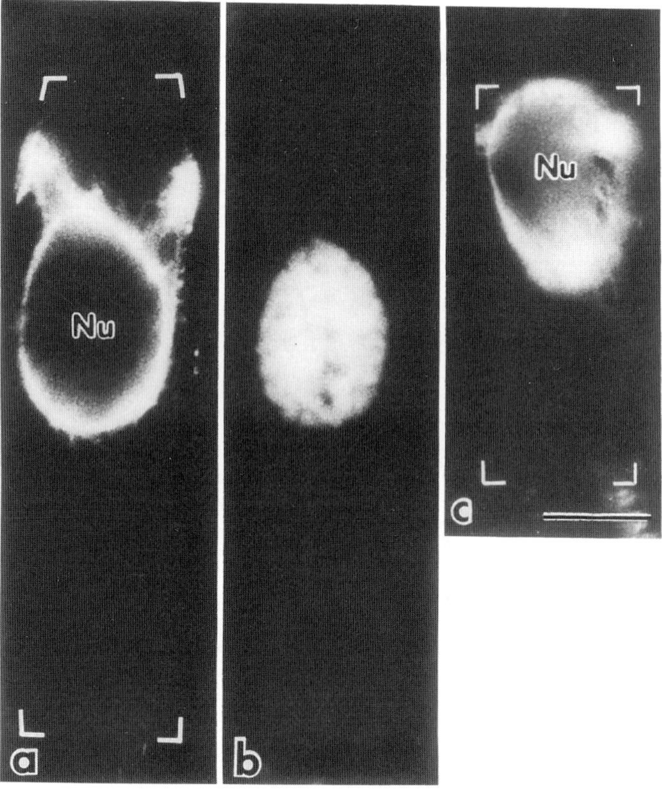

FIG. 3 Asymmetrical PPBs in GMC formation of onion cotyledon epidermis. Tubulin immunofluorescence images of prophase cells with a migrating nucleus (a) and with a nucleus after migration (c) (reproduced with permission from Mineyuki *et al.,* 1991, *J. Plant Physiol.* **138,** p. 645, Fig. 5c). (b) Hoechest fluorescence of the prophase nucleus in a. Note that the nuclear migration in this cell takes place in midprophase. The longitudinal axis of the cotyledon is vertically oriented in the micrographs and the apical end of the cell is positioned toward the top of the page. Brackets mark the cell limits. Note MTs that appear to link the perinuclear MTs with the narrow PPB in the migrating nucleus (a) and that the distal pole is appressed to the apical wall when the nucleus has reached the distal end of the cell (c). Nu, nucleus. Scale bar = 10 μm. See detail in Mineyuki and Palevitz (1990) and Mineyuki *et al.* (1991a).

The process of determination of the division site is also quite different between animals and plants. Figure 4 shows differences between a sand dollar egg and a *Tradescantia* stamen hair cell after experimentally displacing the mitotic apparatus to the distal region of the cell. Rappaport's (1986a,b) experiment shows that in sand dollar eggs, a mitotic apparatus can cause a cleavage furrow to form any region of the cell cortex, and the furrowing site is restricted only by the geometrical position of the mitotic

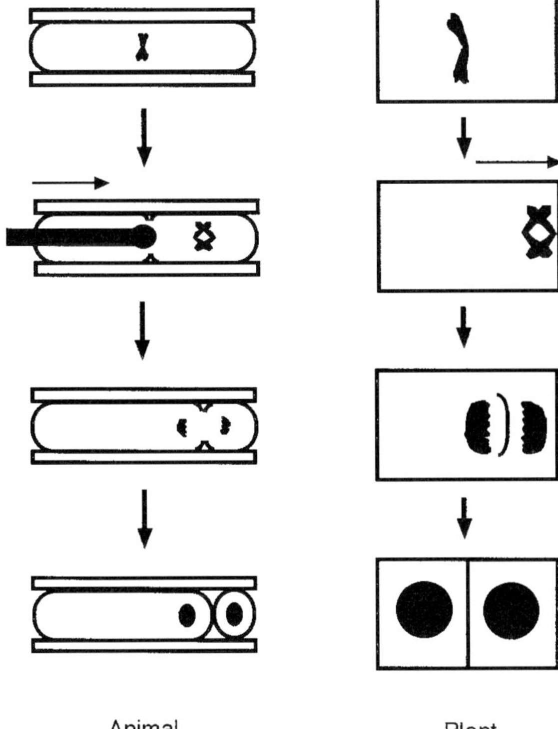

FIG. 4 Effects of the displacement of mitotic apparatus on the position of the division plane in the animal and plant cell. (Left) An animal case and (right) a plant case. Each panel shows a sequence of events from metaphase (top), through anaphase just after the experimental relocation of the mitotic apparatus and telophase, to the end of cytokinesis (bottom). (Animal) Experiments with sand dollar (*Echinarachnius parma*) eggs (Rappaport and Rappaport, 1985; this diagram is redrawn from Rappaport, 1986a, Fig. 2). When a sand dollar egg is confined in a capillary (82-μm inner diameter) so that it is reshaped into a cylinder and the mitotic apparatus is oriented parallel to the capillary axis, the sand dollar egg can divide normally and a furrow forms midway between the asters. However, when the mitotic apparatus at early anaphase is relocated by pushing the cell pole inward using a microneedle (direction is shown by a transversely oriented arrow), the furrow that appears before experimental manipulation disappears and the new furrow begins midway between asters of the relocated mitotic apparatus, and the cleavage furrow completes there. (Plant) Experiments with *Tradescantia* stamen hair cell (Ôta, 1961). A stamen hair cell divides in the middle of the cell in the normal condition. When a stamen hair cell is centrifuged (3550*g* or less for 15 min) to displace the mitotic apparatus in metaphase or early anaphase (right arrow shows the direction of the centrifugation), cell plate formation starts between the daughter nuclei; however, the daughter nuclei and the cell plate gradually move to the central part of the cell. Because the movement of cell plate edge is faster than the nuclear movement, the growing cell plate is often U-shaped, and finally the edge of the cell plate meets the site where it would have been inserted if there had been no centrifugal treatment.

apparatus (at the limited stage of anaphase) within the cell. On the contrary, Ôta's (1961) experiment suggests that the division site is predetermined before displacement of the mitotic apparatus, i.e., before metaphase.

III. Features of the PPB

A. The PPB Is Positioned at the Ultimate Division Site

Documentation on positional consistency between PPBs and the division site has accumulated for a variety of cell types. Figure 5 shows spatial

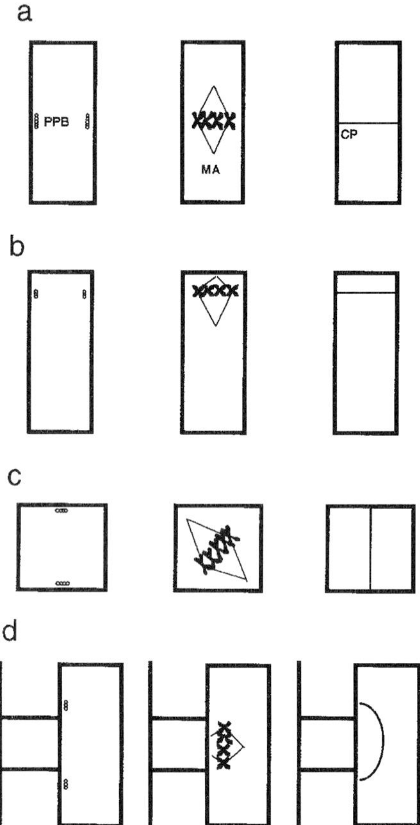

FIG. 5 Spatial relationship among PPBs (left), metaphase spindles (center), and cell plates (right). (a) Symmetrical transverse division. (b) Asymmetrical transverse division. (c) Symmetrical longitudinal division of onion GMC. (d) Asymmetrical division to produce a grass SC. PPB, preprophase band; MA, mitotic apparatus; CP, cell plate.

relationship among PPBs, metaphase spindles, and cell plates in four different cell types. Because the orientation of an equatorial plane does not always match that of a cell plate, a question arose whether a PPB predicts the position of a metaphase equatorial plane or that of a cell plate. Using onion GMCs, in which a cell plate is inserted longitudinally while the spindle initially orients obliquely, Palevitz and Hepler (1974a) clearly demonstrated that the PPB predicts the position of the cell plate but not the position of the equatorial plane (Fig. 5c).

A partial inconsistency between the PPB and the final cell plate arrangement has been found in divisions for triangular and lens-shaped subsidiary cell and leaf hair cell formation of grasses (Galatis *et al.,* 1983, 1984b; Cho and Wick, 1989), in *Marchantia* superficial thallus cells with incomplete PPBs (Apostolakos and Galatis, 1985a,b), and in floating stomata of *Anemia* (Galatis *et al.,* 1986). However, their cell plate divergence can be explained by the disturbance of premitotic polarity and/or space limitation, and thus the generalization that the PPB site predicts where the attachment site of the cell plate with the parental walls will be remains true.

B. PPB Orientation under Altered Cell Polarity

Mispositioning and/or misorientation of cell plates occurs in cells whose PPB formation is interrupted by experimental (Mineyuki and Palevitz, 1990; Mineyuki *et al.,* 1991b; Murata and Wada, 1991a) or genetic (Traas *et al.,* 1995) manipulations, indicating that PPBs are prerequisite for the correct positioning of cell plates. However, it remained an open question whether cell divisions with irregularly oriented division planes have PPBs. An answer has been provided recently by means of two different approaches. Stomatogenesis in the hypocotyl of dark-grown *Cucumis* can be induced by a pulsed red light treatment. Although the GMC does divide, the orientation of its division is altered and irregularly oriented stomata are produced. In this GMC division, PPBs are formed but their orientation is also irregular, suggesting that maloriented GMC division is due to preceding abnormal orientation of PPBs (Kazama and Mineyuki, 1993, 1997). Abnormal orientation of PPBs and division plane is also seen in a maize mutant *tangled 1* (L. Smith, personal communication).

C. Fine Structure of the PPB

The structure of a PPB is illustrated in Figure 6. The width of a mature PPB is usually 2–4 μm, but broader PPBs have been reported (e.g., suspension and protoplast culture cells: Simmonds *et al.,* 1983; Simmonds and

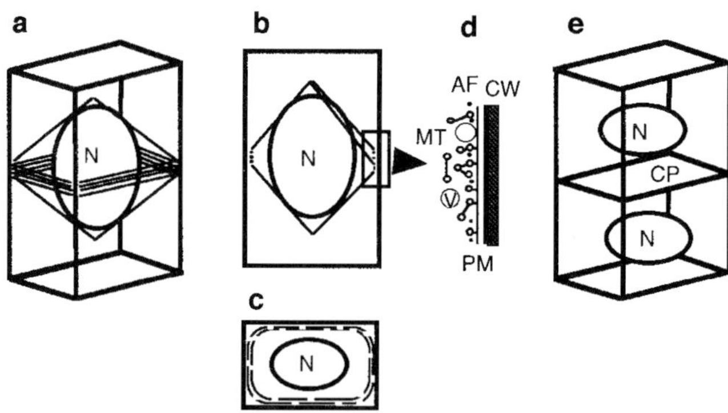

FIG. 6 A diagram illustrating a PPB. (a) A late prophase cell. (b) A midlongitudinal section of a. (c) A midtransverse section of a. (d) The enlarged PPB regions (enclosed by a rectangle) in b. (e) The cell just after cytokinesis. Cell plate attaches to the former PPB region. N, nucleus; CP, cell plate; CW, cell wall; AF, actin filament; MT, microtubule; PM, plasma membrane; V, vesicle.

Setterfield, 1986; Falconer and Seagull, 1985; Gorst *et al.,* 1986; GMCs: Busby and Gunning, 1980; Galatis, 1982; Cho and Wick, 1989; Mullinax and Palevitz, 1989; Galatis *et al.,* 1982). The width of mature PPBs is different among cell types even in the same tissue and it can also be changed by the environmental conditions. While a mean width of asymmetrical PPBs to produce GMCs is 4 μm, that of symmetrical PPBs is 6 μm in onion epidermis (Mineyuki and Palevitz, 1990). While the mean width of PPBs in the dark-induced cell division of *Adiantum* protonemata is 13.3 μm, that in blue light-induced cell division is 9.5 μm (Murata and Wada, 1989). While protoplasts of *Helianthus* cultured in liquid have a narrow PPB, those in agarose beads have only broad PPBs (Caumont *et al.,* 1997).

The MT number in a cross section of a PPB varies from a single layer with $<$20 MTs (epidermal cells in capsule of *Funaria*; Sack and Paolillo, 1985) to $>$10 MT layers with $>$250 MTs (onion root; Nogami *et al.,* 1996). The depth of PPBs in a cross section varies from $<$0.15 μm to 2.6 μm. A comparative study on MT numbers of PPBs among *Triticum* species of different ploidy levels suggests that the number of MTs increases with the number of chromosomes (Eleftheriou, 1985b). Heavy water can induce the formation of additional MTs in a PPB (Burgess and Northcote, 1969; Hardham and Gunning, 1978).

A PPB consists predominantly of overlapping relatively short MTs (average length, 2.7 μm) (Hardham and Gunning, 1977). C- or S-shaped MT profiles are reported in transverse section of a PPB (Burgess and Northcote, 1967; Galatis *et al.,* 1984b). Depletion and the shortening of MTs can be induced by MT drugs, high pressure, or cold treatment. Although C-shaped termination is not frequent in the short MTs induced by colchicine or high pressure, cold treatment produces more C-shaped MT terminations (Hardham and Gunning, 1977).

While the majority of MTs in PPBs follows the contours of the plasma membrane (PM) closely, others curve into the cytoplasm (Pickett-Heaps and Northcote, 1966a; Burgess, 1970b). MTs between perinuclear MTs and PPBs are reported by EM (Burgess, 1970b; Bakhuizen *et al.,* 1985; Eleftheriou, 1985a, 1996; Mineyuki and Furuya, 1986; Schnepf, 1984) and by IFM (Tiwari *et al.,* 1985; Wick and Duniec, 1983, 1984; Mineyuki and Palevitz, 1990; Mineyuki *et al.,* 1991a; Nogami *et al.,* 1996; Baluška *et al.,* 1996).

MT–MT and MT–PM cross bridges in PPBs are seen (Pickett-Heaps and Northcote, 1966a; Burgess and Northcote, 1967; Galatis and Apostolakos, 1977; Hardham and Gunning, 1978; Schnepf, 1973). In leguminous stomatogenesis, MT–PM linkage is often seen in GMCs (Galatis *et al.,* 1982), whereas it is infrequent in the nondifferential and early divisions (Galatis and Mitrakos, 1979). In a number of cases, a cross bridge either to the PM or to an adjacent MT occurred on the terminating profile of MTs. Occasionally, two bridges connected a single MT profile to PM, and cross bridges between MTs and vesicles are also reported (Hardham and Gunning, 1978).

Smooth endoplasmic reticulum (ER) may occur at the periphery of the PPB, but no structural link between ER and MTs has been reported (Burgess and Northcote, 1967, 1968; Galatis and Mitrakos, 1979; Galatis *et al.,* 1982). In *Triticum* roots, smooth ER is seen in the early PPB stage but it is almost absent in the later stage (Burgess and Northcote, 1968). Small electron-dense vesicles, which sometimes appear to fuse with the PM, are seen in a PPB (Burgess and Northcote, 1968; Packard and Stack, 1976; Gunning *et al.,* 1978a; Galatis and Mitrakos, 1979; Galatis, 1982; Galatis *et al.,* 1982; Eleftheriou, 1996). The vesicles are positive to the periodic acid–thiocarbohydrazide–silver proteinate reaction, the reaction against polysaccaride (Galatis, 1982), and in onion the density of vesicle contents corresponds closely to the density of the cell walls (Packard and Stack, 1976). These vesicles and the smooth ER still exist when MTs are depleted by colchicine treatment (Burgess and Northcote, 1969). Coated vesicles in a PPB are reported only in leguminous GMCs (Galatis and Mitrakos, 1979; Galatis *et al.,* 1982).

IV. Occurrence of the PPB

In order to determine whether PPBs appear only in certain specialized types of cells, Gunning *et al.* (1978a) examined all types of cell divisions in *Azolla* roots and concluded that PPBs exist in every category of cell division in the meristem. Observations summarized in Table I imply that PPBs occur in meristems and meristemoids during vegetative growth and in wound-induced cell divisions of higher plants.

PPBs are absent in microsporogenesis (Van Lammeren *et al.*, 1985; Hogan, 1987), in the first asymmetrical mitosis (Heslop-Harrison, 1968; Terasaka and Niitsu, 1990), in other mitosis in pollen grain development (Burgess, 1970a; Terasaka and Niitsu, 1989; Palevitz and Cresti, 1989), and throughout megasporogenesis and in embryo sac development (Bednara *et al.*, 1988; Willemse and Van Lammeren, 1988), but PPBs are reinstated as early as the first zygotic division (Webb and Gunning, 1991). Endosperm also has no cortical MT arrays during syncytial division, cellularization, and the subdivision of starchy endosperm (Bajer and Molè-Bajer, 1972; De Mey *et al.*, 1982), but interphase cortical MTs (ICMs) and PPBs appear during aleurone layer development (Brown *et al.*, 1994). A recessive mutation of the maize gene *Ameiotic* causes the replacement of meiosis I with a synchronized mitotic division. Because PPBs are observed in this ameiotic division, this *ameiotic* gene may encode a product that eliminates the PPB (Staiger and Cande, 1992). This implies that a PPB-containing mitotic cell cycle is the default condition for plant cell division.

PPBs are also seen in certain suspension cultured cells, in cultured protoplasts, and in callus. The "PPB index" (Gorst *et al.*, 1986) or "PPB value" (Wang *et al.*, 1989b) describes the frequency of PPBs in cell cultures. Often a low PPB index reflects the inability of cells to undergo morphogenesis. In *Petunia* callus, a three to five fold increase in the PPB index marks the transition to organized growth (Traas *et al.*, 1990). Cells having the potential for morphogenesis had a higher PPB index than nonembryonic cells, but high PPB index is not obligatorily coupled to embryogenesis or other organized growth (Gorst *et al.*, 1986). Although high frequencies of PPBs are reported in soybean protoplast culture (Wang *et al.*, 1989b) and in suspension cultures of tobacco BY-2 (Katsuta *et al.*, 1990) and *Spartina* (Hogan, 1988), these cultures do not regenerate in the reported culturing conditions.

Most of the meristematic tissues in Pteridophytes have PPBs, but they are absent in the first asymmetrical cell division in *Onoclea* spores (Bassel *et al.*, 1981), in meiosis (Brown and Lemmon, 1985b), and in spermatogenesis (Marc and Gunning, 1986) except for the early steps of antheridia development (Schraudolf, 1993). Some cells in Bryophytes have a PPB, but these may be atypical in terms of the width and number of MTs and are thought to be primitive (Apostolakos and Galatis, 1992). Some types of cells in Brophytes apparently lack a PPB. Moss protonemata do not develop a PPB when the tip cell divides or a side branch is formed (Schmiedel and Schnepf, 1979a; Schmiedel *et al.*, 1981; Doonan *et al.*, 1985, 1987). However,

TABLE I

Occurrence of PPB MTs

Seed plants

Shoot apical meristems: *Coleus* (EM: Lehmann and Schultz, 1976); *Hedera* (IF: Marc and Hackett, 1989)

Root apical meristems (*including root caps*): (Monocot) *Allium* (EM: Deysson and Benbadis, 1968; IF: Wick *et al.*, 1981); *Bouteloua, Chloris,* and *Cyperus* (IF: Cleary and Hardham, 1988); *Hordeum* (IF: Mineyuki *et al.*, 1996); *Lolium* (IF: Cleary and Hardham, 1988); *Ornithogalum* (IF: Hogan, 1987); *Phleum* (EM: Burgess and Northcote, 1967); *Potamogeton and Tradescantia* (IF: Cleary and Hardham, 1988); *Triticum* (EM: Pickett-Heaps and Northcote, 1966a; IF: Marc and Gunning, 1988); *Typha* (IF: Cleary and Hardham, 1988); *Zea* (EM: Hardham and Gunning, 1978; IF: Cleary and Hardham, 1988); *dv* mutant (IF: Staiger and Cande, 1990); (dicot) *Arabidopsis* (IF: Traas *et al.*, 1995); *Brassica* (EM: Gunning and Steer, 1975); *Chrysanthemum* (IF: Mineyuki *et al.*, 1996); *Cucumis* (IF: Woo and Wick, 1995); *Cyperus* (EM: Busby and Gunning, 1980; IF: Gunning and Wick, 1985); *Datura* (IF: Gorst *et al.*, 1986); *Daucus* (IF: Cleary and Hardham, 1988), *Glycine* (IF: Liu *et al.*, 1993); *Hibiscus* (IF: Mineyuki *et al.*, 1996); *Lepidium* Root caps (EM: Hensel, 1984); *Lycopersicon* (IF: Hogan, 1987); *Medicago* (IF: Cleary and Hardham, 1988); *Nicotiana* (EM: Ding *et al.*, 1991; IF: Gorst *et al.*, 1986); *Pisum* (EM: Bakhuizen *et al.*, 1985; IF: Wick, 1985); *Vigna* (IF: Mizuno *et al.*, 1985); *Vicia* (IF: Brown *et al.*, 1989); *Zinnia* (IF: Cleary and Hardham, 1988); (gymnosperm) *Pinus* and *Zamia* (IF: Fowke, 1993)

Vascular bundle development

Root protophloem mother cells: *Aegilops* (EM: Eleftheriou and Tsekos, 1982); *Triticum* (EM: Eleftheriou, 1985a)

Root protophloem sieve elements: *Triticum* (EM: Eleftheriou, 1996)

Root procambial cells: *Triticum* (EM: Eleftheriou, 1985b)

Cambial cells in coleoptile: *Triticum* (EM: Pickett-Heaps and Northcote, 1966a)

Fusiform initials and ray initials: *Ulmus* and *Tilia,* ray cell (EM: Evert and Deshpande, 1970)

Leaf development

Mesophyll cells: (Monocot) *Commelina* (EM: Pickett-Heaps, 1969b); *Hordeum* (IF: Wick *et al.*, 1989); *Triticum* (EM: Pickett-Heaps, 1969a); (dicot) *Nicotiana* (EM: Cronshaw and Esau, 1968); expanding leaf blade (EM: Esau and Gill, 1969); *Spinacia* (IF: Wick *et al.*, 1989)

Spongy parenchyma cells: *Spinacia* (IF: Wick *et al.*, 1989)

Cotyledonary cells: *Allium* (IF: Mineyuki *et al.*, 1991a); *Pisum* (IF: Doonan *et al.*, 1987)

Stomatogenesis (*including nondifferential epidermal division*); (Monocot) *Avena* (EM: Kaufman *et al.*, 1970; IF: Mullinax and Palevitz, 1989); *Allium* (EM: Palevitz and Hepler, 1974a; IF: Mineyuki *et al.*, 1988a); *Commelina* (EM: Pickett-Heaps, 1969b); *Hordeum* (IF: Cho and Wick, 1989); *Lolium* (IF: Cleary and Hardham, 1989); *Saccharum* (EM: Singh *et al.*, 1977); *Secale* (IF: Cho and Wick, 1989); *Triticum* (EM: Pickett-Heaps and Northcote, 1966b; IF: Cho and Wick, 1989); a case that two SC formations are induced by two successively aligned GMCs in an epidermal cell (EM: Galatis *et al.*, 1983); *Zea* (EM: Srivastava and Singh, 1972); (dicot) *Cucumis* (IF: Kazama and Mineyuki, 1997); *Sinapis* (EM: Landré, 1972); *Vigna* (EM: Galatis and Mitrakos, 1979); *Phaseolus, Trifolium, Medicago, Calycotome, Vicia, Pisum, Cicer, Spartium, Robinia, Melilotus, Coronilla, Caratonia, Cercis,* and *Acacia* (EM: Galatis *et al.*, 1982)

(*continues*)

TABLE I (*continued*)

Seed development
- Zygote and proembryogenesis: *Arabidopsis* (IF: Webb and Gunning, 1991)
- Aleurone layer development: *Hordeum* (IF: Brown *et al.,* 1994)

Other specialized cells
- Tapetal cells: *Zea* (IF: Staiger and Cande, 1992)
- Stamen hair and filament cells: *Tradescantia* (EM: Busby and Gunning, 1980; LC: Zhang *et al.,* 1990)
- Leaf hair cells: *Triticum* (EM: Galatis *et al.,* 1984b)
- Ameiotic division: *Zea, am1* mutant (IF: Staiger and Cande, 1992)

Wound-induced cell divisions: (Monocot) *Tradescantia* epidermal cells (IF: Goodbody and Lloyd, 1990); (dicot) *Datura,* periderm (IF: Flanders *et al.,* 1990); *Nautilocalyx,* leaf explants (EM: Venverloo *et al.,* 1980; IF: Goodbody *et al.,* 1991); *Nicotiana* (IF: Wilms and Derksen, 1988); *Pisum,* roots (EM: Hardham and McCully, 1982; IF: Hush *et al.,* 1990)

Suspension culture cells: (Monocot) *Spartina* (IF: Hogan, 1988); (dicot) *Arabidopsis* (IF: Liu *et al.,* 1994); *Datura* (IF: Gorst *et al.,* 1986); *Medicago* (IF: Meijer and Simmonds, 1988); *Nicotiana* (IF: Gorst *et al.,* 1986); BY-2 (field emission SEM: Vesk *et al.,* 1996; IF: Kakimoto and Shibaoka, 1987); *Vicia* (IF: Simmonds *et al.,* 1983); *Zinnia* (IF: Falconer and Seagull, 1985); (gymnosperm) *Picia,* embryonic, and *Pinus,* nonembryonic (IF: Tautorus *et al.,* 1992)

Cultured protoplasts; (Dicot) *Glycine* (IF: Wang *et al.,* 1989a); *Helianthus,* liquid medium, agarose embedding (IF: Caumont *et al.,* 1997); *Medicago* and *Nicotiana,* budding protopast (IF: Meijer and Simmonds, 1988); BY-2 (IF: Sonobe, 1990); *Vicia* (IF: Simmonds, 1986); (gymnosperm) *Picia* (IF: Fowke *et al.,* 1990)

Callus: *Petunia* (IF: Traas *et al.,* 1990)

Pteridophytes

Sporophyte roots: *Adiantum* (IF: Panteris *et al.,* 1991); *Azolla,* all categories of divisions (EM: Gunning *et al.,* 1978a); *Ceratopteris,* apical cell (EM: Gunning, 1982); *Dryopteris,* adventitious roots (EM: Burgess, 1970b); *Isoetes* (EM: Brown and Lemmon, 1984; IF: Cleary *et al.,* 1992b); *Selaginella* (EM: Brown and Lemmon, 1984)

Stomatogenesis: *Azolla,* stomata of a single unspecialized annular gaurd cell with two nuclei (EM: Busby and Gunning, 1984); *Anemia,* a "floating" stomata (EM: Galatis *et al.,* 1986); *Selaginella,* monoplastid GMC (EM: Brown and Lemmon, 1985b; IF: Cleary *et al.,* 1992a)

Trichome development: *Salvinia* (EM: Busby and Gunning, 1980)

Tricoblast and atrichoblast formation: *Azolla* (EM: Gunning *et al.,* 1978b); *Hydrocharis* (EM: Gunning *et al.,* 1978b)

Gametophyte
- Protonemata: *Adiantum* (EM: Wada *et al.,* 1980; IF: Murata and Wada, 1989); *Athyrium* (EM: Jenni *et al.,* 1990)
- Antheridia development: *Anemia,* very early steps of antheridium development (EM: Schraudolf, 1993); *Blechum* (EM: Busby as cited in Gunning, 1982)

Bryophytes

Mosses: *Funaria,* nonstomatal epidermal cells in capsule (EM: Sack and Paolillo, 1985); gametophore tissue (EM: Schnepf in Gunning, 1982); Tmema cell mother cell (EM and IF: Sawidis *et al.,* 1991); *Physcomitrella,* apices of leafy shoot (IF: Doonan *et al.,* 1987); *Sphagnum,* leaflet (EM: Schnepf, 1973)

TABLE I

Hornworts: *Phaeoceros,* meristematic portion of young sporophyte (EM and IF: Brown and Lemmon, 1988)
Liverworts: *Marchantia,* thallus cells (EM: Fowke and Pickett-Heaps, 1978); initial aperture cells (EM: Apostolakos and Galatis, 1985a); differential division of mucilage papillae and other scale cells, inner thallus cells, photosynthetic filament cells and their mother cells (EM: Galatis and Apostolakos, 1977; IF: Apostolakos and Galatis, 1992, 1993); *Reboulia,* archegoniophore stalk cells (IF: Brown and Lemmon, 1990); *Conocephalum,* sporophyte (IF: Shimamura *et al.,* 1998)

Note. Cell types and genus names in which PPBs are observed by electron microscopy (EM), by immunoflurescent microscopy (IF), or by the direct MT observation in living cells (LC) are listed and a reference is cited for each item.

PPBs are found in the formation of tmema cells of *Funaria* protonemata (Sawidis *et al.,* 1991) and in leaflets of mosses (Schnepf, 1973; Doonan *et al.,* 1987). They are not found in the *Funaria* GMC division (Sack and Paolillo, 1985), in mitosis of moss sporogonial initials (Gambardella and Alfano, 1990), in the young sporophyte of hornworts (Brown and Lemmon, 1985a), and in archesporial cells of *Monoclea* (Brown and Lemmon, 1992). PPBs are not found in algae or fungi (Pickett-Heaps, 1974).

V. Temporal Aspects of PPB Development

Early EM observations on synchronously induced cell division of *Adiantum* protonemata showed that MTs reorganize in the G_2 phase and that the PPB is located along the nucleus around the G_2/M boundary (Wada *et al.,* 1980). Microspectrophotometric examination indicates that all of the onion root tip cells with a PPB had the G_2 level of nuclear DNA (Mineyuki *et al.,* 1988b), and pulse-chase experiments with tritiated thymidine (Gunning and Wick, 1985; Mineyuki *et al.,* 1988b) or with 5-bromo-2-deoxyuridine (Gunning and Sammut, 1990) revealed that in some cells PPBs start to form at the end of S phase. However, there is a population of G_2 cells that do not have a PPB but do have an ICM (see Fig. 4 in Mineyuki *et al.,* 1988b; Utrilla *et al.,* 1993).

PPB development and events of the nuclear cycle run in parallel sequences, but they are not fully coupled. When onion seedlings are exposed to DNA synthesis inhibitors, the mitotic index decreases to near zero and very few prophase PPBs are seen. However, the population of cells with a PPB does not decrease remarkably because of the increment of interphase PPBs. In these root tips, some interphase cells with a PPB are in S phase or even in G_1 phase (Mineyuki *et al.*, 1988b). The partial uncoupling of PPB formation and the nuclear cycle is also confirmed using other experimental systems (Utrilla and De la Torre, 1991; Wang *et al.*, 1991). On the contrary, PPB formation as well as the progression of S phase of cultured tobacco BY-2 cells are inhibited by aphidicolin, and PPBs start to form after the drug is removed (Katsuta *et al.*, 1990). PPBs are not formed when cells are reexposed to aphidicolin before the majority of the cultured BY-2 cells have finished S phase (Mizutani *et al.*, 1993). In soybean protoplast cultures, uncoupling of the PPB formation from the nuclear cycle is seen in the presence of aphidicolin; however, not only DNA synthesis but also PPB formation are totally inhibited by a high concentration of aphidicolin. Early PPBs are dominant in aphidicolin-treated cells and very few PPBs are present in later stages of maturation. This suggests tight coupling of the later stage of PPB development and the nuclear cycle (Wang *et al.*, 1991). Although ca. 10% of onion root tip cells have a PPB in the presence of 5-aminouracil, PPBs can be synchronously induced after the removal of 5-aminouracil. Most PPBs in 5-aminouracil-treated cells are broad but narrow; mature PPBs appear when the nuclear cycle progresses after the removal of the 5-aminouracil. This indicates that although the initiation of PPBs is uncoupled from the nuclear cycle, the late stage of PPB development is tightly coupled (C. Okushima and Y. Mineyuki, unpublished data).

VI. Preparation of the Division Site

A. Changes in the MT Array during PPB Development

Changes in MT distribution during development and disappearance of PPBs in onion root tip cells are illustrated in Fig. 7. The first indication of PPB formation is that ICMs (Fig. 7a) gather in the midregion surrounding a nucleus to form a broad MT band (Fig. 7b). MTs in a broad PPB gradually gather to form a narrow PPB. Most prophase PPBs are narrow (Figs. 7d–7f). Double MT bands (Figs. 7c and 8a) are sometimes seen in the G_2/prophase transition stage; they may be in a transitory state shifting from a broad PPB into a narrow PPB (Wick and Duniec, 1983; Utrilla *et al.*, 1993). In cultured BY-2 cells under a specific condition, double MT bands are frequently observed and remain throughout prophase (Hasezawa *et al.*, 1994). Because similar double MT bands are inducible by a high concentration of cycloheximide (Fig. 8b) or by hydroxyurea (Zhang *et al.*, 1996) in onion root tip cells, these double MT bands are PPBs in which MT bundling processes are somehow interrupted.

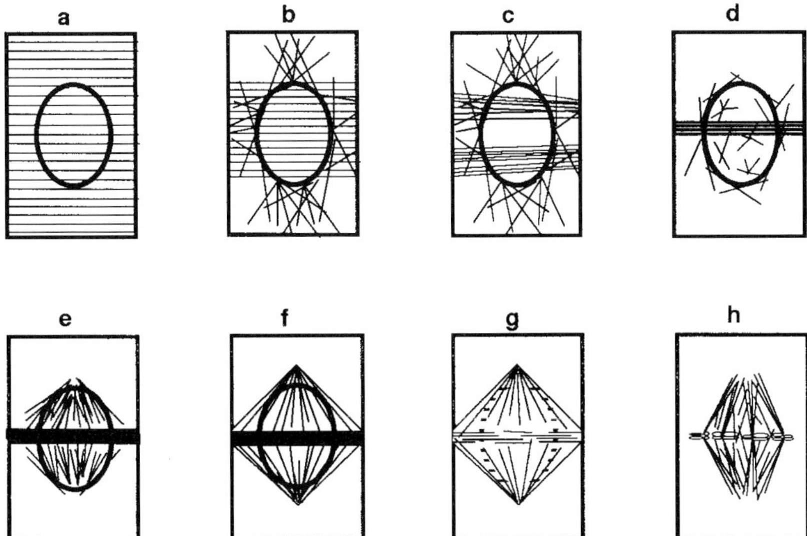

FIG. 7 Changes in MT distribution during the progression of PPB and spindle development in onion root tip cells. The ellipse in the center of the cell represents the nucleus. MTs are shown as thin black lines. The cell cycle stage in each MT stage is as follows: (a) G_1, S, or G_2 phases; (b–d) G_2 or prophase; (e and f) prophase; (g) late prophase or prometaphase; (h) metaphase. These diagrams are made based on observation from Wick *et al.* (1981), Wick and Duniec (1983, 1984), Mineyuki *et al.* (1988b), Mineyuki (1993), and Nogami *et al.* (1996).

PPB fluorescence in late prophase is much brighter than that in the early narrow PPB (Fig. 9). EM study of onion PPBs shows that the number of MTs increases during prophase and the distance between adjacent MTs becomes less as the MT number increases (Nogami *et al.*, 1996). The nar-

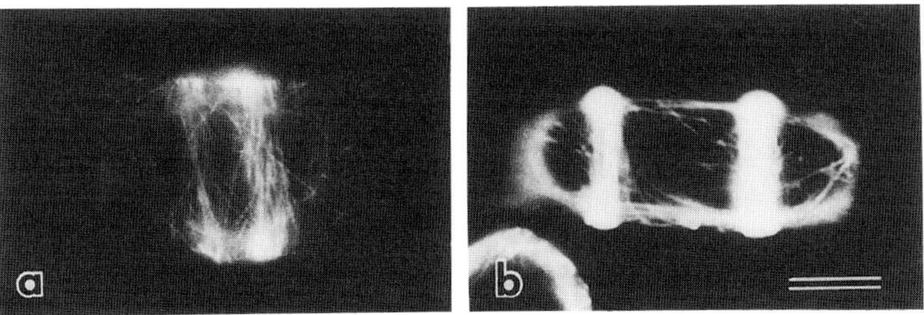

FIG. 8 Tubulin immunofluorescence images of double MT band in onion root tip cells. (a) A double MT band seen in the normal condition. This is thought to be a transition stage from a broad PPB to a narrow PPB. (b) An irregular double MT band induced by the treatment of 36 μM CHM for 2 h. Note that the distance between two MT bands is irregularly long and each MT band is well packed. Scale bar = 10 μm (photograph by A. Nogami).

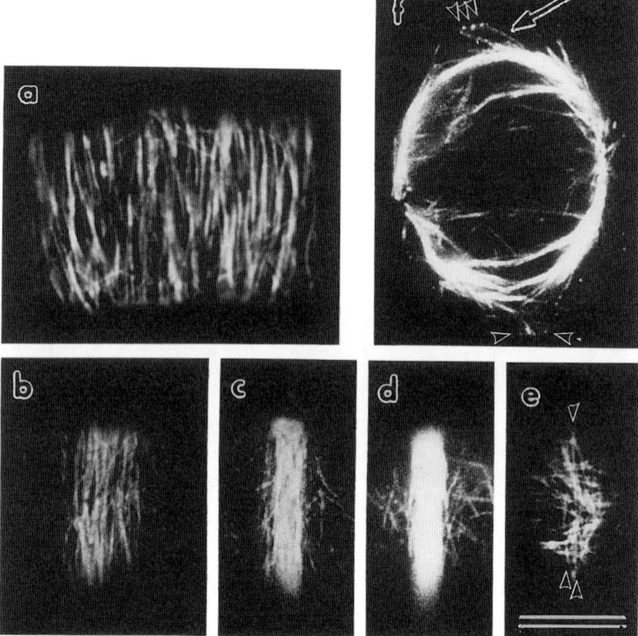

FIG. 9 MT bundling processes on the cell cortex in onion root tip cells. (a) Interphase cortical MTs. (b) A broad PPB. (c) A narrow PPB, which is almost bundled. (d) A mature PPB; each MT in the PPB cannot be distinguished. (e) Loosening of MTs in the PPB at prometaphase. (a–e) Surface view. (f) Midoptical section of the cell in e. MT connection between spindles and the PPB (arrow) are seen. Arrowheads, MTs in a PPB. Scale bar = 10 μm.

rowing of MT bands and loss of the resolution of individual MTs is confirmed by direct observation of the behavior of microinjected fluorescent-labeled tubulin in *Tradescantia* stamen hair cells (Cleary *et al.*, 1992c).

Perinuclear MTs appear concomitantly with the appearance of a broad PPB (Wick and Duniec, 1983). The population of MTs in the cytoplasm thins out slightly and MT foci from which many MTs initiate on the nuclear surface appear when the PPB becomes a narrow band (Fig. 7d). The more developed MT foci gradually gather to the nuclear polar region (Fig. 7e) and MTs orient parallel to the spindle axis to form a bipolar spindle (Fig. 7f). In this stage, MT linkages between PPB and the spindle pole region are clearly seen (arrowheads in Fig. 1a; Nogami *et al.*, 1996). MTs in a PPB remain even after the bipolar spindle is established, but they disappear in late prophase or early prometaphase (Wick and Duniec, 1984). Observations on MT dynamics in living cells show that PPB MTs disappear 5–10 min prior to nuclear envelope breakdown (Cleary *et al.*, 1992c). Figure 9 indicates that although MTs gradually gather to make a narrow band during PPB development, the MT band becomes loosened, keeping its width constant during PPB disappearance. Connections between PPBs and spindle poles remain until the MT band totally disappears (Fig. 9f).

B. PPB Development and Determination of the Division Site

Because a PPB is one of the earliest intracellular structures that predicts the future division site, a question arises whether the division site is determined during PPB formation or before the PPB starts to form. A hint to the answer to this question comes from IFM observations of MT reorganization during stomatogenesis. Figure 10 summarizes changes in MT organization during PPB formation in various types of cell division. Because the orientation of the division plane is the same as that of the preceding division in most cells, ICMs in these cells are already oriented parallel to the future division plane (Fig. 10a). In GMCs, however, the division plane must reorient perpendicular to the former division plane. In onion GMCs (Fig. 10b), interphase MTs are randomly oriented in the cytoplasm and the first indication of the new orientation of the division plane is a broad PPB (Mineyuki *et al.,* 1989). Graminean GMCs (Fig. 10c) have "interphase MT bands" that orient perpendicular to PPBs (Singh, 1977; Busby and Gunning, 1980; Galatis, 1982). This transverse interphase MT band precedes a radial MT array, which is then replaced by a PPB prior to the longitudinal cell division (Mullinax and Palevitz, 1989; Cho and Wick, 1989; Cleary and Hardham, 1989). Because shifts in division plane orientation occur around the time of formation of the broad PPB, broad PPB formation emerges as a process in which the orientation of the division plane is fixed.

Although the broad PPB predicts the orientation of the division plane and its appropriate location in the cell cortex, it cannot precisely define the division site. In asymmetrical division of onion epidermis, ICMs do not always orient transversely, especially in basal regions of young cotyledons. However, all cortical MTs are oriented transversely during broad PPB formation (Fig. 10d). Then, narrowing of the PPB occurs to define the final division site (Mineyuki *et al.,* 1989). Cytochalasin does not affect formation of the broad PPB nor the progression of the cell cycle, but it inhibits PPB narrowing. The persistent broad PPBs in the presence of cytochalasin are associated with subsequent changes in the position of the mitotic apparatus and the cell plate. The number of apically located small GMC-like cells decreases, indicating the importance of the PPB-narrowing step for determination of the division site (Mineyuki and Palevitz, 1990).

PPBs in graminean subsidiary mother cells (SMCs) (Fig. 10e) first appear as fan-shaped arrays of MTs in the paradermal cortex that focus on two

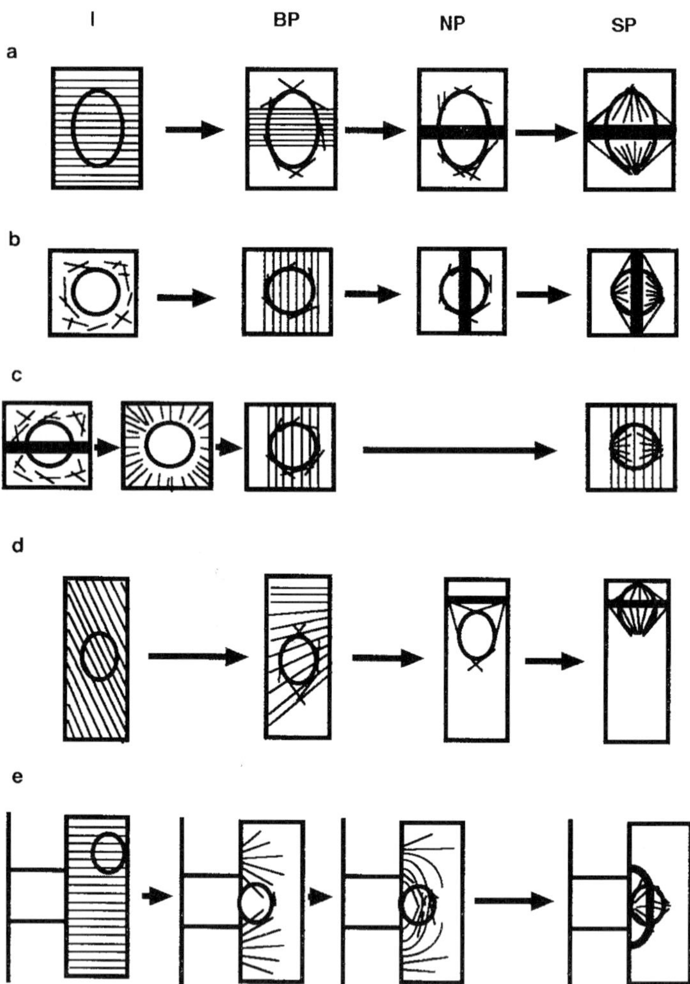

FIG. 10 Changes in MT arrangement during PPB formation. (a) Symmetrical cell division commonly seen in root meristems or epidermal cells (Wick and Duniec, 1983, 1984; Mineyuki *et al.*, 1991a). (b) Onion GMC division (Mineyuki *et al.*, 1989). (c) Graminacean GMC division (Cho and Wick, 1989; Mullinax and Palevitz, 1989). Many graminacean plants (*Avena*, wheat, and maize) maintain a broad PPB until it disappears, but a narrow PPB is reported in *Lolium* (Cleary and Hardham, 1989). (d) Asymmetrical cell division of onion cotyledons (Mineyuki and Palevitz, 1990). (e) Graminacean SMC divisions (Cho and Wick, 1989; Cleary and Hardham, 1989; Mullinax and Palevitz, 1989). The longitudinal axis of the cell is vertically oriented and the apical end of the organ is positioned toward the top of the page. Thick lines show the outline of cells, and the thin lines show MTs. I, interphase; BP, broad PPB stage; NP, narrow PPB stage; SP, a narrow PPB stage with a bipolar spindle.

relatively broad regions along the edges of the SMCs. This arrangement may reflect the transverse cell polarity that is governed by the GMCs. Then, interaction of MTs within and between these arrays gives rise to a curved, narrow band that defines the ultimate division site (Mullinax and Palevitz, 1989; Cho and Wick, 1989; Cleary and Hardham, 1989). These steps may correspond to the successive stages of formation of a broad PPB and a band narrowing stage, as seen in other cell types.

The division plane orientation in a cell can be influenced by nearly mechanical injury. The first indication of the new division polarity is a reorientation of cortical actin filaments (AFs) that occurs within 30 min after wounding (Goodbody and Lloyd, 1990). Reorientation of ICMs to the future division plane is also reported to occur 2–6 h after wounding, i.e., long before PPB formation (Wilms and Derksen, 1988; Hush *et al.*, 1990). However, in *Nautilocalyx* explants, if a second wound is inflicted before cell division, reorientation of the division plane is inducible until the early stages of PPB formation (Venverloo, 1990). This means that although the orientation of the division plane can be determined early before PPB formation, it is reversible until the beginning (or early stages) of the PPB formation. Coexistence of differently oriented MT bands in a cell is reported in *Azolla* and pea root tips (Gunning *et al.*, 1978a; Gunning and Wick, 1985) and in the shoot apical meristem of *Hedera* (Marc and Hackett, 1989). This may be explained as a transitional stage from an old division plane axis to a new axis.

This evidence supports our hypothesis (Mineyuki *et al.*, 1989) that formation of a broad PPB is a necessary step to fix the axis of division polarity (division plane orientation) in a cell and the subsequent narrowing of the MT band determines the ultimate site where the cell plate will join the parental walls. In some cases, cell polarity is predetermined long before the former divisions, but it can be adjusted by environmental factors until the beginning of PPB formation. The division polarity is fixed during broad PPB formation and the precise division site is determined by the narrowing of the MT bands, with submicrometer accuracy.

C. Other Signs That Predict the Future Division Site

1. Phragmosomes in Highly Vacuolated Cells

In highly vacuolated cells, the position of the future division site is often predictable in terms of the location of cytoplasm in a cell. For examples, divisions of vacuolated cells can be induced by wounding. The nucleus, which has been flattened against the wall at interphase, becomes round and starts to move to the center of the cell before cell division. In the early stage of nuclear migration, the nucleus, now separated from the cortex, is suspended across the vacuole by a few thin cytoplasmic strands. Later the nucleus is tethered in a central position by more numerous, thicker cytoplasmic strands. These cytoplasmic strands tend to become aggregated into a more or less continuous diaphragm known as a phragmosome (PS). This organization persists from prophase throughout nuclear division, and the expanding cell plate follows exactly the course of the plane of the PS (Sinnott and Bloch 1940, 1941; Venverloo *et al.,* 1980; Venverloo and Libbenga, 1987). Laser microsurgery in leaf epidermal cells of *Nautilocalyx* demonstrates that the cytoplasmic strands seen during the premitotic nuclear migration and positioning are under tension, and that the tension is likely to influence the alignment of strands as well as the position of the nucleus during division plane formation (Goodbody *et al.,* 1991).

The PS contains MTs (Goosen-de Roo *et al.,* 1984; Bakhuizen *et al.,* 1985; Flanders *et al.,* 1990; Katsuta *et al.,* 1990; Goodbody *et al.,* 1991; Lloyd *et al.,* 1992) and AFs (Trass *et al.,* 1987; Lloyd and Trass, 1988; Kakimoto and Shibaoka, 1987; Katsuta *et al.,* 1990; Goodbody and Lloyd, 1990). A PPB and a PS form simultaneously and "bunching up" of the PPB MTs and the realignment of the radial MT strands into a PS occur concomitantly (Flanders *et al.,* 1990; Lloyd *et al.,* 1992). All stages leading to PS formation, including the formation of the cytoplasmic strands, can be inhibited by colchicine, oryzalin, and cytochalasin. These anticytoskeletal drugs also attack mature prophase PS to cause loss of integrity of PSs within 30 min, suggesting that both AFs and MTs are involved in the formation and maintenance of prophase PSs (Venverloo and Libbenga, 1987).

Although MTs in the PS disappear after prometaphase, AFs remain, connecting the nucleus to the cell cortex. Chloroisopropyl phenylcarbamate treatment results in multiple nuclei and branched phragmoplasts (Clayton and Lloyd, 1984). In these cells, AFs are seen connecting the leading edges of the abnormal phragmoplast to the cortex, but their cortical sites cannot be predicted by PPBs (Lloyd and Traas, 1988). Polar strands, which also contain AFs and connect the mitotic apparatus to the cell cortex, also occur. Here these AFs tend to align at right angle to the plane of the PS. Polar strands seem to play an important role in the remigration of the mitotic apparatus after centrifugation (Ôta, 1961). Weakening of polar strands causes the displacement of nucleus within the PS in *Nautilocalyx* explants (Venverloo and Libbenga, 1987).

PSs are not clear or not detectable in some types of cell divisions (e.g., fern gametophytes; Wada, 1941), although PS-like structures can be induced by centrifugal treatment of some cells that normally do not have a PS (Ôta, 1961; Murata and Wada, 1997). Doubts about the universality of the PS have been expressed. It may be that its cytoskeletal components are present in both vacuolated and nonvacuolated cells but that the cytoplasmic organization is visible in vacuolated cells only.

2. Premitotic Nuclear Migration

Premitotic nuclear migration is a prerequisite for divisions in vacuolated cells with PSs and in asymmetrical cell divisions. Migration of the nucleus in SMC divisions (Pickett-Heaps, 1969b; Kennard and Cleary, 1997) and cell divisions with a PS (Venverloo and Libbenga, 1987; Katsuta *et al.,* 1990) precede PPB formation. Premitotic nuclear migration in *Adiantum* protonemata ceases at the G_2/M boundary when PPBs appear (Wada *et al.,* 1980; Mineyuki and Furuya, 1980). In other cases nuclear migration in some asymmetrical divisions takes place in midprophase after a PPB at the asymmetrical location has been formed (Fig. 3; Mineyuki and Palevitz, 1990; Gunning *et al.,* 1978a).

Migration of the nucleus from the periphery to the center of the cell is inhibited by colchicine, but not by cytochalasin, in *Nautilocalyx* explants (Venverloo and Libbenga, 1987) and in cultured tobacco cells (Katsuta *et al.,* 1990). However, AFs and not MTs are required for traumatotactic migration (Schnepf and von Taitteur, 1973; Goodbody and Lloyd, 1990) and for the nuclear migration in *Tradescantia* SMCs that occurs in G_1 phase (Kennard and Cleary, 1997). On the contrary, in the asymmetrical division of onion GMC formation, MTs connecting the migrating nucleus and the asymmetrical PPB are seen (Fig. 3), and colchicine (Y. Mineyuki, unpublished observation) and cytochalasin (Mineyuki and Palevitz, 1990) inhibit the nuclear migration to the apical part of the cell. Premitotic nuclear migration is also seen in spore germination of a fern and in side branch formation of moss caulonemata, in which the existence of PPBs has not been reported (Schmiedel and Schnepf, 1979a; Schmiedel *et al.,* 1981; Bassel *et al.,* 1981; Jensen, 1981; Doonan *et al.,* 1985, 1986). MTs but not AFs are somehow involved in premitotic nuclear migration in these cases (Bassel *et al.,* 1981; Vogelmann *et al.,* 1981; Schmiedel and Schnepf, 1979b; Doonan *et al.,* 1985, 1986).

VII. Spatial Aspects of PPB Development

A. Origin of the PPB MTs

Where do PPB MTs come from? Are they nucleated at the site of incipient PPB formation? Because MTs can translocate over a substratum with the help of motor proteins and cofactors *in vitro,* the idea of rearrangement, translocation, and reutilization of intact MTs cannot be discounted (Palevitz, 1991). The number of MTs in a PPB is similar to that of the preceding ICMs (Schnepf, 1973; Hardham and Gunning, 1979). Moreover, PPBs can still form in cells whose *de novo* protein or RNA synthesis is inhibited by drugs (Benbadis *et al.,* 1974; Olszewska *et al.,* 1990; Utrilla and De la Torre, 1991; Mineyuki *et al.,* 1994; Nogami *et al.,* 1996), indicating that a PPB can form without new tubulin synthesis.

The idea that a PPB is formed by condensation or "bunching up" of existing ICMs has been proposed (Pickett-Heaps, 1969c; Wick and Duniec, 1983; Doonan *et al.,* 1987), but this apparently cannot be applicable to cells whose PPB formation requires the redirection of cell polarity. The appearance of perinuclear MTs before PPB formation in some suspension culture cells (Flanders *et al.,* 1990) and the alignment of perinuclear MTs parallel to the anticipated PPB in wheat protophloem (Eleftheriou, 1985a) support another idea—that the nuclear surface has an MTOC activity in this stage and that tubulins are first polymerized on the nuclear surface and the MTs and then move to the cell cortex. However, in many cell types, perinuclear MTs appear concomitantly with a PPB (e.g., Wick and Duniec, 1983; Mineyuki *et al.,* 1989; Mineyuki and Palevitz, 1990) and in soybean protoplast cultures they appear after the PPB is well developed (Wang *et al.,* 1989b). MTs can polymerize in the PPB region after cold treatment even when the nucleus is displaced far away from the PPB site (Murata and Wada, 1991b). The results of this experiment are clearly in contrast to the idea that perinuclear MTs can move to the PPB site.

Although rearrangement of preexisting MTs is a reasonable hypothesis, observations of fixed cells cited previously do not deny another hypothesis—that preexisting MTs depolymerize and MTs are newly polymerized in the PPB. Recovery experiments using cells with depolymerized MTs show that MTs can be polymerized from cytoplasmic tubulin (Cleary and Hardham, 1988; Galatis and Apostolakos, 1991; Murata and Wada, 1991b). The failure of PPBs to form in preprophasic cells in the presence of taxol also supports the idea (Panteris *et al.,* 1995). Direct observation of fluorescent-labeled tubulin in living stamen hair cells shows that tubulin subunits exchange rapidly in MTs (half-time of MT turnover, 62.1 s; Hush *et al.,* 1994), and also shows the rapid appearance of an incipient PPB from few cortical MTs (Cleary *et al.,* 1992c). That MTs can first organize randomly in the cell cortex and then reorient is shown in wheat protophloem (Eleftheriou, 1985a) and in *Adiantum* protonemata recovering from cold treatment (Murata and Wada, 1991b).

Coexistence of an incipient PPB and ICMs in cotyledonary cells suggests that ICMs are not triggered to depolymerize as soon as the PPB begins to form (Doonan *et al.,* 1987). In *Adiantum* protonemata arrested at early G_1 phase under red light, a transversely oriented cortical MT band exists in the subapical region. This MT band is thought to be involved in the maintenance of the cylindrical shape (Wada and O'Brien, 1975). A PPB is not formed by the direct displacement of this subapical MT band (Wada *et al.,* 1980; Murata and Wada, 1989). In blue light-induced cell division, an early PPB coexists with ICMs. Disruption of ICMs starts at ca. 120 μm from the nucleus (and PPB) and spreads toward the tip and base, but cortical MTs far away from the nucleus (more than 300–400 μm from the tip) remain (Murata and Wada, 1989). In prophase of dark-induced cell division, tip growth ceases and tubulins in the nuclear region are used as PPB and perinuclear MTs. The tip growth can be reinduced in the preprophasic cells by irradiation with red light at the appropriate time. In this condition, many prophase cells have neither a PPB nor subapical cortical MTs, although perinuclear and cytoplasmic MTs are seen. Competition for tubulins between PPB and the subapical MT band for tip growth may be the reason why only perinuclear and cytoplasmic MT systems remain and the cortical MT band cannot develop (Mineyuki *et al.,* 1991b).

B. MTOCs

Are there any specific MTOCs in the PPB? Packard and Stack (1976) pointed out that the PM is a candidate for a MT-nucleating site, claiming that PPB MTs often appear to terminate in or on the PM. Hardham and Gunning (1978) could not confirm such patterns of termination in PPBs. Gunning *et al.* (1978b) proposed that initiation sites for the PPB are situated along the cell edges. This possibility is also raised by work on other cell types (Galatis, 1982; Apostolakos and Galatis, 1992). The observation that MTs are better organized at cell edges than cell surfaces in broad PPBs that appear during recovery from colchicine for 12 h further supports the idea (Galatis and Apostolakos, 1991). However, there have been no reports on the specific localization of candidate MTOC molecules at cell edges.

A survey of MTOCs in the PPB using autoimmune serum 5051, which recognizes animal centriolar materials, failed to reveal any positive signs of MTOCs in PPBs (Clayton *et al.,* 1985; Wick *et al.,* 1985; Palevitz, 1988). However, the specificity of this serum to plant MTOCs is questionable (Harper *et al.,* 1989). Anti-centrin also fails to stain the PPB (Del Vecchio *et al.,* 1997). Colchicine-induced paracrystals often occupy the cortical cytoplasmic zone where a PPB is expected to be assembled, suggesting the possible existence of MTOCs (Apostolakos *et al.,* 1990; Karagiannidou *et al.,* 1995). Antibodies that recognize an EF-1α homolog reveal PPBs (Hasezawa and Nagata, 1993). Although these antibodies were originally raised against sea urchin centrosomes, immunologically cross-reactive peptides are colocalized with all categories of MTs in plants and animals (Ohta *et al.*, 1988, 1990). This peptide may therefore be a MAP-like protein rather than a candidate for an MTOC. Anti-γ-tubulin signal is seen in both broad and narrow PPBs but declines while MTs in the PPB are still present. Its distribution is punctate (Liu *et al.*, 1993, 1994). As another MTOC candidate in a PPB, some monoclonal antibodies raised against fern spermatozoids are reported to detect PPBs (Marc and Gunning, 1988).

C. PM–MT Connection

Electron-dense cross-bridges exist between MTs and the PM in both PPB and ICMs (see Section III,C), and these cortical MTs disappear when lysed protoplasts are treated with ATP under specific conditions (Sonobe, 1990). There are no differences between ICM and PPB MTs in terms of the sensitivity to MT drugs (Cleary and Hardham, 1988) or in MT turnover rates (Hush *et al.*, 1994). However, there is a remarkable difference in stability against cycloheximide between ICMs and PPB MTs. PPB MTs remain in the cell cortex, whereas ICMs disappear following 2 h of cycloheximide treatment, suggesting that some components involved in the connection of MTs to the PM are different between these two MT arrays (Mineyuki *et al.*, 1994). ICMs control the direction of cell expansion, which may change from hour to hour depending on the environment (e.g., direction of light). If ICMs stay in one position for a long time, it may be that the cells cannot change their growth direction quickly. Rapid turnover of proteins that are involved in the connection of ICMs to the PM may provide a mechanism for rapid growth adjustment in changing environments. By comparison, determination of the new division plane is not necessarily a quick response. One cell cycle in normal plant cells usually takes more than half a day. Slow turnover of PM–MT bridging proteins in PPBs may reflect this. Two monoclonal antibodies, *Pas1D3* and *Pas5F4,* which are raised against a fern spermatozoid, detect some diffuse cytoplasmic materials that follow the distribution of perinuclear and PPB MTs, but they do not colocalize with ICMs (Marc and Gunning, 1988). This observation could also suggest differences between PM–MT connections in ICMs and PPBs. Further characterization of molecules that are involved in PM–MT connection will be necessary.

D. Nuclear Position and PPB Development

The question of whether the nucleus, or the position of the nucleus, influences the determination of the PPB site has been investigated. In *Adiantum* protonemata, nuclear position apparently influences the site of early PPB formation (Murata and Wada, 1991a). A PPB forms around a basipetally displaced nucleus when a protonema cell is centrifuged before the time of PPB formation. However, double MT bands are formed, one in the correct position and another around the displaced nucleus, when the nucleus is displaced in an early stage of PPB formation (Figs. 11a–11c). Similarly, when binuclei induced by caffeine are separated using centrifugation, two PPBs are formed, one around each nucleus (Murata and Wada, 1993). Counter to the centrifugation experiments in *Adiantum,* other pieces of evidence indicate that the position of PPBs is not particularly influenced by the nucleus or nuclei, especially in the organized tissue. First, in some asymmetrical divisions, a PPB is formed asymmetrically although the nucleus has not yet migrated (e.g., Fig. 3a; Mineyuki and Palevitz, 1990). Second, centrifugation experiments on cell division in stomatal development show that PPBs remain at the correct position (Pickett-Heaps, 1969c; Galatis *et al.,* 1984a). Third, in cases in which two SCs are induced by two successively aligned GMCs in a wheat epidermal cell, four PPB sites are seen in a longitudinal anticlinal wall (Galatis *et al.,* 1983), although the induced cell has only one nucleus. Finally, in division of caffeine-induced binucleate cells in organized tissues, the PPB position is suggested to be

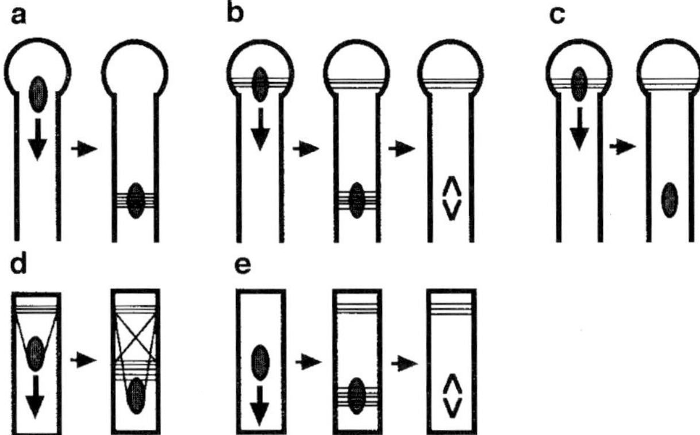

FIG. 11 A comparison of the centrifugal experiments between protonemal division of *Adiantum* (a–c; Murata and Wada, 1991a) and asymmetrical divisions of onion cotyledons (d and e; Mineyuki *et al.,* 1991b). Vertical arrows mean basipetal centrifugation to displace nuclei from the expected PPB site. (a–c) The effect of the timing of the centrifugation on the PPB positioning. A nucleus is displaced just before PPB formation (a), in the early PPB stage (b), or in a late PPB stage (c), respectively. (d) A cell with asymmetrical PPBs is centrifuged to displace the nucleus (left) and fixed immediately (right). (e) A cell is centrifuged far before PPB formation (left), and the nucleus remains in the basipetal region even when it is in the preprophase stage (middle). The cells shown on the right in b and e are in prometaphase or later stages. Although PPBs are expected to disappear before this stage without centrifugation, a PPB far from the nucleus still remains there.

determined not by the position or number of nuclei but by stimuli related to the establishment of cell polarity as well as the plane of division (Pickett-Heaps, 1969a; Apostolakos and Galatis, 1987; Manandhar *et al.,* 1996a,b). However, in contrast to Pickett-Heaps's (1996c) observation, Burgess and Northcote (1968) report the existence of PPB in the displaced nucleus of asymmetrical divisions.

In interpreting some of the previous observations, it is necessary to pay careful attention to the stage of the cell cycle at the time of the experiment. For example, Mineyuki *et al.* (1991a) report that when a cell with an asymmetrically located PPB is centrifuged basipetally, a second band-like MT array appears in the central part of the cell, near the displaced nucleus, in addition to the apical PPB, suggesting that a new MT band is formed during the basipetal centrifugation (Fig. 11d). When interphase nuclei are centrifuged as far down to the basal end of the cell as possible and then the tissue is incubated for a time sufficient to allow entry into prophase, a cell whose nucleus has not yet recovered from the centrifugation and is still located toward the basal end of the cell will develop two MT bands, one near the apical end where a PPB is normally found and another located around the displaced nucleus (Fig. 11e). Although the MT band around the nucleus is broader and coarser than the normal PPB, this extra MT band apparently formed under the influence of the nucleus. In conclusion, PPB formation is influenced not only by intercellular morphogenetic stimuli but also internally by the nucleus. However, in some cases, especially at the time of asymmetrical division, intercellular morphogenetic stimuli are stronger than the nuclear signal.

Interestingly, as illustrated in Figs. 11b and 11e, the MT band located remotely from the nucleus remains even after prometaphase, although the MT band around the nucleus disappears (Mineyuki *et al.,* 1991a; Murata and Wada, 1991a). The MT band remote from the displaced nucleus can be disrupted when the displaced nucleus and the surrounding cytoplasm are moved again to the original site by another centrifugation (Murata and Wada, 1992). These results support the idea that the endoplasm surrounding the nucleus contains factors that are important for the disappearance of PPBs (cyclins? see Section VIII,B), and that the PPB is a source of tubulin/MT for spindle development.

VIII. Molecular Aspects of PPB Organization

A. Actin

Cortical actin bands encompassing the PPB (Fig. 12c) have been reported in various cell types (Palevitz, 1987, 1988; Trass *et al.,* 1987; Kakimoto and

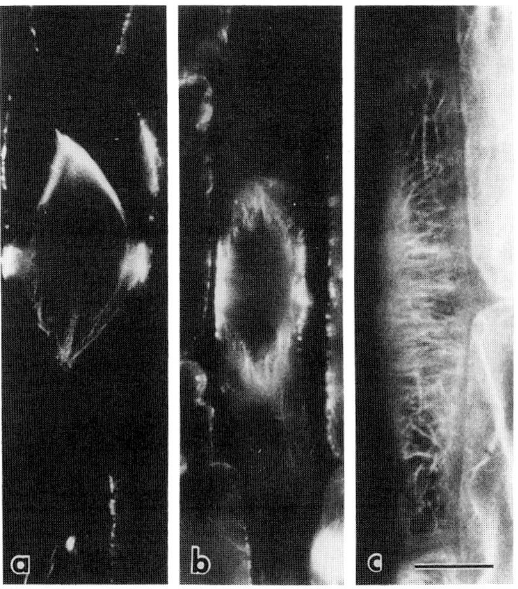

FIG. 12 PPB MTs and actin in onion cotyledon. All three cells are in late prophase judged from the nuclear staining (data not shown). (a, b) Tubulin immunofluorescence images and (c) an image of AFs stained with rhodamine phalloidin. Cells are treated with (b) or without (a, c) 20 μM cytochalasin D. Note that in control, PPB MT is narrow but AF band is broad. Scale bar = 10 μm. See detail in Mineyuki and Palevitz (1990).

Shibaoka, 1987; Lloyd and Traas, 1988; McCurdy *et al.,* 1988; Mineyuki and Palevitz, 1990; McCurdy and Gunning, 1990; Katsuta *et al.,* 1990; Liu and Palevitz, 1992; Eleftheriou and Palevitz, 1992; Panteris *et al.,* 1992; Cleary *et al.,* 1992a, c; Cleary, 1995; Cleary and Mathesius, 1996; Baluška *et al.,* 1997) using fluorescent phalloidin derivatives, immunostaining, or microinjection. The existence of microfilaments in the PPB is also confirmed by EM (Ding *et al.,* 1991). The width of the actin band always exceeds the area covered by the MT band (compare Figs. 12a and 12c), and it sometimes occupies the whole of the cell surface. While formation of the actin band precedes the incipient PPB formation in onion and *Tradescantia* roots (Liu and Palevitz, 1992), it appears sometime after a broad PPB formation in wheat roots (McCurdy and Gunning, 1990) and *Tradescantia* stamen hair cells (Cleary *et al.,* 1992c). While the actin band disappears when PPB MTs are destroyed by MT inhibitors (Palevitz, 1987; Trass *et al.,* 1987; Mineyuki and Palevitz, 1990; Katsuta *et al.,* 1990; Panteris *et al.,* 1992), it remains in other cases (Lloyd and Traas, 1988; McCurdy and Gunning, 1990). Also, while the actin band disappears before the breakdown of PPB MTs in some systems (Liu and Palevitz, 1992; McCurdy and Gunning, 1990; Cleary *et al.,* 1992c), in others it persists, even in metaphase (Lloyd and Traas, 1988; Mineyuki and Palevitz, 1990; Fig. 12 in Cleary, 1995). Recovery experiments from cytoskeletal inhibitors indicate that MTs are necessary for the maintenance of the actin band (McCurdy and Gunning, 1990; Panteris *et al.,* 1992).

While some actin persists in the cell cortex throughout mitosis, it disappears from the division site after the disappearance of the actin band. This "actin-depleted zone" was found independently by IFM of root tips (Liu and Palevitz, 1992) and by microinjection of rhodamin–phalloidin into *Tradescantia* stamen hair cells (Cleary *et al.,* 1992c); its existence was confirmed in other systems (Cleary, 1995; Baluška *et al.,* 1997). Although it is not conspicuous, some photographs presented in the earlier works on actin band also suggest the existence of actin-depleted zone. For example, I present a picture of Fig. 12c as a broad PPB actin, but if the image is carefully examined, the fluorescent signals in the central part of the actin band appear to be weak. If we regard this as an actin-depleted zone, transversely aligned cortical AFs are not a PPB actin. Reexamination of actin bands described in the early work will be necessary. Because observation of living cells indicates that cell plates are inserted accurately in the area of the cortex defined by the actin-depleted zone, the actin-depleted zone may have some essential roles for the memory of the division site (see Section IX).

Cytochalasin D interferes with the narrowing of the PPB MTs (Fig. 12b) (Mineyuki and Palevitz, 1990; Eleftheriou and Palevitz, 1992). PPBs become wide within 15 min after cytochalasin treatment, indicating that the cytochalasin treatment also leads to rewidening of the MT band. PPB MTs do not bundle to form a narrow band in GMCs of winter rye in normal conditions (Fig. 10c). The abscence of actin bands in these cells (Cho and Wick, 1990, 1991) supports the idea of involvement of actin bands in PPB bundling, although it is not certain that the procedures used were capable of detecting the delicate actin (Clearly and Mathesius, 1996). Although PPBs in GMCs of *Selaginella krausiana* never become narrow, an actin band is observed (Cleary *et al.,* 1992a). Whether the difference between the presence and absence of actin bands between these GMCs with a broad PPB is a result of the phylogenetical difference between a grass and a fern ally remains unsolved.

B. Cyclin-Dependent Kinases

Cyclin-dependent kinases (Cdks) are eukariotic protein kinases that bind to cyclins and activate proteins that are necessary for the entry into S or M phase. Since the first finding that the mitosis-promoting factor (MPF) is a complex of a cdc2 kinase (= cdk1) and a cyclin B, several types of cdks and cyclins have been reported in animal cells (Martin-Castellanos and Moreno, 1997). The existence of cdc2 homologs in a PPB has been shown by IFM in a variety of cell types and species (Mineyuki *et al.,* 1991c, 1996; John *et al.,* 1993; Colasanti *et al.,* 1993; Bögre *et al.,* 1997; Mews *et al.,* 1997). One of the functions of cdc2 may be a promotion of disassembly of PPB MTs because microinjection of active MPF into *Tradescantia* stamen hair cells causes rapid disassembly of PPB MTs (Hush *et al.,* 1996). This idea is supported by experiments with general kinase inhibitors, K-252a and staurosporin (Katsuta and Shibaoka, 1992).

Maize cdc2-specific antibody recognizes only 10% of the prophase PPBs (Colasanti *et al.,* 1993) and these PPBs are late PPBs (Mews *et al.,* 1997). On the contrary, anti-PSTAIR, the antibody against a peptide of a well-conserved domain of cdks, recognizes very wide developmental stages of PPBs (Mineyuki *et al.,* 1991c), even in maize roots (Mineyuki *et al.,* 1996). This raises the possibility that other anti-PSTAIR cross-reactive molecules exist in the PPB. A study using phosphorylation-state specific antibody against the ATP-binding domain of cdks suggests that there are different phosphorylation states of cdks in onion PPBs (Okushima *et al.,* 1996). Two types of cyclins are reported in the PPB: Cyclin Ib appears transiently in the late PPBs, and cyclin II associates with developing PPBs (Mews *et al.,* 1997). This observation also suggests our hypothesis.

The width of a PPB is broad in onion root tips treated with cycloheximide (Nogami *et al.,* 1996) or those treated with staurosporin, and MTs could not bundle in the presence of another protein kinase inhibitor, 6-dimethylaminopurine (Nogami *et al.,* 1994). These findings suggest that cdk homologs are involved in MT bundling during PPB development. Double staining with anti-PSTAIR and anti-tubulin showed that the width of anti-PSTAIR cross-reactive materials is always narrower than that of the PPB MTs (Y. Mineyuki, unpublished data) and that they disappear when MTs are disassembled by oryzaline (Colasanti *et al.,* 1993) or colchicine. Some MTs in a PPB are stable against cold treatment; anti-PSTAIR cross-reactive materials colocalize with these MT bundles (Y. Mineyuki, unpublished data). These data suggest that anti-PSTAIR cross-reactive materials are accumulated in the bundled part of a PPB.

Whether cdks in a PPB directly phosphorylate tubulins or modify MT arrangement via some MT-associated proteins is not known. The distance between adjacent MTs in a PPB cannot be shortened in the presence of cycloheximide (Nogami *et al.,* 1996). Because cycloheximide presumably inhibits cdk activity by inhibiting cyclin synthesis, there is a possibility that some MAPs are involved in the cdk-induced MT bundling in PPBs. A kinesin-related protein, TKRP125, which locates in the PPB and has a site of phosphorylation by cdc2 (Asada *et al.,* 1997), is a candidate substrate for cdks in the PPB. $P13^{suc1}$ is an essential cell cycle protein that binds to cdc2, but it does not associate with the PPB (Hepler *et al.,* 1994). A monoclonal antibody, MPM-2, has been used to identify a family of proteins that becomes phosphorylated at mitosis in a range of eukaryotic cells (Davis *et al.,* 1983). Although MPM-2 fails to stain the PPB, the increase in MPM-2 fluorescence in a cell apparently correlates with PPB formation (Trass *et al.,* 1992; Young *et al.,* 1994). Effects of phosphatase inhibitors on PPBs are also reported. Hasezawa and Nagata (1992) show inhibition of the PPB formation by okadaic acid, whereas Zhang *et al.* (1992) observe that cells incubated with okadaic acid are accumulated in G_2 and that the PPB develops and disassembles with the same timing as controls, although the treated cells do not progress to prometaphase. More examination may be necessary to understand these inconsistent results. However, it is clear that some kinases (including cdc2) and phosphatases act in concert to control PPB development and degradation.

C. Other Molecules Associated with the PPB

Several other molecules are reported to localize in the PPB using IFM. Some antibodies that recognize animal intermediate filaments or plant cytoplasmic bundles are known to detect molecules in the PPB (Dawson *et al.,* 1985; Hargreaves *et al.,* 1989; Fairbairn *et al.,* 1994) although others fail to detect PPBs (Parke *et al.,* 1987). Localization of plant MT-associated proteins in the PPB has also been reported (Jiang and Sonobe, 1993; Chan *et al.,* 1996). In an attempt to examine the localization of motor proteins in the PPB, Parke *et al.* (1986) failed to find myosin. Kinesin-like proteins are reported to be in the PPB (Asada *et al.,* 1997; Bowser and Reddy, 1997), whereas antibodies against the MT-binding domain of *Arabidopsis* kinesin-like protein stain PPBs very weakly (Liu *et al.,* 1996). Calmodulins (Del Vecchio *et al.,* 1997) and molecules that react with an antibody against rabbit skeletal muscle troponin T (Lim *et al.,* 1986) are also detected in the PPB. *Pas5D8,* a monoclonal antibody against a fern spermatozoid, stains whole cell cortex except for the PPB region (Marc and Gunning, 1988). Because this antibody is an IgM and similar results are obtained by other IgM antibodies such as Amersham anti-actin N-350 (McCurdy *et al.,* 1988), whether this PPB excluding zone is meaningful or just the result of steric exclusion from the PPB site by close packing of the MTs remains unsolved.

IX. Operation of the PPB (Division) Site

As discussed in Section III,A, cell plates fuse to the parental walls at the former PPB site, without exception. Cell plate formation includes two separate phases (Mineyuki *et al.*, 1991b): (i) the appearance of cell plate fragments between the daughter nuclei and (ii) expansion of cell plates to the former PPB site. In cells whose equatorial planes are not parallel to the future division plane (Fig. 5c), location of the incipient phragmoplast is independent of the PPB site and it is restricted by the position of daughter nuclei. Reorientation of cell plates to the division site occurs in the second phase. Experimental obliteration of the PPB in *Adiantum* protonemata (Mineyuki *et al.*, 1991b; Murata and Wada, 1991a) causes incorrect insertion of cell plate. In these cases, the appearance of the phragmoplast between the daughter nuclei looks normal, but phragmoplast expansion toward the correct division site is affected by the treatment (Figs. 13a and 13b). Malorientation of the cell plate is also reported in large cells with PSs following colchicine treatment during preprophase (Venverloo and Libbenga, 1987). This indicates the involvement of the PPB site in the guidance of phragmoplast expansion. Because cytochalasin and phalloidin also inhibit the cell plate expansion process, actin must somehow be involved in this process (Palevitz and Hepler, 1974b; Palevitz, 1980, 1986; Mineyuki and Gunning, 1990; Cho and Wick, 1990). In cells with a PS, it is proposed that actin (see Section VI,C,1) guides the expanding phragmoplast to the division site (Lloyd and Traas, 1988), but this does not seem applicable to meristematic cells, for which evidence to the contrary has accumulated (Cleary, 1995). However, Valster and Hepler (1997) recently showed that AFs link the expanding edge of the phragmoplast with the cortical division site and

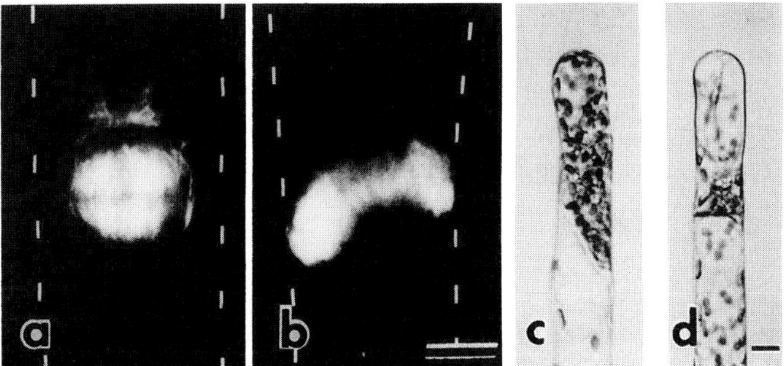

FIG. 13 Cell plate formation in altered cell division of *Adiantum* protonemata. (a, b) Tubulin immunofluorescence of a development of phragmoplasts in cells whose PPB formation is obliterated by the experimental manipulation with red light (see detail in Mineyuki *et al.*, 1991b). (a) Early telophase. (b) Late telophase. Note that while the incipient phragmoplast appearance seems normal, the expansion of the phragmoplast becomes abnormal. (c, d) Cell plate in the dark-induced cell division in the presence (c) or absence (d) of 2% dimethyl sulfoxide (DMSO). Note that although cell plate is inserted transversely in control, DMSO-treated cell plate is inserted obliquely. Scale bars = 10 μm.

cortical actin patches within the actin-deplet zone in *Tradescantia* stamen hair cells using microinjection with a high phalloidine concentration. Observation of living onion GMCs shows that the rotational movement is also inhibited by blue light or by sodium azide (Palevitz, 1986).

The newly formed cell wall that forms just after the cell plate attaches to the PPB site is fluid and wrinkled, but then it becomes stiff and flat in *Tradescantia* stamen hair cells. This may be one process of the cell plate maturation. When cell plates fail to reach the correct division site by the experimental manipulations, cell plate maturation is not completed. This observation suggests that cell plate maturation is a function of the PPB site (Mineyuki and Gunning, 1990).

PPBs disappear before metaphase and cell plate formation takes place in their absence. Hence, some positional information must be inserted at the division site when a PPB MT locates there, later helping to guide the edges of the centrifugally extending phragmoplast to the correct site and helping in the process of cell plate maturation. Experimental manipulation of the division site after PPB disappearance shows that the positional information resists centrifugation at 3550*g* or less for 15 min (Ôta, 1961) but is disrupted by making minute wounds in the division site of mitotic cells with a microneedle (Gunning and Wick, 1985). Small vesicles associate with PPB MTs and fuse with PM (see Section III,C). Cell walls at the division site are sometimes differentially thickened (Packard and Stack, 1976; Galatis and Mitrakos, 1979; Galatis, 1982). These observations suggest promotion of localized cell wall deposition by the PPB, but molecules involved in this process have not been identified. Because the actin-depleted zone is seen at the division site (see Section VIII,A), one of the candidates may be molecules that block binding of actin with the PM. When and how this information is accumulated is not known. When *Adiantum* protonemata are grown under 2% dimethyl sulfoxide or 1% methanol, the frequency with which cell plates are inserted obliquely increased (Fig. 13c). This suggests that the PM is important for the accumulation of positional information.

X. Other Functions of the PPB

A. Premitotic Nuclear Positioning

As discussed in Section VI,C,2, while some premitotic nuclear migrations occur after the PPB is formed, those of some other cell types take place long before the PPB formation and in some cells nuclear migration occurs entirely without PPBs. Premitotic nuclear migration may therefore be a response to cellular polarization which can be completed without involving a PPB.

Although the path of the nuclear migration in *Adiantum* protonemata is different among the experimental conditions, nuclei cease their migration at the time of G_2/M boundary to determine the ultimate division site (Mineyuki and Furuya, 1980; Furuya, 1984, Fig. 2). A nucleus that has finished its migration shows enhanced resistance to centrifugation (Pickett-Heaps, 1969c; Mineyuki and Furuya, 1986) and the apparent cytoplasmic viscosity in the PPB–nucleus region increases, based on observation of changes in organelle movement (Mineyuki *et al.,* 1984). They are sensitive to colchicine but not to cytochalasin, indicating that PPB MTs and MTs connecting PPB and the nucleus are involved in the anchoring of the preprophasic nucleus (Mineyuki and Furuya, 1986). MT linkages between PPBs and the perinuclear MTs have been reported (see Section III,C) and the observation that the PPB and the nucleus remain associated when cells are broken (Wick and Duniec, 1983, 1984; Tiwari *et al.,* 1985) also indicates a strong connection between the nucleus and the PPB. Therefore, PPBs and endoplasmic MTs between PPBs and the nuclear surface are involved in premitotic nuclear anchoring.

B. Bipolar Spindle Formation and Orientation

Because the ultimate prophase spindle axis has been found to be perpendicular to the PPB in all cases reported to date, involvement of the PPB in prophase spindle orientation has been suggested (Wick and Duniec, 1984). This has further been confirmed by observation of GMCs (Figs. 10b and 10c; Mineyuki *et al.,* 1988a; Cho and Wick, 1989; Cleary and Hardham, 1989). In onion GMCs, the prophase spindle orients transversely, perpendicular to the PPB, but it gradually changes its orientation during prometaphase and it finally orients obliquely in meta and anaphase (Mineyuki *et al.,* 1988a). In *Haemanthus* endosperm, a cell type which lacks PPBs, three well-developed spindle poles are often formed in late prophase but are transformed into normal bipolar spindles during prometaphase (Schmit *et al.,* 1983, 1985; Smirnova and Bajer, 1994). On the contrary, cells with a PPB never show well-developed tripoles, suggesting that the PPB provides a reference plane to which spindle poles can be established unambiguously. Clear MT-linkage between spindle poles and the PPB have been seen in late prophase (Figs. 1a and 7f; Mineyuki *et al.,* 1991a; Nogami *et al.,* 1996). When the narrowing of a PPB is inhibited by cytochalasin (Mineyuki and Palevitz, 1990) or cycloheximide (Nogami *et al.,* 1996), bipolar spindle formation is also inhibited. These results indicate the possible involvement of the PPB on bipolar spindle organization and/or spindle orientation.

Against these observations, there is evidence that spindle poles can be established prior to PPB formation in lower land plants (e.g., Brown and Lemmon, 1990). Relationships between spindles and the PPB may differ between higher plants and lower land plants.

XI. Conclusion

The determination of the division site and correct insertion of a cell plate at the predetermined division site contain the following elementary processes: (i) Orientation of the division plane is fixed in a cell in terms of the parallel alignment of cortical MTs to the future division plane orientation (broad PPB formation); (ii) the precise position at which the cell plate will fuse with the parental wall is determined by narrowing of the broad MT band (maturation of a PPB); (iii) accumulation of some information at the PPB site (division site) is necessary in the final stage of cytokinesis; (iv) disappearance of the PPB to utilize MTs/tubulin for spindle formation at the prophase/prometaphase transition stage; (v) appearance of cell plate fragments between daughter nuclei at the anaphase/telophase transition stage; (vi) centrifugal expansion of the cell plate to the division site; and (vii) cell plate maturation after the cell plate edges attach to the parental cell walls. The PPB is somehow involved in steps i–iv, and the positional information accumulated at step iii may be involved in steps vi and vii. As discussed in Section VI,B, the PPB is not a determinant of the division plane orientation but a determinant of the ultimate division site. The most important function of a PPB may be the accumulation of positional information at the division site that may be necessary for the correct insertion and maturation of the cell plate (see Section IX). Thus, it is important to know the molecular mechanism of PPB narrowing for the ultimate division site determination and finding molecules which are accumulated at the PPB site may be the next crucial step for studying the function of PPBs.

Acknowledgments

I thank Dr. Masashi Tazawa (Fukui Institute of Technology) for the encouragement to write this chapter and to Dr. Takashi Murata (University of Tokyo) and Professor Brian E. S. Gunning (Australian National University) for critical reading of the manuscript. I also thank colleagues both inside and outside Japan for their helpful comments on the manuscript and for generously providing their unpublished information, micrographs, and figures. I especially thank Ms. Akiko Nogami and Chikage Okushima for their help in preparing the manuscript. This work was supported by the Ito Science Foundation, the Sumitomo Foundation, and a research grant from the Ministry of Education, Science and Culture in Japan.

References

Apostolakos, P., and Galatis, B. (1985a). Studies on the development of the air pores and air chambers of *Marchantia paleacea.* III. Microtubule organization in preprophase–prophase initial aperture cells—Formation of incomplete preprophase microtubule bands. *Protoplasma* **128,** 120–135.

Apostolakos, P., and Galatis, B. (1985b). Studies on the development of the air pores and air chambers of *Marchantia paleacea.* IV. Cell plate arrangement in initial aperture cells. *Protoplasma* **128,** 136–146.

Apostolakos, P., and Galatis, B. (1987). Induction polarity and spatial control of cytokinesis in some abnormal subsidiary mother cells of *Zea mays. Protoplasma* **140,** 26–42.

Apostolakos, P., and Galatis, B. (1992). Patterns of microtubule organization in two polyhedral cell types in the gametophyte of the liverwort *Marchantia paleacea* Bert. *New Phytol.* **122,** 165–178.

Apostolakos, P., and Galatis, B. (1993). Interphase and preprophase microtubule organization in some polarized cell types of the liverworts *Marchantia paleacea* Bert. *New Phytol.* **124,** 409–421.

Apostolakos, P., and Galatis, B., Katsaros, C., and Schnepf, E. (1990). Tubulin conformation in microtubule-free cells of *Vigna sinensis.* An immunofluorescent and electron microscope study. *Protoplasma* **154,** 132–143.

Asada, T., Kuriyama, R., and Shibaoka, H. (1997). TKRP125, a kinesin-related protein involved in the centrosome-independent organization of the cytokinetic apparatus in tobacco BY-2 cells. *J. Cell Sci.* **110,** 179–189.

Bajer, A., and Molè-Bajer, J. (1972). Spindle dynamics and chromosome movements. *Int. Rev. Cytol. Suppl.* **3,** 1–271.

Bakhuizen, R., Spronsen, P. C. van, Sluiman-den Hertog, F. A. J., Venverloo, C. J., and Goosen-de Roo, L. (1985). Nuclear envelope radiating microtubules in plant cells during interphase mitosis transition. *Protoplasma* **128,** 43–51.

Baluška, F., Barlow, P. W., Parker, J. S., and Volkmann, D. (1996). Symmetric reorganizations of radiating microtubules around pre- and post-mitotic nuclei of dividing cells organized within intact root meristems. *J. Plant Physiol.* **149,** 119–128.

Baluška, F., Vitha, S., Barlow, P. W., and Volkmann, D. (1997). Rearrangement of F-actin arrays in growing cells of intact maize root apex tissues: A major developmental switch occurs in the postmitotic transition region. *Eur. J. Cell Biol.* **72,** 113–121.

Bassel, A. R., Kuehnert, C. C., and Miller, J. H. (1981). Nuclear migration and asymmetric cell division in *Onoclea sensibilis* spores: An ultrastructural and cytochemical study. *Am. J. Bot.* **68,** 350–360.

Bednara, J., Van Lammeren, A. A. M., and Willemse, M. T. M. (1988). Microtubular configurations during meiosis and megasporogenesis in *Gasteria verrucosa* and *Chamaenerion angustifolium. Sex Plant Reprod.* **1,** 164–172.

Benbadis, M.-C., Levy, F., and Deysson, M. G. (1974). Interruption de la mitose en prophase avec persistance de la membrane nucléar sous l'influence du cycloheximide: Étude ultrastructurale. *C. R. Acad. Sci. Paris D* **278,** 1353–1355.

Bögre, L., Zwerger, K., Meskiene, I., Binarova, P., Csizmadia, V., Planck, C., Wagner, E., Hirt, H., and Heberle-Bors, E. (1997). The cdc2Ms kinase is differently regulated in the cytoplasm and in the nucleus. *Plant Physiol.* **113,** 841–852.

Bowser, J., and Reddy, A. S. N. (1997). Localization of a kinesin-like calmodulin-binding protein in dividing cells of *Arabidopsis* and tobacco. *Plant J.* **12,** 1429–1437.

Brown, R. C., and Lemmon, B. E. (1984). Plasmid apportionment and preprophase microtubule bands in monoplastidic root meristem cells of *Isoetes* and *Selaginella. Protoplasma* **123,** 95–103.

Brown, R. C., and Lemmon, B. E. (1985a). Preprophasic establishment of division polarity in monoplastidic mitosis of hornworts. *Protoplasma* **124,** 175–183.

Brown, R. C., and Lemmon, B. E. (1985b). A cytoskeletal system predicts division plane in meiosis of *Selaginella. Protoplasma* **127,** 101–109.

Brown, R. C., and Lemmon, B. E. (1988). Preprophasic microtubule systems and development of the mitotic spindle in hornworts (Bryophyta). *Protoplasma* **143,** 11–21.

Brown, R. C., and Lemmon, B. E. (1990). Polar organizers mark division axis prior to preprophase band formation in mitosis of the hepatic *Reboulia hemisphaerica* (Bryophyta). *Protoplasma* **156,** 74–81.

Brown, R. C., and Lemmon, B. E. (1992). Polar organizers in monoplastidic mitosis of hepatics (Bryophyta). *Cell Motil. Cytoskel.* **22,** 72–77.

Brown, R. C., Lemmon, B. E., and Mullinax, J. B. (1989). Immunofluorescent staining of microtubules in plant tissues: Improved embedding and sectioning techniques using polyethylene glycol (PEG) and Steedman's wax. *Bot. Acta* **102,** 54–61.

Brown, R. C., Lemmon, B. E., and Olsen, O.-A. (1994). Endosperm development in barley: Microtubule involvement in the morphogenetic pathway. *Plant Cell* **6,** 1241–1252.

Bünning, E. (1952). Morphogenesis in plants. *Surv. Biol. Progr.* **2,** 105–140.

Bünning, E., and Biegert, F. (1953). Die Bildung der Spaltöffnungsinitiation bei *Allium cepa. Z. Bot.* **41,** 17–39.

Burgess, J. (1970a). Cell shape and mitotic spindle formation in the generative cell of *Endymion non-scriptus. Planta* **95,** 72–85.

Burgess, J. (1970b). Interactions between microtubules and the nuclear envelope during mitosis in a fern. *Protoplasma* **71,** 77–89.

Burgess, J., and Northcote, D. H. (1967). A function of the preprophase band of microtubules in *Phleum pratense. Planta* **75,** 319–326.

Burgess, J., and Northcote, D. H. (1968). The relationship between the endoplasmic reticulum and microtubular aggregation and disaggregation. *Planta* **80,** 1–14.

Burgess, J., and Northcote, D. H. (1969). Action of colchicine and heavy water on the polymerization of microtubules in wheat root meristem. *J. Cell Sci.* **5,** 433–451.

Busby, C. H., and Gunning, B. E. S. (1980). Observations on pre-prophase bands of microtubules in uniseriate hairs, stomatal complexes of sugar cane, and *Cyperus* root meristems. *Eur. J. Cell Biol.* **21,** 214–223.

Busby, C. H., and Gunning, B. E. S. (1984). Microtubules and morphogenesis in stomata of the water fern *Azolla*: An unusual mode of guard cell and pore development. *Protoplasma* **122,** 108–119.

Caumont, C., Petitprez, M., Woynaroski, S., Barthou, H., Brière, C., Kallerhoff, J., Borin, C., Souvré, A., and Alibert, G. (1997). Agarose embedding affects cell wall regeneration and microtubule organization in sunflower hypocotyl protoplasts. *Physiol. Plant.* **99,** 129–134.

Chan, J., Rutten, T., and Lloyd, C. (1996). Isolation of microtubule-associated proteins from carrot cytoskeletons: A 120 kDa map decorates all four microtubule arrays and the nucleus. *Plant J.* **10,** 251–259.

Cho, S.-O., and Wick, S. M. (1989). Microtubule orientation during stomatal differentiation in grasses. *J. Cell Sci.* **92,** 581–594.

Cho, S.-O., and Wick, S. M. (1990). Distribution and function of actin in the developing stomatal complex of winter rye (*Secale cereale* cv. Puma). *Protoplasma* **157,** 154–164.

Cho, S.-O., and Wick, S. M. (1991). Actin in the developing stomatal complex of winter rye: A comparison of actin antibodies and Rh-phalloidin labeling of control and CB-treated tissues. *Cell Motil. Cytoskel.* **19,** 25–36.

Clayton, L., and Lloyd, C. W. (1984). The relationship between the division plane and spindle geometry in *Allium* cells treated with CIPC and griseofulvin: An anti-tubulin study. *Eur. J. Cell Biol.* **34,** 248–253.

Clayton, L., Black, C. M., and Lloyd, C. W. (1985). Microtubule nucleating sites in higher plant cells identified by an auto-antibody against pericentriolar material. *J. Cell Biol.* **101,** 319–324.

Cleary, A. L. (1995). F-actin redistributions at the division site in living *Tradescantia* stomatal complexes as revealed by microinjection of rhodamine-phalloidin. *Protoplasma* **185,** 152–165.

Cleary, A. L., and Hardham, A. R. (1988). Depolymerization of microtubule arrays in root tip cells by oryzalin and their recovery with modified nucleation patterns. *Can. J. Bot.* **66,** 2353–2366.

Cleary, A. L., and Hardham, A. R. (1989). Microtubule organization during development of stomatal complexes in *Lolium rigidum. Protoplasma* **149,** 67–81.

Cleary, A. L., and Mathesius, U. (1996). Rearrangements of F-actin during stomatogenesis visualized by confocal microscopy in fixed and permeabilised *Tradescantia* leaf epidermis. *Bot. Acta* **109,** 15–24.

Cleary, A. L., Brown, R. C., and Lemmon, B. E. (1992a). Establishment of division plane and mitosis in monoplastidic guard mother cells of *Selaginella. Cell Motil. Cytoskel.* **23,** 89–101.

Cleary, A. L., Brown, R. C., and Lemmon, B. E. (1992b). Microtubule arrays during mitosis in monoplastidic root tip cells of *Isoetes. Protoplasma* **167,** 123–133.

Cleary, A. L., Gunning, B. E. S., Wasteneys, G. O., and Hepler, P. K. (1992c). Microtubule and F-actin dynamics at the division site in living *Tradescantia* stamen hair cells. *J. Cell Sci.* **103,** 977–988.

Colasanti, J., Cho, S.-O., Wick, S. M., and Sundaresan, V. (1993). Localization of the functional $p34^{cdc2}$ homolog of maize in root tip and stomatal complex cells: Association with predicted division sites. *Plant Cell* **5,** 1101–1111.

Cronshaw, J., and Esau, K. (1968). Cell division in leaves of *Nicotiana. Protoplasma* **65,** 1–24.

Davis, F. M., Tsao, T. Y., Fowler, S. K., and Rao, P. N. (1983). Monoclonal antibodies to mitotic cells. *Proc. Natl. Acad. Sci. USA* **80,** 2926–2930.

Dawson, P. J., Hulme, J. S., and Lloyd, C. W. (1985). Monoclonal antibody to intermediate filament antigen cross-reacts with higher plant cells. *J. Cell Biol.* **100,** 1793–1798.

De Mey, J., Lambert, A.-M., Bajer, A. S., Moeremans, M., and De Brabander, M. (1982). Visualization of microtubules in interphase and mitotic plant cells of *Haemanthus* endosperm with the immuno-gold staining (IGS) method. *Proc. Natl. Acad. Sci. USA* **79,** 1898–1902.

Del Vecchio, A. J., Harper, J. D. I., Vaughn, K. C., Baron, A. T., Salisbury, J. L., and Overall, R. L. (1997). Centrin homologues in higher plants are prominently associated with the developing cell plate. *Protoplasma* **196,** 224–234.

Deysson, G., and Benbadis, M. C. (1968). Etude ultrastructurale de la mise en placede de l'appareil fusorial dans les cellules méristématiques de végétaux supérieurs. *C. R. Biol. D* **162,** 601–604.

Ding, B., Turgeon, R., and Parthasarathy, M. V. (1991). Microfilaments in the preprophase band of freeze substituted tobacco root cells. *Protoplasma* **165,** 209–211.

Doonan, J. H., Cove, D. J., and Lloyd, C. W. (1985). Immunofluorescence microscopy of microtubules in intact cell linages of the moss, *Physcomitrella patens.* I. Normal and CIPC-treated tip cells. *J. Cell Sci.* **75,** 131–147.

Doonan, J. H., Jenkins, G. I., Cove, D. J., and Lloyd, C. W. (1986). Microtubules connect the migrating nucleus to the prospective division site during side branch formation in the moss *Physcomitrella patens. Eur. J. Cell Biol.* **41,** 157–164.

Doonan, J. H., Cove, D. J., Corke, F. M. K., and Lloyd, C. W. (1987). Pre-prophase band of microtubules, absent from tip-growing moss filaments, arises in leafy shoots during transition to intercalary growth. *Cell Motil. Cytoskel.* **7,** 138–153.

Eleftheriou, E. P. (1985a). Microtubules and root protophloem ontogeny in wheat. *J. Cell Sci.* **75,** 165–179.

Eleftheriou, E. P. (1985b). Abundance of microtubules in preprophase bands of some *Triticum* species. *Planta* **163,** 175–182.

Eleftheriou, E. P. (1996). Developmental features of protophloem sieve elements in roots of wheat (*Triticum aestivum* L.). *Protoplasma* **193,** 204–212.

Eleftheriou, E. P., and Palevitz, B. A. (1992). The effect of cytochalasin D on preprophase band organization in root tip cells of *Allium. J. Cell Sci.* **103,** 989–998.

Eleftheriou, E. P., and Tsekos, I. (1982). Development of protophloem in roots of *Aegilops comosa* var. *thessalica.* I. Differential divisions and pre-prophase bands of microtubules. *Protoplasma* **113,** 110–119.

Esau, K., and Gill, R. H. (1969). Structural relations between nucleus and cytoplasm during mitosis in *Nicotiana tabacum* mesophyll. *Can. J. Bot.* **47,** 581–591.

Evert, R. F., and Deshpande, B. P. (1970). An ultrastructural study of cell division in the cambium. *Am. J. Bot.* **57,** 942–961.

Fairbairn, D. J., Goodbody, K. C., and Lloyd, C. W. (1994). Simultaneous labelling of microtubules and fibrillar bundles in tobacco BY-2 cells by the anti-intermediate filament antibody, ME101. *Protoplasma* **182,** 160–169.

Falconer, M. M., and Seagull, R. W. (1985). Immunofluorescent and calcofluor white staining of developing tracheary elements in *Zinnia elegans* L. suspension cultures. *Protoplasma* **125,** 190–198.

Flanders, D. J., Rawlins, D. J., Shaw, P. J., and Lloyd, C. W. (1990). Nucleus-associated microtubules help determine the division plane of plant epidermal cells: Avoidance of four-way junctions and the role of cell geometry. *J. Cell Biol.* **110,** 1111–1122.

Fowke, L. C. (1993). Microtubules in dividing root cells of the conifer *Pinus radiata* and the cycad *Zamia furfuracea. Cell Biol. Int.* **17,** 143–151.

Fowke, L. C., and Pickett-Heaps, J. D. (1978). Electron microscope study of vegetative cell division in two species of *Marchantia. Can. J. Bot.* **56,** 467–475.

Fowke, L. C., Attree, S. M., Wang, H., and Dunstan, D. I. (1990). Microtubule organization and cell division in embryogenic protoplast cultures of white spruce (*Picea glauca*). *Protoplasma* **158,** 86–94.

Furuya, M. (1984). Cell division patterns in multicellular plants. *Annu. Rev. Plant Physiol.* **35,** 349–373.

Galatis, B. (1982). The organization of microtubules in guard cell mother cells of *Zea mays. Can. J. Bot.* **60,** 1148–1166.

Galatis, B., and Apostolakos, P. (1977). On the fine structure of differentiating mucilage papillae of *Marchantia. Can. J. Bot.* **55,** 772–795.

Galatis, B., and Apostolakos, P. (1991). Patterns of microtubule reappearance in root cells of *Vigna sinensis* recovering from a colchicine treatment. *Protoplasma* **160,** 131–143.

Galatis, B., and Mitrakos, K. (1979). On the differential divisions and preprophase microtubule bands involved in the development of stomata of *Vigna sinensis* L. *J. Cell Sci.* **37,** 11–37.

Galatis, B., Apostolakos, P., Katsaros, C., and Loukari, H. (1982). Pre-prophase microtubule band and local wall thickening in guard cell mother cells of some Legminosae. *Ann. Bot.* **50,** 779–791.

Galatis, B., Apostolakos, P., and Katsaros, C. (1983). Synchronous organization of two preprophase microtubule bands and final cell plate arrangement in subsidiary cell mother cells of some *Triticum* species. *Protoplasma* **117,** 24–39.

Galatis, B., Apostolakos, P., and Katsaros, C. (1984a). Experimental studies on the function of the cortical cytoplasmic zone of the preprophase microtubule band. *Protoplasma* **122,** 11–26.

Galatis, B., Apostolakos, P., and Katsaros, C. (1984b). Positional inconsistency between preprophase microtubule band and final cell plate arrangement during triangular subsidiary cell and atypical hair cell formation in two *Triticum* species. *Can. J. Bot.* **62,** 343–359.

Galatis, B., Apostolakos, P., and Palafoutas, D. (1986). Studies on the formation of "floating" guard cell mother cells in *Anemia. J. Cell Sci.* **80,** 29–55.

Gambardella, R., and Alfano, F. (1990). Monoplastidic mitosis in the moss *Timmiella barbuloides* (Bryophyta). *Protoplasma* **156,** 29–38.

Goodbody, K. C., and Lloyd, C. W. (1990). Actin filaments line up across *Tradescantia* epidermal cells, anticipating wound-induced division planes. *Protoplasma* **157,** 92–101.

Goodbody, K. C., Venverloo, C. J., and Lloyd, C. W. (1991). Laser microsurgery demonstrates that cytoplasmic strands anchoring the nucleus across the vacuole of premitotic plant cells are under tension. Implications for division plane alignment. *Development* **113,** 931–939.

Goosen-de Roo, L., Bakhuizen, R., Spronsen, P. C. van, and Libbenga, K. R. (1984). The presence of extended phragmosomes containing cytoskeletal elements in fusiform cambial cells of *Fraxinus excelsior* L. *Protoplasma* **122,** 145–152.

Gorst, J., Wernicke, W., and Gunning, B. E. S. (1986). Is the preprophase band of microtubules a marker of organization in suspention cultures? *Protoplasma* **134,** 130–140.

Gunning, B. E. S. (1980). Spatial and temporal regulation of root tip cells of the water fern *Azolla pinnata. Eur. J. Cell Biol.* **23,** 53–65.

Gunning, B. E. S. (1982). The cytokinetic apparatus: Its development and spatial regulation. *In* "The Cytoskeleton in Plant Growth and Development" (C. W. Lloyd, ed.), pp. 229–292. Academic Press, London.

Gunning, B. E. S. (1992). Use of confocal microscopy to examine transitions between successive microtubule arrays in the plant cell division cycle. *In* "Cellular Basis of Growth and Development in Plants" (H. Shibaoka, ed.), Proceedings of the VII International Symposium Awarding International Prize Biology pp. 145–155. Osaka University, Toyonaka.

Gunning, B. E. S., and Hardham, A. R. (1982). Microtubules. *Annu. Rev. Plant Physiol.* **33,** 651–698.

Gunning, B. E. S., and Sammut, M. (1990). Rearrangements of microtubules involved in establishing cell division planes start immediately after DNA synthesis and are completed just before mitosis. *Plant Cell* **2,** 1273–1282.

Gunning, B. E. S., and Steer, M. W. (1975). "Ultrastructure and the Biology of Plant Cells." Arnold, London.

Gunning, B. E. S., and Wick, S. M. (1985). Preprophase bands, phragmoplasts, and spatial control of cytokinesis. *J. Cell Sci. Suppl.* **2,** 157–179.

Gunning, B. E. S., Hardham, A. R., and Hughes, J. E. (1978a). Pre-prophase bands of microtubules in all categories of formative and proliferative cell division in *Azolla* roots. *Planta* **143,** 145–160.

Gunning, B. E. S., Hardham, A. R., and Hughes, J. E. (1978b). Evidence for initiation of microtubules in discrete regions of the cell cortex in *Azolla* root-tip cells, and an hypothesis on the development of cortical arrays of microtubules. *Planta* **143,** 161–179.

Haber, A. H. (1962). Nonessentiality of concurrent cell divisions for degree of polarization of leaf growth. I. Studies with radiation-induced mitotic inhibition. *Am. J. Bot.* **49,** 583–589.

Hardham, A. R., and Gunning, B. E. S. (1977). The length and disposition of cortical microtubules in plant cells fixed in glutaraldehyde-osmium tetroxide. *Planta* **134,** 201–203.

Hardham, A. R., and Gunning, B. E. S. (1978). Structure of cortical microtubule arrays in plant cells. *J. Cell Biol.* **77,** 14–34.

Hardham, A. R., and Gunning, B. E. S. (1979). Interpolation of microtubules into cortical arrays during cell elongation and differentiation in roots of *Azolla pinnata. J. Cell Sci.* **37,** 411–442.

Hardham, A. R., and McCully, M. E. (1982). Reprogramming of cells following wounding in pea (*Pisum sativum* L.) roots. I. Cell division and differentiation of new vascular elements. *Protoplasma* **112,** 143–151.

Hargreaves, A. J., Dawson, P. J., Butcher, G. W., Larkins, A., Goodbody, K. C., and Lloyd, C. W. (1989). A monoclonal antibody raised against cytoplasmic fibrillar bundles from carrot cells, and its cross-reaction with animal intermediate filaments. *J. Cell Sci.* **92,** 371–378.

Harper, J. D., Mitchison, J. M., Williamson, R. E., and John, P. C. L. (1989). Does the autoimmune serum 5051 specifically recognize microtubule organizing centers in plant cells? *Cell Biol. Int. Rep.* **13,** 471–483.

Hasezawa, S., and Nagata, T. (1992). Okadaic acid as a probe to analyze the cell cycle progression in plant cells. *Bot. Acta* **105,** 63–69.

Hasezawa, S., and Nagata, T. (1993). Microtubule organizing centers in plant cells: Localization of a 49kDa protein that is immunologically cross-reactive to 51kDa protein from sea urchin centrosomes in synchrononized BY-2 cells. *Protoplasma* **176,** 64–74.

Hasezawa, S., Sano, T., and Nagata, T. (1994). Oblique cell plate formation in tobacco BY-2 cells originates in double preprophase bands. *J. Plant Res.* **107,** 355–359.

Hensel, W. (1984). Microtubules in statocytes form roots of cress (*Lepidium sativum* L.). *Protoplasma* **119,** 121–134.

Hepler, P. K., and Palevitz, B. A. (1974). Microtubules and microfilaments. *Annu. Rev. Plant Physiol.* **25,** 309–362.

Hepler, P. K., Cleary, A. L., Gunning, B. E. S., Wardsworth, P., Wasteneys, G. O., and Zhang, D. H. (1993). Cytoskeletal dynamics in living plant cells. *Cell Biol. Int.* **17,** 127–142.

Hepler, P. K., Sek, F. J., and John, P. C. L. (1994). Nuclear concentration and mitotic dispersion of the essential cell cycle protein, $p13^{suc1}$, examined in living cells. *Proc. Natl. Acad. Sci. USA* **91,** 2176–2180.

Heslop-Harrison, J. (1968). Synchronous pollen mitosis and the formation of the generative cell in massulate orchids. *J. Cell Sci.* **3,** 457–466.

Hogan, C. J. (1987). Microtubule patterns during meiosis in two higher plant species. *Protoplasma* **138,** 126–136.

Hogan, C. J. (1988). Preprophase bands in a suspension culture of the monocot *Spartina pectinata. Exp. Cell Res.* **175,** 216–222.

Hush, J. M., Hawes, C. R., and Overall, R. L. (1990). Interphase microtubule re-orientation predicts a new cell polarity in wounded pea roots. *J. Cell Sci.* **96,** 47–61.

Hush, J. M., Wadsworth, P., Callaham, D. A., and Hepler, P. K. (1994). Quantification of microtubule dynamics in living plant cells using fluorescence redistribution after photobleaching. *J. Cell Sci.* **107,** 775–784.

Hush, J. M., Wu, L., John, P. C. L., Hepler, L. H., and Hepler, P. K. (1996). Plant mitosis promoting factor disassembles the microtubule preprophase band and accelerates prophase progression in *Tradescantia. Cell Biol. Int.* **20,** 275–287.

Jarosch, R. (1989). Lateral hydrodynamic effects of rotating filaments. *Plant Syst. Evol.* **164,** 285–322.

Jarosch, R. (1990). Lateral hydrodynamic effects of rotating cytoskeletal components as possible determining factors for cytomorphogenesis in plant cells. *Protoplasma* **157,** 38–51.

Jenni, V., Cattelan, H., and Roos, U. (1990). Immunofluorecence and ultrastructure of mitosis and cell division in the fern *Athyrium filix* Femina. *Bot. Helvetica* **100,** 101–120.

Jensen, L. C. W. (1981). Division, growth, and branch formation in protonema of the moss *Physcomitrium turbinatum:* Studies of sequential cytological changes in living cells. *Protoplasma* **107,** 301–317.

Jiang, C.-J., and Sonobe, S. (1993). Identification and preliminary characterization of a 65kDa higher-plant microtubule associated-protein. *J. Cell Sci.* **105,** 891–901.

John, P. C. L., Zhang, K., and Dong, C. (1993). A $p34^{cdc2}$-based cell cycle: Its significance in monocotyledonous, dicotyledonous and unicellular plants. *In* "Molecular and Cell Biology of the Plant Cell Cycle" (J. C. Ormrod and D. Francis, eds.), pp. 9–34. Kluwer, Dordrecht.

Kakimoto, T., and Shibaoka, H. (1987). Actin filaments and microtubules in the preprophase band and phragmoplast of tobacco cells. *Protoplasma* **140,** 151–156.

Karagiannidou, T. H., Eleftheriou, E. P., Tsekos, I., Galatis, B., and Apostolakos, P. (1995). Colchicine-induced paracrystals in root cells of wheat (*Triticum aestivum* L.). *Ann. Bot.* **76,** 23–30.

Katsuta, J., and Shibaoka, H. (1992). Inhibition by kinase inhibitors of the development and the disappearance of the preprophase band of microtubules in tobacco BY-2 cells. *J. Cell Sci.* **103,** 397–405.

Katsuta, J., Hashiguchi, Y., and Shibaoka, H. (1990). The role of the cytoskeleton in positioning of the nucleus in premitotic tobacco BY-2 cells. *J. Cell Sci.* **95,** 413–422.

Kaufman, P. B., Petering, L. B., Yokcum, C. S., and Baic, D. (1970). Ultrastructural studies on stomata development in internodes of *Avena sativa. Am. J. Bot.* **57,** 33–49.

Kazama, H., and Mineyuki, Y. (1993). Division plane control in photo-induced stomatal differentiation. *In* "XV International Botanical Congress," p. 72. Yokohama, Japan. [Abstract]

Kazama, H., and Mineyuki, Y. (1997). Alteration of division polarity and preprophase band orientation in stomatogenesis by light. *J. Plant Res.* **110,** 489–493.

Kennard, J. L., and Cleary, A. L. (1997). Pre-mitotic nuclear migration in subsidiary mother cells of *Tradescantia* occurs in G1 of the cell cycle and requires F-actin. *Cell Motil. Cytoskel.* **36,** 55–67.

Landré, P. (1972). Oligine et development de epidermes cotylédonares et foliaires de la moutarde (*Sinapis alba* L.). Differentiation ultrastructurale de stomat. *Ann. Sci. Nat. Bot. Ser. 12* **13,** 247–322.

Lehmann, H., and Schultz, D. (1976). "Die Pflanzenzelle." Ulmer, Stuttgart.

Lim, S.-S., Hering, G. E., and Borisy, G. G. (1986). Widespread occurrence of anti-troponin T crossreactive components in non-muscle cells. *J. Cell Sci.* **85,** 1–19.

Liu, B., and Palevitz, B. A. (1992). Organization of cortical microfilaments in dividing root cells. *Cell Motil. Cytoskel.* **23,** 252–264.

Liu, B., Marc, J., Joshi, H. C., and Palevitz, B. A. (1993). A γ-tubulin-related protein associated with the microtubule arrays of higher plants in a cell cycle-dependent manner. *J. Cell Sci.* **104,** 1217–1228.

Liu, B., Joshi, H. C., Wilson, T. J., Silflow, C. D., Palevitz, B. A., and Sunstad, D. P. (1994). γ-Tubulin in *Arabidopsis:* Gene sequence, immunoblot, and immunofluorescence studies. *Plant Cell* **6,** 303–314.

Liu, B., Cyr, R. J., and Palevitz, B. A. (1996). A kinesin-like protein, KatAp, in the cells of *Arabidopsis* and other plants. *Plant Cell* **8,** 119–132.

Lloyd, C. W., and Trass, J. A. (1988). The role of F-actin in determining the division plane of carrot suspension cells. Drug studies. *Development* **102,** 211–222.

Lloyd, C. W., Venverloo, C. J., Goodbody, K. C., and Shaw, P. J. (1992). Confocal laser microscopy and three-dimensional reconstruction of nucleus-associated microtubules in the division plane of vacuolated plant cells. *J. Microsc.* **166,** 99–109.

Lyndon, R. F. (1990). "Plant Development: The Cellular Basis," Topics in Plant Physiology, Vol. 3. Unwin Hyman, London.

Mabuchi, I. (1986). Biochemical aspects of cytokinesis. *Int. Rev. Cytol.* **101,** 175–213.

Manandhar, G., Apostolakos, P., and Galatis, B. (1996a). Cell division of binuclear cells induced by caffeine: Spindle organization and determination of division plane. *J. Plant Res.* **109,** 265–275.

Manandhar, G., Apostolakos, P., and Galatis, B. (1996b). Nuclear and microtubular cycles in heterophasic multinuclear *Triticum* root-tip cells induced by caffeine. *Protoplasma* **194,** 164–176.

Marc, J., and Gunning, B. E. S. (1986). Immunofluorescent localization of cytoskeletal tubulin and actin during spermatogenesis in *Pteridium aquilinum* (L.) Kuhn. *Protoplasma* **134,** 163–177.

Marc, J., and Gunning, B. E. S. (1988). Monoclonal antibodies to a fern sermatozoid detect novel components of the mitotic and cytokinetic apparatus in higher plant cells. *Protoplasma* **142,** 15–24.

Marc, J., and Hackett, W. P. (1989). A new method for immunofluorescent localization of microtubules in surface cell layers. Application to the shoot apical meristem of *Hedera. Protoplasma* **148,** 70–79.

Martin-Castellanos, C., and Moreno, S. (1997). Recent advances on cyclins. CDKs and CDK inhibitors. *Trends Cell Biol.* **7,** 95–98.

McCurdy, D. W., and Gunning, B. E. S. (1990). Reorganization of cortical actin microfilaments and microtubules at preprophase and mitosis in wheat root-tip cells: A double label immunofluorescence study. *Cell Motil. Cytoskel.* **15,** 76–87.

McCurdy, D. W., Sammut, M., and Gunning, B. E. S. (1988). Immunofluorescent visualization of arrays of transverse cortical actin microfilaments in wheat root-tip cells. *Protoplasma* **147,** 204–206.

Meijer, E. G. M., and Simmonds, D. H. (1988). Microtubule organization during the development of the mitotic apparatus in cultured mesophyll protoplasts of higher plants—An immunofluorescence microscopic study. *Physiol. Plant.* **74,** 225–232.

Mews, M., Sek, F. J., Moore, R., Volkmann, D., Gunning, B. E. S., and John, P. C. L. (1997). Mitotic cyclin distribution during maize division: Implications for the sequence diversity and function of cyclins in plants. *Protoplasma* **200,** 128–145.

Mineyuki, Y. (1993). Analysis of microtubule distribution in root tip cells of *Allium cepa* L. using a confocal laser scanning microscope. *Plant Morphol.* **5,** 127–133.

Mineyuki, Y., and Furuya, M. (1980). Effect of centrifugation on the development and timing of premitotic positioning of the nucleus in *Adiantum* protonemata. *Dev. Growth Differ.* **22,** 867–874.

Mineyuki, Y., and Furuya, M. (1986). Involvement of colchicine-sensitive cytoplasmic element in premitotic nuclear positioning of *Adiantum* protonemata. *Protoplasma* **130,** 83–90.

Mineyuki, Y., and Gunning, B. E. S. (1990). A role for preprophase bands of microtubules in maturation of new cell walls, and a general proposal on the function of preprophase band sites in cell division in higher plants. *J. Cell Sci.* **97,** 527–537.

Mineyuki, Y., and Palevitz, B. A. (1990). Relationship between preprophase band organization, F-actin and the division site in *Allium.* Fluorescence and morphometric studies on cytochalasin-treated cells. *J. Cell Sci.* **97,** 283–295.

Mineyuki, Y., Takagi, M., and Furuya, M. (1984). Changes in organelle movement in the nuclear region during the cell cycle of *Adiantum* protonema. *Plant Cell Physiol.* **25,** 297–308.

Mineyuki, Y., Marc, J., and Palevitz, B. A. (1988a). Formation of the oblique spindle in dividing guard mother cells of *Allium. Protoplasma* **147,** 200–203.

Mineyuki, Y., Wick, S. M., and Gunning, B. E. S. (1988b). Preprophase bands of microtubules and the cell cycle: Kinetics and experimental uncoupling of their formation from the nuclear cycle in onion root-tip cells. *Planta* **174,** 518–526.

Mineyuki, Y., Marc, J., and Palevitz, B. A. (1989). Development of the preprophase band from random cytoplasmic microtubules in guard mother cells of *Allium* cepa L. *Planta* **178,** 291–296.

Mineyuki, Y., Marc, J., and Palevitz, B. A. (1991a). Relationship between the preprophase band, nucleus and spindle in dividing *Allium* cotyledon cells. *J. Plant Physiol.* **138,** 640–649.

Mineyuki, Y., Murata, T., and Wada, M. (1991b). Experimental obliteration of the preprophase band alters the sites of cell division, cell plate orientation and phragmoplast expansion in *Adiantum* protonemata. *J. Cell Sci.* **100,** 551–557.

Mineyuki, Y., Yamashita, M., and Nagahama, Y. (1991c). p34^{cdc2} kinase homologue in the preprophase band. *Protoplasma* **162,** 182–186.

Mineyuki, Y., Iida, H., and Anraku, Y. (1994). Loss of microtubules in the interphase cells of onion (*Allium cepa* L.) root tips from the cell cortex and their appearance in the cytoplasm after treatment with cycloheximide. *Plant Physiol.* **104,** 281–284.

Mineyuki, Y., Aioi, H., Yamashita, M., and Nagahama, Y. (1996). A comparative study on stainability of preprophase bands by the PSTAIR antibody. *J. Plant Res.* **109,** 185–192.

Mizuno, K., Sek, F., Perkin, J., Wick, S. M., Duniec, J., and Gunning, B. E. S. (1985). Monoclonal antibodies specific to plant tubulin. *Protoplasma* **129,** 100–108.

Mizutani, T., Katsuta, J., and Shibaoka, H. (1993). Nuclear-cycle dependence of the development of the preprophase band in tobacco BY-2 cells. *Plant Cell Physiol.* **34,** 215–219.

Mullinax, J. B., and Palevitz, B. A. (1989). Microtubule reorganization accompanying preprophase band formation in guard mother cells of *Avena sativa* L. *Protoplasma* **149,** 89–94.

Murata, T., and Wada, M. (1989). Re-organization of microtubules during preprophase band development in *Adiantum* protonemata. *Protoplasma* **151,** 73–80.

Murata, T., and Wada, M. (1991a). Effects of centrifugation on preprophase-band formation in *Adiantum* protonemata. *Planta* **183,** 391–398.

Murata, T., and Wada, M. (1991b). Re-formation of the preprophase band after cold-induced depolymerization of microtubules in *Adiantum* protonemata. *Plant Cell Physiol.* **32,** 1145–1151.

Murata, T., and Wada, M. (1992). Cell cycle-specific disruption of the preprophase band of microtubules in fern protonemata: Effects of displacement of the endoplasm by centrifugation. *J. Cell Sci.* **101,** 93–98.

Murata, T., and Wada, M. (1993). Cell division in caffeine-induced binucleate protonemal cells of *Adiantum.* II. Formation of the preprophase band and cell division in centrifuged and non-centrifuged cells. *J. Plant Res.* **106,** 313–318.

Murata, T., and Wada, M. (1997). Formation of a phragmosome-like structure in centrifuged protonemal cells of *Adiantum capillus-veneris* L. *Planta* **201,** 273–281.

Newcomb, E. H. (1969). Plant microtubules. *Annu. Rev. Plant Physiol.* **20,** 253–288.

Nogami, A., Yamashita, M., Nagahama, Y., and Mineyuki, Y. (1994). Inhibition of bundling of microtubules in preprophase bands by a protein kinase inhibitor, 6-dimethylaminopurine (6-DMAP). *Cell Struct. Funct.* **19,** 487.

Nogami, A., Suzaki, T., Shigenaka, Y., Nagahama, Y., and Mineyuki, Y. (1996). Effects of cycloheximide on preprophase bands and prophase spindles in onion (*Allium cepa* L.) root tip cells. *Protoplasma* **192,** 109–121.

O'Brien, T. P. (1983). The preprophase band of microtubules: Does it block cleavage? *Cytobios* **37,** 101–105.

Ohta, K., Toriyama, M., Hamaguchi, Y., Maekawa, S., Endo, S., and Sakai, H. (1988). Characterization of monoclonal antibodies against the mitotic apparatus-associated 51-kDa protein and the effect of the microinjection on cell division in sand dollar eggs. *Protoplasma Suppl.* **1,** 113–121.

Ohta, K., Miyazaki, M., Murofushi, H., Hosoda, S., Endo, S., and Sakai, H. (1990). The mitotic apparatus-associated 51-kDa protein from sea urchin eggs is a GTP-binding protein and is immunologically related to yeast polypeptide elongation factor 1. *J. Biol. Chem.* **265,** 3240–3247.

Okushima, C., Yokota, E., and Mineyuki, Y. (1996). Analysis of cdc2 homologs in PPBs using an antibody against a cdc2 peptide containing phosphorylation site, T14 and Y15. *Cell Struct. Funct.* **21,** 609.

Olszewska, M. J., Marciniak, K., and Kuran, H. (1990). The timing of synthesis of proteins required for mitotic spindle and phragmoplast in partially synchronized root meristems of *Vicia faba* L. *Eur. J. Cell Biol.* **53,** 89–92.

Ôta, T. (1961). The role of cytoplasm in cytokinesis of plant cells. *Cytologia* **26,** 428–447.

Packard, M. J., and Stack, S. M. (1976). The preprophase band: Possible involvement in the formation of the cell wall. *J. Cell Sci.* **22,** 403–411.

Palevitz, B. A. (1980). Comparative effects of phalloidin and cytochalasin B on motility and morphogenesis in *Allium. Can. J. Bot.* **58,** 773–785.

Palevitz, B. A. (1986). Division plane determination in guard mother cells of *Allium:* Video time-lapse analysis of nuclear movements and phragmoplast rotation in the cortex. *Dev. Biol.* **117,** 644–654.

Palevitz, B. A. (1987). Actin in the preprophase band of *Allium cepa. J. Cell Biol.* **104,** 1515–1519.

Palevitz, B. A. (1988). Cytochalasin-induced reorganization of actin in *Allium* root cells. *Cell Motil. Cytoskel.* **9,** 283–298.

Palevitz, B. A. (1991). Potential significance of microtubule rearrangement, translocation and reutilization in plant cells. *In* "The Cytoskeletal Basis of Plant Growth and Form" (C. W. Lloyd, ed.), pp. 45–55. Academic Press, London.

Palevitz, B. A., and Cresti, M. (1989). Cytoskeletal changes during generative cell division and sperm formation in *Tradescantia virginiana. Protoplasma* **150,** 54–71.

Palevitz, B. A., and Hepler, P. K. (1974a). The control of the plane of division during stomatal differentiation in *Allium.* I. Spindle reorientation. *Chromosoma* **46,** 297–326.

Palevitz, B. A., and Hepler, P. K. (1974b). The control of the plane of division during stomatal differentiation in *Allium.* II. Drug studies. *Chromosoma* **46,** 327–341.

Panteris, E., Galatis, B., and Apostolakos, P. (1991). Patterns of cortical and perinuclear microtubule organization in meristematic root cells of *Adiantum capillus-veneris. Protoplasma* **165,** 173–188.

Panteris, E., Apostolakos, P., and Galatis, B. (1992). The organization of F-actin in root tip cells of *Adiantum capillus-veneris* throughout the cell cycle. A double label fluorescence microscopy study. *Protoplasma* **170,** 128–137.

Panteris, E., Apostolakos, P., and Galatis, B. (1995). The effect of taxol on *Triticum* preprophase root cells: Preprophase microtubule band organization seems to depend on new microtubule assembly. *Protoplasma* **186,** 72–78.

Parke, J., Miller, C. C. J., and Amderton, B. J. (1986). Higher plant myosin heavy-chain identified using a monocronal antibody. *Eur. J. Cell Biol.* **41,** 9–13.

Parke, J. M., Miller, C. C. J., Cowell, I., Dodson, A., Dowing, A., Downes, M., Duckett, J. G., and Anderton, B. J. (1987). Monoclonal antibodies against plant proteins recognize animal intermediate filaments. *Cell Motil. Cytoskel.* **8,** 312–323.

Pickett-Heaps, J. D. (1969a). Preprophase microtubule bands in some abnormal mitotic cells of wheat. *J. Cell Sci.* **4,** 397–420.

Pickett-Heaps, J. D. (1969b). Preprophase microtubules and stomatal differentiation in *Commelina cyanea. Aust. J. Biol. Sci.* **22,** 375–391.

Pickett-Heaps, J. D. (1969c). Preprophase microtubules and stomatal differentiation; Some effects of centrifugation on symmetrical and asymmetrical cell division. *J. Ultrastruct. Res.* **27,** 24–44.

Pickett-Heaps, J. D. (1974). Plant microtubules. *In* "Dynamic Aspects of Plant Ultrastructure" (A. W. Roberts, ed.), pp. 219–255. McGraw-Hill, London.

Pickett-Heaps, J. D., and Northcote, D. H. (1966a). Organization of microtubules and endoplasmic reticulum during mitosis and cytokinesis in wheat meristems. *J. Cell Sci.* **1,** 109–120.

Pickett-Heaps, J. D., and Northcote, D. H. (1966b). Cell division in the formation of the stomatal complex of young leaves of wheat. *J. Cell Sci.* **1,** 121–128.

Rappaport, R. (1986a). Mitotic apparatus–surface interaction and cell division. *Int. J. Invertebr. Reprod. Dev.* **9,** 263–277.

Rappaport, R. (1986b). Establishment of the mechanism of cytokinesis in animal cells. *Int. Rev. Cytol.* **105,** 245–281.

Rappaport, R., and Rappaport, B. N. (1985). Surface contractile activity associated with isolated asters in cylindrical sand dollar eggs. *J. Exp. Zool.* **235,** 217–226.

Roberts, K., Burgess, J., Roberts, I. N., and Linstead, P. (1985). Microtubule rearrangements during plant growth and development: An immunofluorescence study. *In* "Botanical Microscopy" (A. W. Roberts, ed.), pp. 105–127. Oxford Univ. Press, New York.

Sack, F. D., and Paolillo, D. J., Jr. (1985). Incomplete cytokinesis in *Funaria* stomata. *Am. J. Bot.* **72,** 1325–1333.

Sawidis, T., Quader, H., Bopp, M., and Schnepf, E. (1991). Presence and absence of the preprophase band of microtubules in moss protonemata: A clue to understanding its function? *Protoplasma* **163,** 156–161.

Schmiedel, G., and Schnepf, E. (1979a). Side branch formation and orientation in the caulonema of the moss, *Funaria hygrometrica:* Normal development and fine structure. *Protoplasma* **100,** 367–383.

Schmiedel, G., and Schnepf, E. (1979b). Side branch formation and orientation in the caulonema of the moss, *Funaria hygrometrica:* Experiments with inhibitors and with centrifugation. *Protoplasma* **101,** 47–59.

Schmiedel, G., Reiss, H., and Schnepf, E. (1981). Association between membranes and microtubules during mitosis and cytokinesis in caulonema tip cells of the moss *Funaria hygrometrica. Protoplasma* **108,** 173–190.

Schmit, A.-C., Vantard, M., De Mey, J., and Lambert, A. M. (1983). Aster-like microtubule centers establish spindle polarity during interphase–mitosis transition in higher plant cells. *Plant Cell Rep.* **2,** 285–288.

Schmit, A. C., Vantard, M., and Lambert, A. M. (1985). Microtubules and F-actin rearrangement during the initiation of mitosis in acentriolar higher plant cells. *In* "Cell Motility: Mechanism and Regulation" (H. Ishikawa, S. Hatano, and H. Sato, eds.), pp. 415–433. Univ. of Tokyo Press, Tokyo.

Schnepf, E. (1973). Mikrotubulus-Anordnung und -Umordnung, Wandbildung und Zellmorphogenese in jungen *Sphagnum*-Blättchen. *Protoplasma* **78,** 145–173.

Schnepf, E. (1984). Pre- and postmitotic reorientation of microtubule arrays in young *Sphagnum* leaflets: Transitional stages and initiation sites. *Protoplasma* **120,** 100–112.

Schnepf, E., and von Traitteur, R. (1973). Über die traumatotaktische Bewegung der Zellkerne in *Tradescantia*-Blättern. *Z. Pflanzenphysiol.* **69,** 181–184.

Schraudolf, H. (1993). Cytoskeleton elements in spermatocytes and spermatozoids of schizaceous ferns. *Crypt. Bot.* **3,** 133–138.

Shimamura, M., Mineyuki, Y., and Deguchi, H. (1998). Diversity and evolution of cell division in lower land plants. *Proc. Jpn. Soc. Plant Taxon.* **14,** 13–29.

Simmonds, D. H. (1986). Prophase bands of microtubules occur in protoplast cultures of *Vicia hajastana* Grossh. *Planta* **167,** 469–472.

Simmonds, D. H., and Setterfield, G. (1986). Aberrant microtubule organization can result in genetic abnormalities in protoplast cultures of *Vicia hajastana* Grossh. *Planta* **167,** 460–468.

Simmonds, D., Setterfield, G., and Brown, D. L. (1983). Organization of microtubules in dividing and elongating cells of *Vicia hajastana* Grossh in suspension culture. *Eur. J. Cell Biol.* **32,** 59–66.

Singh, A. P. (1977). The subcellular organization of stomatal initials in sugarcane leaves: The guard and subsidiary mother cells. *Can. J. Bot.* **55,** 2801–2809.

Singh, A. P., Shaw, M., and Hollins, G. (1977). Preprophase bands of microtubules in developing stomatal complexes of sugarcane. *Cytologia* **42,** 611–620.

Sinnott, E. W. (1960). "Plant Morphogenesis." McGraw-Hill, New York.

Sinnott, E. W., and Bloch, R. (1940). Cytoplasmic behavior during division of vacuolate plant cells. *Proc. Natl. Acad. Sci. USA* **26,** 223–227.

Sinnott, E. W., and Bloch, R. (1941). Division in vacuolate plant cells. *Am. J. Bot.* **28,** 225–232.

Smirnova, E. A., and Bajer, A. S. (1994). Microtubule organizing centers and reorganization of the interphase cytoskeleton and the mitotic spindle in higher plant *Haemanthus. Cell Motil. Cytoskel.* **27,** 219–233.

Smith, L. G., Hake, S., and Sylvester, A. W. (1996). The *tangled-1* mutation alters cell division orientations throughout maize leaf development without altering leaf shape. *Development* **122,** 481–489.

Sonobe, S. (1990). ATP-dependent depolymerization of cortical microtubules by an extract in tobacco BY-2 cells. *Plant Cell Physiol.* **31,** 1147–1153.

Srivastava, L. M., and Singh, A. P. (1972). Stomatal structure in corn leaves. *J. Ultrastruct. Res.* **39,** 345–363.

Staiger, C. J., and Cande, W. Z. (1990). Microtubule distribution in *dv,* a maize meiotic mutant defective in the prophase to metaphase transition. *Dev. Biol.* **138,** 231–242.

Staiger, C. J., and Cande, W. Z. (1992). *Ameiotic,* a gene that controls meiotic chromosome and cytoskeletal behavior in maize. *Dev. Biol.* **154,** 226–230.

Tautorus, T. E., Wang, H., Fowke, L. C., and Dunstan, D. I. (1992). Microtubule pattern and the occurrence of pre-prophase bands in embryogenic cultures of black spruce (*Picea mariana* Mill.) and non-embryogenic cultures of jack pine (*Pinus banksiana* Lamb.). *Plant Cell Rep.* **11,** 419–423.

Terasaka, O., and Niitsu, T. (1989). Peculiar spindle configuration in the pollen tube revealed by the anti-tubulin immunofluorescence method. *Bot. Mag.* **102,** 143–147.

Terasaka, O., and Niitsu, T. (1990). Unequal cell division and chromatin differentiation in pollen grain cells. II. Microtubule dynamics associated with the unequal cell division. *Bot. Mag.* **103,** 133–142.

Tiwari, S. C., Wick, S. M., Williamson, R. E., and Gunning, B. E. S. (1985). Cytoskeleton and interaction of cellular function in cells of higher plants. *J. Cell Biol.* **99,** 63s–69s.

Traas, J. A., Doonan, J. H., Rawlins, D. J., Shaw, P. J., Watts, J., and Lloyd, C. W. (1987). An actin network is present in the cytoplasm throughout the cell cycle of carrot cells and associates with the dividing nucleus. *J. Cell Biol.* **105,** 387–395.

Traas, J. A., Renaudin, J. P., and Teyssendier de la Serve, B. (1990). Changes in microtubular organization mark the transition to organized growth during organogenesis in *Petunia hybrida. Plant Sci.* **68,** 249–256.

Traas, J. A., Beven, A. F., Doonan, J. H., and Shaw, P. J. (1992). Cell cycle-dependent changes in labelling of specific phosphoproteins by the monoclonal antibody MPM-2 in plant cells. *Plant J.* **2,** 723–732.

Traas, J., Bellini, C., Nacry, P., Kronenberger, J., Bouchez, D., and Caboche, M. (1995). Normal differentiation patterns in plants lacking microtubular preprophase bands. *Nature* **375,** 676–677.

Utrilla, L., and De la Torre, C. (1991). Loss of microtubular orientation and impaired development of prophase bands upon inhibition of RNA synthesis in root meristem cells. *Plant Cell Rep.* **9,** 492–495.

Utrilla, L., Giménez-Abián, M. I., and De la Torre, C. (1993). Timing of the phase of the microtubule cycles involved in cytoplasmic and nuclear divisions in cells of undisturbed onion root meristems. *Biol. Cell* **78,** 235–241.

Valster, A. H., and Hepler, P. K. (1997). Caffeine inhibition of cytokinesis: Effect on the phragmoplast cytoskeleton in living *Tradescantia* stamen hair cells. *Protoplasma* **196,** 155–166.

Van Lammeren, A. A. M., Keijzer, C. J., Willemse, M. T. M., and Kieft, H. (1985). Structure and function of the microtubular cytoskeleton during pollen development in *Gasteria verrucosa* (Mill.) H. Duval. *Planta* **165,** 1–11.

Venverloo, C. J. (1990). Regulation of the plane of cell division in vacuolated cells. II. Wound-induced changes. *Protoplasma* **155,** 85–94.

Venverloo, C. J., and Libbenga, K. R. (1987). Regulation of the plane of cell division in vacuolated cells. I. The function of nuclear positioning and phragmosome formation. *J. Plant Physiol.* **131,** 267–284.

Venverloo, C. J., Hovenkamp, P. H., Weeda, A. J., and Libbenga, K. R. (1980). Cell division in *Nautilocalyx* explants. I. Phragmosome, preprophase band and plane of division. *Z. Pflanzenphysiol.* **100,** 161–174.

Vesk, P. A., Vesk, M., and Gunning, B. E. S. (1996). Field emission scanning electron microscopy of microtubule arrays in higher plant cells. *Protoplasma* **195,** 168–182.

Vogelmann, T. C., Bassel, A. R., and Miller, J. H. (1981). Effects of microtubule-inhibitors on nuclear migration and rhizoid differentiation in germinating fern spores (*Onoclea sensibilis*). *Protoplasma* **109,** 295–316.

Wada, B. (1941). Über die Spindelfigur bei der somatischen Mitose der Prothalliumzellen von *Osmunda japonica* Thunb. *in vivio. Cytologia* **11,** 353–368.

Wada, M., and O'Brien, T. P. (1975). Observations on the structure of the protonema of *Adiantum capillus-veneris* L. undergoing cell division following white-light irradiation. *Planta* **126,** 213–227.

Wada, M., Mineyuki, Y., Kadota, A., and Furuya, M. (1980). The changes of nuclear position and distribution of circumferentially aligned cortical microtubules during the progression of cell cycle in *Adiantum* protonemata. *Bot. Mag.* **93,** 237–245.

Wang, H., Cutler, A. J., and Fowke, L. C. (1989a). Preprophase bands in cultured multinucleate soybean protoplasts. *Protoplasma* **150,** 110–116.

Wang, H., Cutler, A. J., and Fowke, L. C. (1989b). High frequencies of preprophase bands in soybean protoplast cultures. *J. Cell Sci.* **92,** 575–580.

Wang, H., Cutler, A. J., and Fowke, L. C. (1991). DNA replication and the development of preprophase bands in soybean protoplast cultures. *Physiol. Plant.* **82,** 150–156.

Webb, M. C., and Gunning, B. E. S. (1991). The microtubular cytoskeleton during development of the zygote, proembryo and free-nuclear endosperm in *Arabidopsis thaliana* (L.) Heynh. *Planta* **184,** 187–195.

Wick, S. M. (1985). The higher plant mitotic apparatus: Redistribution of microtubules, calmodulin and microtubule initiation material during its establishment. *Cytobios* **43,** 285–294.

Wick, S. M. (1991). The preprophase band. *In* "The Cytoskeletal Basis of Plant Growth and Form" (C. W. Lloyd, ed.), pp. 231–244. Academic Press, London.

Wick, S. M., and Duniec, J. (1983). Immunofluorescence microscopy of tubulin and microtubule arrays in plant cells. I. Pre-prophase band development and concomitant appearance of nuclear envelope-associated tubulin. *J. Cell Biol.* **97,** 235–243.

Wick, S. M., and Duniec, J. (1984). Immunofluorescence microscopy of tubulin and microtubule arrays in plant cells. II. Transition between the pre-prophase band and the mitotic spindle. *Protoplasma* **122,** 45–55.

Wick, S. M., Seagull, R. W., Osborn, M., Weber, K., and Gunning, B. E. S. (1981). Immunofluorescence microscopy of organized microtubule arrays in structurally stabilized meristematic plant cells. *J. Cell Biol.* **89,** 685–690.

Wick, S. M., Cho, S.-O., and Mundelius, A. R. (1989). Microtubule deployment within plant tissues: Fluorescence studies of sheets of intact mesophyll and epidermal cells. *Cell Biol. Int. Rep.* **13,** 95–106.

Willemse, M. T. M., and Van Lammeren, A. A. M. (1988). Structure and function of the microtubular cytoskeleton during megasporogenesis and embryo sac development in *Gasteria verrucosa* (Mill.) H. Duval. *Sex. Plant Reprod.* **1,** 74–82.

Wilms, F. H. A., and Derksen, J. (1988). Reorganization of cortical microtubules during cell differentiation in tobacco explants. *Protoplasma* **146,** 127–132.

Woo, Y. M., and Wick, S. M. (1995). Effects of Benlate 50 DF on microtubules of cucumber root tip cells and on growth of cucumber seedlings. *Am. J. Bot.* **82,** 496–503.

Young, T., Hyams, J. S., and Lloyd, C. W. (1994). Increased cell cycle-dependent staining of plant cells by the antibody MPM-2 correlates with preprophase band formation. *Plant J.* **5,** 279–284.

Zhang, D., Wadsworth, P., and Hepler, P. K. (1990). Microtubule dynamics in living dividing plant cells. Confocal imaging of microinjected fluorescent brain tublin. *Proc. Natl. Acad. Sci. USA* **87,** 8820–8824.

Zhang, H. Q., Li, Y. Q., Kuraś, M., Bednara, J., and Cresti, M. (1996). Influence of hydroxyurea on cell divisions and microtubular cytoskeleton in *Allium cepa* root meristem. *Acta Soc. Bot. Pol.* **65,** 179–185.

Zhang, K., Tsukitani, Y., and John, P. C. L. (1992). Mitotic arrest in tobacco caused by the phosphoprotein phosphatase inhibitor okadaic acid. *Plant Cell Physiol.* **33,** 677–688.

Molecular Characteristics of the Centrosome

Søren S. L. Andersen
Department of Molecular Biology, Princeton University, Princeton, New Jersey 08540-1014

As an organizer of the microtubule cytoskeleton in animals, the centrosome has an important function. From the early light microscopic observation of the centrosome to examination by electron microscopy, the centrosome field is now in an era of molecular identification and precise functional analyses. Tables compiling centrosomal proteins and reviews on the centrosome are presented here and demonstrate how active the field is. However, despite this intense research activity, many classical questions are still unanswered. These include those regarding the precise function of centrioles, the mechanism of centrosome duplication and assembly, the origin of the centrosome, and the regulation and mechanism of the centrosomal microtubule nucleation activity. Fortunately, these questions are becoming elucidated based on experimental data discussed here. Given the fact that the centrosome is primarily a site of microtubule nucleation, special focus is placed on the process of microtubule nucleation and on the regulation of centrosomal microtubule nucleation capacity during the cell cycle and in some tissues.

KEY WORDS: Centrosome, Microtubule-organizing center, Centriole, Assembly, Cell cycle, Microtubule, Mitotic spindle, Spindle pole body, Microtubule nucleation.

I. Introduction

No element of the cell has aroused a wider interest of late than the remarkable body known as the *centrosome,* which is now generally regarded as the especial organ of cell-division and in this sense the *dynamic centre* of the cell—Wilson (1896, p. 36). More than 100 years later this organelle

0074-7696/99 $30.00

still "arouses" as much interest as ever before, as evident by the fact that an international meeting held in 1997 was entirely devoted to the centrosome (Stearns and Winey, 1997).

It is interesting to briefly review the historical development of the centrosome field (Fulton, 1971). Initially, light microscopists observed centrosomes (Fig. 1) until the advent of electron microscopy in the 1950s. Studies during these early phases were mostly descriptive, and the lack of techniques to complete the analysis led to a field characterized by much philosophical and mystical thinking. In particular, the replication process of this organelle was a matter of speculation. Ever since its discovery, it was described as an "autonomous organelle" which implies that it is capable of self-reproduction. The ideas regarding centrosomal self-replication were fueled by the discovery of the DNA double helix, and evidence for the direct involvement of either DNA or RNA as templates during centriole duplication has often been reported (Hall *et al.,* 1989; Heidemann *et al.,* 1977; Huang, 1990; Johnson and Rosenbaum, 1990). However, careful stud-

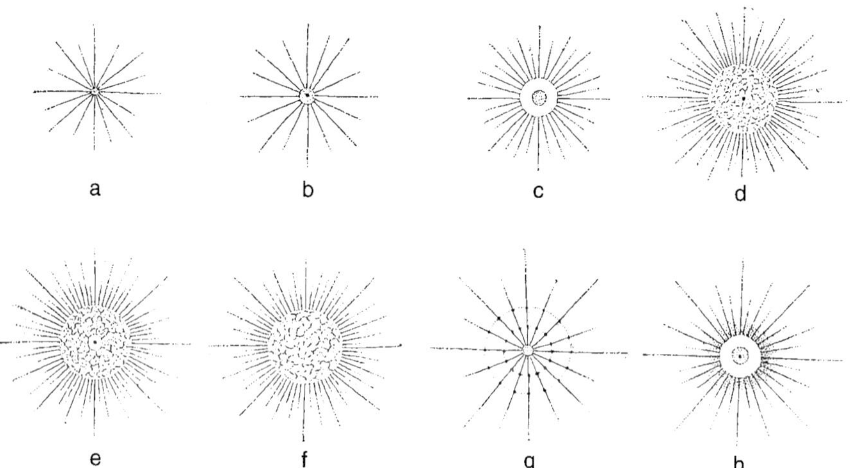

FIG. 1 Diagrams illustrating various descriptions of the centrosome and centrosphere. (a) Simplest type; only a minute centrosome at the focus of the rays (sperm-aster in many forms). (b) Rays proceeding directly from a centrosome of considerable size within which a central granule resides. (c) Rays proceeding from a clear centrosphere, enclosing a centrosome like that in b but with no central granule. (d) An extremely minute centrosome lying in the middle of a large reticulated centrosphere. (e) Like that in d, but with a small spherical body surrounding the centrosome. (f) No centrosome as distinguished from the reticulated centrosphere. (g) The centrosome contains a central granule or centriole (cf. b); outside this is a clear zone (medullary zone of Van Beneden), and outside this is a vaguely defined granular zone, probably corresponding to Van Beneden's cortical zone. (h) The same as g, only according to Boveri (reproduced from Wilson, 1896).

ies on somatic cells have shown that the pericentriolar matrix is practically devoid of nucleic acids (Dirksen, 1991). Today, we can with certainty say that centrioles do not contain DNA or RNA required for their duplication (Dirksen, 1991).

The nomenclature of the centrosome has also been heavily debated through the years (Wilson, 1896, pp. 36 and 232; see also legend to Fig. 1 and Bornens, 1992). The nomenclature adopted here, with which most people would agree, is the following (see also Section II and Figs. 2–6): First, a *centrosome* consists of a pair of centrioles with an associated proteinaceous *pericentriolar matrix* (PCM). Second, a *centriole* is a microscopic structure composed of nine groups of microtubules (MTs) (9+0 configuration). Third, a *basal body* is a centriole bearing a flagellum or cilium (at the transition zone between basal body and flagella two extra MTs originate at the centriole center giving rise to the so-called 9+2 configuration of the flagellum; Fulton, 1971; Gibbons and Grimstone, 1960). Fourth, a MT-organizing center (MTOC) is a structure capable of organizing and nucleating MTs (Pickett-Heaps, 1969; Porter, 1966). In fungi the centrosome is called the "spindle pole body" (SPB) and is the primary MT-nucleating center. All centrosomes are MTOCs, but MTOCs that are not centrosomes are frequently encountered. It is now clear that an MTOC alone can organize the MTs required for cell division (Heald *et al.,* 1997; Lambert, 1993). Thus, a centrosome is not essential for cell division but when present it is dominant over other MTOCs (Heald *et al.,* 1997; González *et al.,* 1998). Therefore, although the centrosome is important it is not "the organ of cell division *par excellence*" (Wilson, 1896, p. 85).

The descriptive phase lasted approximately 100 years until the molecular-biological revolution at the beginning of the 1970s. The past 10 years in particular have been the decade of molecular identification. With increasingly powerful molecular and genetic techniques, more proteins have been described as "centrosomal." The number of centrosomal proteins now form an impressive list (Table I). The latest approach consists of purification of the organelle, separation of associated proteins by gel electrophoresis, and identification by mass spectroscopy via computer-linked gene databases (Wigge *et al.,* 1998). With such techniques new proteins have been identified and added to the already impressive long list of centrosomal proteins (Table I; because this review is not on the yeast SPB only very few SPB components are included in Table I).

The centrosome equivalent in yeast is called the SPB. Because of the ease with which genetics can be performed, yeast has proven very powerful at identifying components and regulatory mechanisms for SPB duplication. It is certain that yeast will continue to yield important information on SPB components and duplication. These discoveries may be applicable to the vertebrate centrosome. At least three aspects of similarities can be compared. First, it is certain that the process of MT nucleation is conserved between the SPB and the centrosome (i.e., the components γ-tubulin, Tub4, Spc97, and Spc98; Table I; see Section III; Stearns and Winey, 1997). Second, the machinery that regulates SPB duplication also shares components with the centrosome duplication machinery just like the components that regulate the cell cycle in yeast and animal cells share components with each other. Third, the biggest differences are probably between the proteins actively duplicating and assembling SPBs and centrioles. Because comparing the SPB and the centrosome is a review in itself, and also because there are recent reviews in this series and elsewhere on the topic (Balczon, 1996; Table II), this review will primarily focus on the vertebrate centrosome.

The most important function of the centrosome is to be a focal dominant point for MT polymerization. In most cells the centrosome is localized at the perinuclear region during interphase. During mitosis the duplicated centrosomes form the poles of the spindle (see Fig. 12). When talking about MTs, it is important to grasp the special polymerization properties of MTs, described by a concept called "dynamic instability." In the dynamic instability model (Erickson and O'Brien, 1992; Kirschner and Mitchison, 1986), MTs polymerize with a certain "growth rate" (v_g) and depolymerize with a certain "shrinkage rate" (v_s). The transition from growth to shrinkage is called a "catastrophe" (McIntosh, 1984) and occurs with a certain frequency (f_{cat}). The opposite event, from shrinkage to growth, is called a "rescue" and occurs with a frequency (f_{res}) (Walker *et al.,* 1988). By modulating these four parameters, and in particular the catastrophe frequency, the cell manages to very efficiently control the amount of MT polymerization across the cell cycle (Verde *et al.,* 1992; Belmont *et al.,* 1990). Recently, the importance of control of MT polymerization via "treadmilling" in addition to dynamics has been demonstrated (Rodionov and Borisy, 1997; Margolis and Wilson, 1998). For example, treadmilling is a way to regulate dynamics of MTs in the cytoplasm after ejection from the centrosome (Rodionov and Borisy, 1997; Keating *et al.,* 1997; see Section V,C). Moreover, spindle MT flux is also dependent on treadmilling (Waters *et al.,* 1996). However, the most powerful mechanism of MT assembly regulation occurs via regulation of the four parameters in the dynamic instability model. This is because regulation of the catastrophe frequency can drastically determine the final average length of a population of MTs (Verde *et al.,* 1992). Treadmilling, MT dynamics, and their regulation will not be discussed further in this context.

Nucleation of MTs is the first step in the formation of MTs and precedes MT assembly. Thus, regulation of nucleation is another important step in MT formation. Although the "nucleation frequency" is not part of the dynamic instability model, this frequency could actually be included as a fifth parameter. Because of its importance for MT dynamics regulation, much of this review will be devoted to the process of MT nucleation.

It is certain that the centrosome will continue to attract biologists' attention for years to come. Thus, one may ask the following fundamental questions: How does a centrosome assemble? What is the function of the centrioles? How are centrioles duplicated? What is the evolutionary origin of centrosomes and centrioles? How are MTs nucleated by the centrosome? How much is dynamic instability influenced and regulated by the centrosome through regulation of MT minus-end dynamics? These are questions that have puzzled cell biologists since the discovery of the centrosomes by Van Beneden and Boveri 110 years ago (Wilson, 1925). Numerous reviews, some of which are listed here (Table II), have through history summarized results regarding these questions.

In the history of reviews (Table II), this review belongs to the molecular biological era of analyses of centrosome function through the combined power of biochemistry, genetics, and advanced microscopy. This review contains the classical sections encompassing definitions and descriptive morphological studies of the centrosomes. Emphasis is put on centrosomal components, possible molecular mechanisms of centriole and centrosome assembly, and MT nucleation. Readers interested in the *Saccharomyces cerevisiae* SPB should consult other reviews (Table II).

II. On the Origin and Low-Resolution Structure of the Vertebrate Centrosome

Regarding the centrosome structure, it is worth noting that the first studies using light microscopy were rather remarkably accurate (Fig. 1). With electron microscopy it became possible to visualize the centrioles, which are at the heart of the centrosome. Centrioles look like barrels surrounded by proteinatious material, the so-called "pericentriolar matrix" (Fig. 2). In recent years we have gained much insight into the constituents, organization, and regulation of this matrix (see Section III). In Section IV, I discuss how this matrix assembles onto the centrioles and concentrate on centriolar structure.

There are numerous excellent reports on centriole structure, and they are in agreement to a certain level of resolution (Alvey, 1985; Chrétien *et al.,* 1997; Fuller *et al.,* 1995; Kenny *et al.,* 1997; Kuriyama and Borisy, 1981; Paintrand *et al.,* 1992; Rieder and Borisy, 1982; Tournier *et al.,* 1991a; Wilsman and Farnum, 1983). Centrioles are mostly observed in pairs and are never more than 1 μm apart (Fig. 2; Tournier *et al.,* 1991a). The length of the mature centrioles is constant but cell type dependent and in the

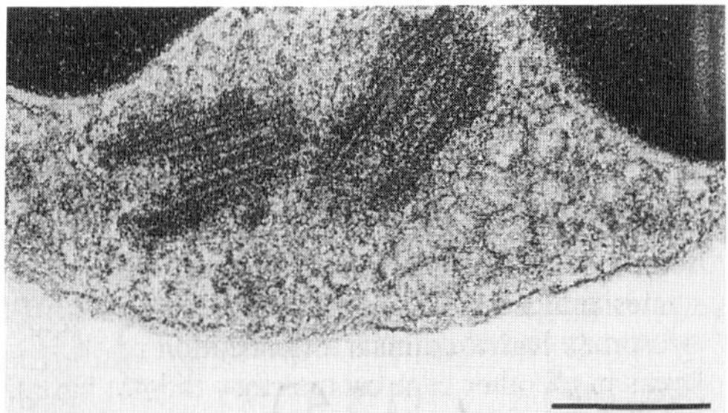

FIG. 2 Electron micrograph showing the structural organization of mouse thymocyte centrioles in centrosomes *in situ.* Scale bar = 0.5 μm. Note the nonlinear arrangement (reproduced from Tournier *et al., The Journal of Cell Biology,* 1991, **113,** 1361–1369 by copyright permission of The Rockefeller University Press).

range of 0.1–0.7 μm (Kenny *et al.,* 1997). The centrioles are linked to each other by filamentous and amorphous material (Figs. 3 and 4). The two centrioles of the pair most frequently adopt a nonlinear conformation with respect to each other. This means that the axis of one centriole is not parallel with the axis of the other centriole (Fig. 2). A centriole is like a barrel, with the sides made of MTs. The diameter of the barrel is 225–335 nm (Chrétien *et al.,* 1997) or 9–13 times the diameter of a single MT. To describe the orientation of the one barrel with respect to the other,

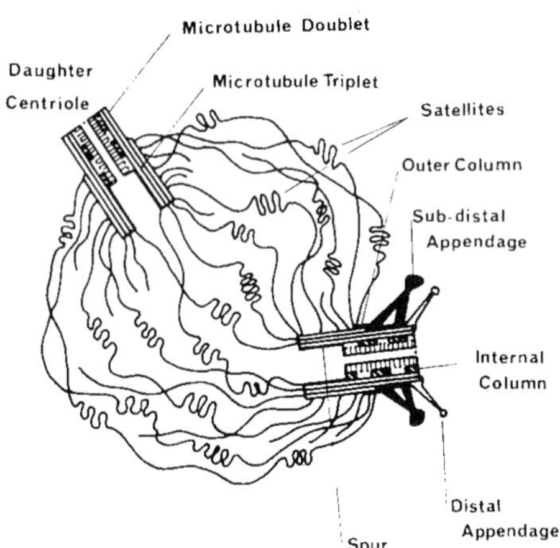

FIG. 3 Schematic representation and nomenclature of centriole organization and architecture. Note the appendages associated with the mother centriole and the filamentous material linking the two centrioles (reproduced from Paintrand *et al.,* 1992, with permission from Academic Press).

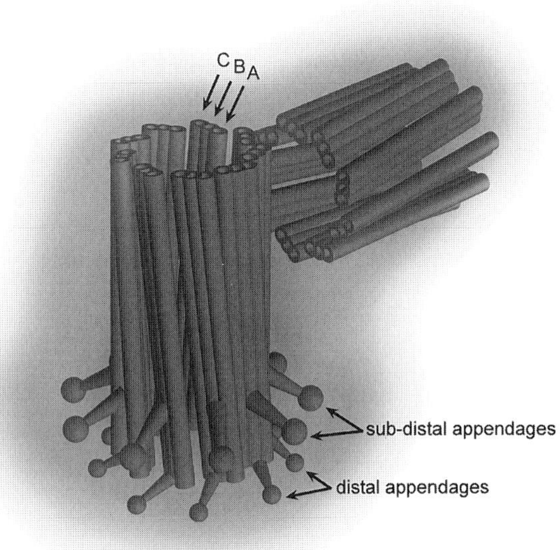

FIG. 4 Schematic simplified three-dimensional representation of the centrioles. Each centriole comprises nine sets of microtubules made of the A–B–C tubules (however, see text and Fig. 5). The mother (mature) centriole carries two sets of appendages at its distal extremity called the distal and subdistal appendages. The daughter (immature) centriole is devoid of appendages. The two centrioles are shown in an orthogonal configuration, with the proximal end of the daughter centriole facing the wall of the mother centriole. The centrioles are surrounded by ill-defined material called the pericentriolar matrix (PCM) (reproduced from Chrétien *et al.*, 1997, with permission from Academic Press).

"proximal" and "distal" have been adopted. Proximal refers to the end of the barrel that is closest to the other centriole (Figs. 2–4). The proximal and distal ends are functionally distinct. Duplication of centrioles commences at the proximal end. The flagellum and cilium originate from the distal end of the centriole barrel. The MTs in the wall of the centriole wall do not form a closed barrel, and the arrangement of the MT in the wall is complicated. At the proximal end the wall consists of nine triplet MTs, as shown in the simplified drawings in Figs. 3 and 4. Toward the distal end this organization changes, and the wall here consists of nine *doublets* of MTs (Fig. 5). To keep track of these MTs a special nomenclature has been adopted. The MT closest to the center of the centriole barrel, an ordinary MT with 13 protofilaments, is called "A." The two others further from the center are called "B" and "C," respectively. There seems to be a continuous change from the triplet A–B–C to the doublet A–B organization from the proximal

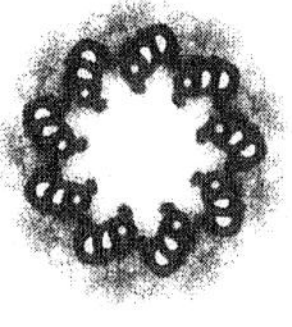

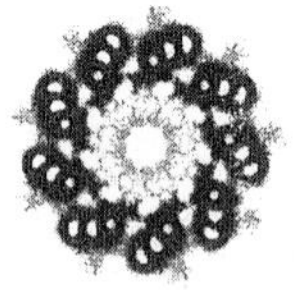

FIG. 5 Electron micrograph showing serial sections of the centriole cylinder from the proximal (top) to the distal (bottom) end. (Top) Note the three A–B–C microtubules as well as their tight connection. (Bottom) Note the doublet structure of the centriole blades and the appendages. Scale bar = 0.2 μm (reproduced from Paintrand *et al.*, 1992, with permission from Academic Press).

to the distal end. This is probably because MTs share walls with each other. In some cases it is observed that the walls of MT C and B are not completely closed and contain 11 and 10 protofilaments, respectively (Tilney *et al.*, 1973). Thus, the centriole cylinder consists of two parts. The proximal part is more like the classical organization of the basal body with nine opened triplets and a distal one which is more reminiscent of the organization of an axoneme with nine doublets. Moreover, the distal end is indeed the end from which a primary cilium and flagellum originate in certain cell types. Thus, the centriole possesses the double organization of basal bodies and cilia.

The triplets/doublets of MTs in the centrioles can be visualized as rafts (Fig. 4). With respect to the tangent to the centriole cylinder, the rafts are tilted like spokes in a wheel. Looking from the distal end, the vector from the A to the B tubule moves clockwise (Fig. 5, top). Interestingly, the angle between the vector and the cylindrical centriole "wall" changes from the proximal to the distal end. At the proximal end this angle is 45°, and it is 20° at the distal end. Thus, there is a twist in the MTs from the proximal to the distal end (Fig. 4). This suggests a strain in the entire centriolar wall. Although not impossible, this seems unlikely. Rather, the change from three to two MTs is the cause of this angular change from the proximal to the distal end.

The next level of organization involves the material associated with the centrioles. This material is often referred to as the "pericentriolar material." The presence of filamentous-like structures has been reported and the name "centrosome matrix" or pericentriolar matrix has for this reason been proposed (Fuller *et al.,* 1995; Paintrand *et al.,* 1992). As discussed in Section IV, this nomenclature seems more up to date than pericentriolar material since a high degree of organization has been observed for some of the proteins present in the matrix (Dictenberg *et al.,* 1998). Moreover, proteinatious appendages are observed with one of the centrioles (the older of the centrioles in the pair). This particular aspect of centrosomes is discussed in detail by Chrétien *et al.* (1997) and Paintrand *et al.* (1992) (see Figs. 3, 4, and 8). Interestingly, it was reported that the distance between the centrioles changed markedly, depending on the presence of Ca^{2+} ions (Paintrand *et al.,* 1992). The calcium-dependent conformational changes may reflect contractile activity in some of the proteins in the centrosome matrix (i.e., centrin) and show that the centriole pair with associated matrix is a dynamic structure.

A particularly puzzling question is the origin of the centrosome. As described in Section IV, this question can be reduced to a question of the origin of the centrioles. Moreover, the question can be addressed both in evolutionary terms and in physiological terms, i.e., having to do with maternal or paternal origin.

With mouse as the only known exception, centrioles are paternally inherited (Schatten, 1994; Simerly *et al.,* 1995). As described in Section IV,B, the pool of maternal proteins complements the paternally delivered sperm centrioles to form a mature centrosome.

If we consider the evolutionary question of centrosome origin, it has been proposed that centrioles were acquired through endosymbiosis much like mitochondria (Sagan, 1967). In Sagan's meticulous theoretical account of the origin of mitosis it is proposed that ingestion of a flagellar-driven motile parasite by an ancestral amoeboid is the origin of eukaryotic flagella. These ancestral ameboids would have gained a selective advantage of an

TABLE I
Centrosomal Proteins

Protein/mass (group)	Source/sequence	Suggested function(s)/comments	Reference
CTR56/350 (2)	Human/no	Ab cross reacts with myosin heavy chain	Bailly *et al.* (1992a)
350 kDa (2)	Dictyostelium/no	Maintenance of the structural integrity of the nucleus-associated body/Ab cross reacts with myosin	Kalt and Schliwa (1996)
MAP1B/325 (1)	Human/yes	Nucleation at the centrosome	Domínguez *et al.* (1994)
Ninein/245 (2)	Mouse/yes	?/Like PCM-1 and centrin, very acidic; extensive coil–coil structure; GTP binding site and EF-hand; PEST sequence	Bouckson Castaing *et al.* (1996)
PCM-1/228 (2)	Human/yes	Centrosome duplication and/or regulation of MT nucleation/not centrosomal during mitosis; very acidic and contains a nucleotide binding site	Balczon *et al.* (1994, 1995)
Pericentrin/220 (1)	Human/yes	MT nucleation/colocalizes with γ-tubulin but has distinct function in pole formation	Dictenberg *et al.* (1998), Doxsey *et al.* (1994)
NuMA/220 (3)	Human *Xenopus*/yes	Pole formation/in a complex with dynein	Merdes *et al.* (1996)
XKLP2/160 (3)	*Xenopus*/yes	Centrosome separation/member of the BimC kinesin family	Boleti *et al.* (1996), Wittmann *et al.* (1998)
LK6/134 (2)	*Drosophila*/yes	Regulation of MTs dynamics/domain ressembling Ca^{2+}; calmodulin-regulated kinases; PEST-rich sequence	Kidd and Raff (1997)

CP190/120 (1) Ident.: Bx63	*Drosophila*/yes	?/In the nucleus during interphase and centrosomal during mitosis; classical N-terminal NLS; central 124-aa MT and centrosomal-binding domain overlap (close to four zinc finger motifs)	Kellogg *et al.* (1989), Oegema *et al.* (1995a, 1997), Whitfield *et al.* (1995)
Basonuclein/120 (1)	Human/yes	Associated with meiotic centrioles, organization/zinc finger motif	Yang *et al.* (1997)
Nuf1p(Spc110p)/110 (1)	*S. cerevisiae*/yes	Essential structural component of the SPB required during spindle formation; spanning the central and inner SPB plaques; central coil–coil and globular ends	Friedman *et al.* (1996)
Spc98p/98 (1) hGCP3 Xgrip109 HsSpc98p	*S. cerevisiae*/yes Human/yes *Xenopus*/yes	γ-TuRC assembly and see Spc97p/mammalian and *Xenopus* homolog	Geissler *et al.* (1996), Murphy *et al.* (1998), Tassin *et al.* (1998), Martin *et al.* (1998)
Spc97p/97 (1) hGCP2 HsSpc97p	*S. cerevisiae*/yes Human/yes	MT nucleation and SPB duplication/ complexed with Spc98p and Tub4p; localized to both inner and outer SPB plaques	Knop *et al.* (1997), Murphy *et al.* (1998), Martin *et al.* (1998)
Cenexin/96 (1)	Human/no	Associated with mature centrioles and perhaps involved in centriole replication	Lange and Gull (1995)
Phosphoinositide reg. subunit/85 (1)	Human/yes	Orienting centrosome during chemotaxis/ direct association with α-, β-, and γ-tubulin stimulated by insulin and PDGF	Kapeller *et al.* (1995)
hsp73/70 (1)	Many/yes	Recovery of centrosome structure and function after heat shock	Brown (1996b)
Tcp1/60 (1)	Mouse, yeast/yes	Chaperone for MT growth of centrosomes/antibodies block centrosomal MT nucleation	Brown (1996a)

(*continues*)

TABLE I (*continued*)

Protein/mass (group)	Source/sequence	Suggested function(s)/comments	Reference
CP60/60 (1)	*Drosophila*/yes	?/Centrosomal during anaphase and telophase; degraded or in the nucleus during interphase; complexed with CP190 and an unknown kinase; MT binding activity	Kellog *et al.* (1995)
Sad1p/58 (1)	*S. cerevisiae*/yes	SPB structure/membrane spanning domain; acidic N terminus	Hagan and Yanagida (1995)
γ-Tubulin/55 (1)	All eucaryotes/yes	MT nucleation and centriole duplication	Fuller *et al.* (1995), Moudjou *et al.* (1996), Pereira and Schiebel (1997)
Tub4p/55 (1)	*S. cerevisiae*/yes	γ-Tubulin homolog	Sobel and Snyder (1995)
Nuf2p/53 (1)	*S. cerevisiae*/yes	SPB separation and duplication/coil–coil structure	Osborne *et al.* (1994)
Tektins/47–55 (1)	Human, molluscan/yes	Link between centrosome and the intermediate filament cytoskeleton, centriole stabilization/centrosomal localization from prophase to anaphase; coil–coil structure	Steffen *et al.* (1994)
Nek2/48 (1)	Human/yes	Regulation of centrosome separation	Fry *et al.* (1998)
pEg2/46 (1)	*Xenopus*/yes	Unknown/Ser/Thr kinase; member of a growing family of protein kinases	Roghi *et al.* (1998)
Actin-RPV or centractin/43 (1)	Human/yes	Unclear: link centrosome to the nucleus; structural; couple dynein motors to vesicles/50% homology with actin; part of dynactin complex	Clark and Meyer (1992), Merdes and Cleveland (1997)

Cdc2/34 (1)	Human/yes	Centrosome architecture/localized to the centrosome throughout the cell cycle	Bailly *et al.* (1989), Pockwinse *et al.* (1997)
PPX/35 (2)	Human/yes	Regulation of the level of centrosomal protein phosphorylation/PP2A homolog	Brewis *et al.* (1993)
CycB1/50 (2)	Human/yes	Regulation of cdc2 kinase and centrosomal protein phosphorylation/centrosomal up to metaphase of each cell cycle	Bailly *et al.* (1992b), Buendia *et al.* (1992)
CycA/50 (2)	Human/yes	Regulation of cdc2 kinase and centrosomal protein phosphorylation/centrosomal from prophase to metaphase	Bailly *et al.* (1992b), Buendia *et al.* (1992)
cAMP RII/49 (1)	Human/yes	?	Keryer *et al.* (1993, 1995)
p23/23 (3)	Chicken/yes	?/Ca^{2+} binding like centrin but no homology	Zhu *et al.* (1995)
Centrin or Cdc31p or caltractin/20 (1)	Human, plant, yeast/yes	SPB duplication and clipping of centrosomal MTs/member of the calmodulun superfamily; has four Ca^{2+}-binding EF hands; 70% homology between algae and human form	Errabolu *et al.* (1994), Levy *et al.* (1996), Paoletti *et al.* (1996)
PCM-1/PCM-2/PCM-3 (2)	Human/only for PCM-1	Regulation of MT nucleation, centriole assembly/proteins recognized by autoimmune sera	Balczon *et al.* (1994, 1995), Balczon and West (1991)
Katanin (3)/81 and 61	Sea urchin/yes	Severing of centrosomal MTs/centrosomal association is MT dependent; ATPase activity; heterooligomeric protein	McNally *et al.* (1996), Hartman *et al.* (1998)

TABLE II
Reviews on the Centrosome

Review	Subjects covered	Reference
	Reviews on specific centrosomal components	
"Centrosome–Microtubule Nucleation"	Mechanism of nucleation of microtubules by γ-tubulin	Pereira and Schiebel (1997)
"Cytoskeleton: Anatomy of an Organizing Center"	Components localized to the SPB	Marschall and Stearns (1997)
"Centrosomes and Microtubules: Wedded with a Ring"	The γ-tubulin ring complex (γ-TuRC)	Raff (1996)
"γ-Tubulin and Microtubule Organization in Plants"	γ-Tubulin in plants	Joshi and Palevitz (1996)
"In Search of a Function for Centrins"	Possible conserved function of centrins between species	Schiebel and Bornens (1995)
"Centrin, Centrosomes, and Mitotic Spindle Poles"	Role of centrin in centrosome dynamics	Salisbury (1995)
	Comprehensive reviews on the centrosome	
"Centrosomes and Microtubule Organization during *Drosophila* Development"	Centrosome and centriolar structure and function during development, role of centrosomes in spindle assembly	González *et al.* (1998)
Structure and Function of the Centriole in Animal Cells: Progress and Questions"	See ref.	Lange and Gull (1996)
"The Centrosome in Animal Cells and Its Functional Homologs in Plant and Yeast Cells"	See ref.	Balczon (1996)
"Regulation of Centrosome Function during Mitosis"	See ref.; in particular on *in vitro* reconstitution of centrosome nucleation capacity	Buendia and Karsenti (1995)

"The Centrosome and Cellular Organization"	Centrosomal components, assembly, duplication, and role for cellular organization	Kellogg *et al.* (1994)
"The Centrosome and Its Mode of Inheritance: The Reduction of the Centrosome during Gametogenesis and Its Restoration during Fertilization"	Compilation of the evidence that the centrosome is paternally inherited during fertilization and of organisms devoid of centrosomes	Schatten (1994)
"Microtubules, Centrosomes and Intermediate Filaments in Directed Cell Movement"	Role of the centrosome in cell locomotion	Schliwa and Höner (1993)
"Unravelling the Tangled Web at the Microtubule-Organizing Center"	Review on centrosomal components notably from yeast	Rose *et al.* (1993)
The Centrosome	Book on centrosome definition, components, and structure from different organisms	Kalnins (1992)
"Spindle Poles in Higher Plant Mitosis"	See ref.	Smirnova and Bajer (1992)
"Centriole and Basal Body Formation during Ciliogenesis Revisited"	Compilation of literature on basal body formation and relevance for centriole formation	Dirksen (1991)
"The Chromosome Cycle and the Centrosome Cycle in the Mitotic Cycle"	Historical review on centrosome research	Mazia (1987)
"Microtubule Organizing Centers"	Examination of the centrosome as MTOC	Brinkley (1985)
"Centrosomes and Mitotic Poles"	Historical review on speculations on centrosome function and assembly	Mazia (1984)
"The Centrosome as an Organizer of the Cytoskeleton"	Review on the status of centrosomal research at the molecular level	McIntosh (1983)

ization signal (NLS). If the NLS is deleted, CP190 associates with the centrosome throughout the cell cycle. The region required for centrosomal localization is centrally located, 124 amino acids long, and overlaps the region required for MT association but has no homology with other MAP-binding domains. Moreover, this sequence contains one of a total of four zinc finger motifs (Whitfield *et al.,* 1995).

CP60 is especially abundant on centrosomes during anaphase and telophase and is apparently degraded at the onset of interphase up to nuclear cycle 12. Then, CP60 is found localized to the nucleus in interphase and to the centrosome during mitosis. CP60 has no homology with other known proteins but contains six cdk consensus phosphorylation sites and a sequence similar to the "destruction box" in cyclins (Kellogg *et al.,* 1995). It is unknown how CP60 localizes to the centrosome but it may do so indirectly through binding to CP190 (Oegema *et al.,* 1997). The function of CP60 and CP190 remain elusive (Moritz *et al.,* 1998).

PCM-1 was identified using serum from patients with systemic sclerosis and Raynaud's phenomenon (Balczon and West, 1991). With these sera, three proteins of 39, 185, and 220 kDa, called PCM-3, PCM-2, and PCM-1 respectively, were identified. Balczon *et al.* (1994) described the cloning of the 228-kDa PCM-1. PCM-1 has no homology to other known proteins but is very acidic and contains a nucleotide consensus ATP/GTP-binding motif. PCM-1 is associated with the centrosome throughout the cell cycle but there is a decrease in staining intensity at the onset of mitosis. This led to the hypothesis that PCM-1 may be an inhibitor of MT nucleation since release of the protein at metaphase could explain the increased MT nucleation at the onset of metaphase. Considering the acidic p*I* of PCM-1, it is possible that it rejects acidic tubulin subunits from the centrosome and thereby inhibits MT nucleation. However, perhaps it is not involved in MT nucleation but rather in centriole duplication (Balczon *et al.,* 1995).

Centrin is a 20-kDa filamentous protein member of the EF-hand Ca^{2+}-binding superfamily with four helix–loop–helix motifs. It was first discovered in the flagella apparatus of unicellular green algae and is concentrated on centrioles. Like γ-tubulin, only a fraction of the total cellular pool of centrin is centrosome associated (Paoletti *et al.,* 1996). Recent reviews summarize knowledge about centrin from different organisms and its possible functional role (Baron and Salisbury, 1992; Salisbury, 1995; Schiebel and Bornens, 1995). The precise function of centrin is unknown but the yeast homolog cdc31p is essential and involved in SPB duplication. Interestingly, centrin contracts upon exposure to Ca^{2+} and it has been suggested to be important for centriole duplication and excision of the flagella associated with the sperm centriole (Sanders and Salisbury, 1994). Centrin may be responsible for the Ca^{2+}-dependent changes in centriolar structure observed by Paintrand *et al.* (1992) (Fig. 3).

Cenexin is a 96-kDa centriole-associated protein (Lange and Gull, 1995). Cenexin associates with the maturing mother centriole at the G_2-M phase transition (Fig. 8). Cenexin has not been cloned but the data on the association to the new mother centriole suggest that this protein may be directly involved in regulating centriole duplication. Note that the possibility that cenexin represents a posttranslational modification cannot be formally excluded. A recent review describes details on centriole duplication with respect to cenexin (Lange and Gull, 1996).

Tektins were discovered by studies on flagellar and ciliary doublet MTs (Chen *et al.,* 1993; Linck and Stephens, 1987; Steffen and Linck, 1988). Because of the relationship between flagellar MTs and the centriole it was reasonable to assume that the tektins may also be present on centrioles. This has indeed been confirmed biochemically and by immunofluorescence (Steffen *et al.,* 1994; Steffen and Linck, 1988). Tektins form filamentous polymers in the walls of ciliary and flagellar MTs. The central region of the tektins forms a coil–coil rod, consisting of four major α-helical regions that are separated by nonhelical linkers. The tektins have a secondary structure and molecular design similar to that of intermediate filament proteins although the primary sequence homology is low. The function of tektins appears to be stabilization and regulation of the species-specific length of centrioles (Steffen and Linck, 1988; Kenny *et al.,* 1997).

Molecular **chaperones** are proteinaceous structures that are thought to facilitate the folding of proteins *in vivo* (Frydman and Hartl, 1996). Tailless complex polypeptide-1 (TCP-1) and hsp73 have been shown to be integral components of the centrosome and to colocalize with pericentrin (Brown *et al.,* 1996a, b; Nelson and Craig, 1992). However, for both proteins, like centrin, γ-tubulin, and Nek2, only a minor fraction is centrosome associated and the remainder is diffusely localized in the cytoplasm. TCP-1 is a cytosolic chaperonin and exists as a 25S complex with at least eight related subunits (the TRiC complex) and this complex has been shown to chaperone tubulin folding (Tian *et al.,* 1996). hsp73 belongs to the heat shock proteins that are produced in increasing amounts after thermal stress probably to either repair and/or replace proteinaceous components. Apparently, TCP-1 and hsp73 participate in different aspects of the modulation of the structure and function of the centrosome. In the case of TCP-1, antibodies to the protein were capable of blocking the initiation of MT growth off the centrosome. hsp73, on the other hand, facilitates the recovery of centrosomal structure and function after heat shock, indicating that hsp73 may participate in the process of centrosomal assembly following centriole duplication during S phase. The function of chaperones in the centrosome may be to facilitate the movement of proteins into and out of the organelle as well as to catalyze spatial changes in the organization of pericentriolar material during the cell cycle. It seems possible that chaperones would be required to facilitate the nucleation of MTs at the centrosome. Nucleation sites, especially during mitosis, are very closely packed and could inhibit each other's function without some external help in organizing their position and spatial orientation.

One of the first biochemical characterizations of the centrosomes consisted of the observation that the phosphorylation level of centrosomal proteins increases at the onset of mitosis, during which the nucleation capacity is also increased (Tash *et al.,* 1980). The regulation of this change in phosphorylation involves cell cycle-regulated **kinases** and **phosphatases.** Several studies have shown localization to and functional effects of cyclin A and cyclin B on centrosomes (Bailly *et al.,* 1989, 1992b; Buendia *et al.,* 1992; Maldonado Codina and Glover, 1992), and cdc2 kinase was recently described as constitutively associated with the centrosome throughout the cell cycle (Pockwinse *et al.,* 1997). Recently, two new kinases, LK6 and pEg2, were reported to be centrosomally associated (Kidd and Raff, 1997; Roghi *et al.,* 1998). A mammalian type 2A phosphatase PPX (Brewis *et al.,* 1993) has also been localized to the centrosome. Genetically lowering the dosage of a type 2A phosphatase not associated with the centrosome was shown to uncouple the nuclear and centrosomal cycles (Snaith *et al.,* 1996; Tournebize *et al.,* 1997). In summary, both kinases and phosphatases have been localized to the centrosome. How the balance in their activities determines the final level of centrosomal protein phosphorylation remains unknown. How phosphorylation affects the nucleation and duplication capacity of centrosomes has to be investigated further (see Section V,C).

Nek2 is a kinase of the NIMA family associated with the centrosome. Apart from cyclin A-dependent cdc2 kinase (Buendia *et al.,* 1992), it is the only kinase for which a centrosomal function has been convincingly shown (Fry *et al.,* 1998; Fry and Nigg, 1995). Nek2 localizes to the centrosome throughout the cell cycle. When cells were incubated with taxol for prolonged periods, mitotic arrest and formation of multiple asters ensued. These asters are examples of self-assembled MTOCs (see Section IV,B). In such cells Nek2 was associated only with the two MTOCs containing centrioles. Depolymerization of MTs did not affect the localization to centrosomes. Moreover, Nek2 was found to associate with purified centrosomes. Therefore, Nek2 is a truly integral centrosomal protein. Only 10% of total Nek2 is centrosome associated as reported for centrin, γ-tubulin, and chaperones. Overexpression of Nek2 had two effects. Using γ-tubulin as a marker Fry *et al.* (1998) observed that centrosomes had separated into two foci and that the pericentriolar matrix in many cases was dispersed from the centrosomes. However, transfection with a catalytically inactive Nek2 mutant also caused dispersal of centrosomal material but did not trigger any centrosome separation. The authors conclude that centrosome separation depends on kinase activity but that dispersal is brought about by a different mechanism. It is suggested that Nek2 activates centrosome separation at the end of G_2 prior to cdc2-activated centrosome migration by motors such as Eg5 and XKLP2 (Blangy *et al.*, 1995; Boleti *et al.*, 1996; Sawin and Mitchison, 1995; Wittmann *et al.*, 1998; see Fig. 10 and Section IV,C).

B. Proteins Concentrated at the Centrosome

Katanin is a heterodimeric (81 and 60 kDa), ATP-dependent, MT-severing protein from sea urchin eggs (McNally *et al.*, 1996). Microtubule-severing proteins promote the disassembly of MTs by generating internal breaks within a MT (Karsenti, 1993; McNally *et al.*, 1996). It is proposed that Katanin is involved in severing MTs at their minus ends and in this way involved in generating spindle MT flux (McNally *et al.*, 1996; Sawin and Mitchison, 1994; Waters *et al.*, 1996). The localization studies showed that MTs are required to localize Katanin to the centrosome. When cells were preextracted prior to immunofluorescence the staining pattern obtained was a large hollow ball of Katanin around the centrosomal area with γ-tubulin at the interior of the ball. These data show that Katanin is not an integral part of the centrosome and can be assigned as a group 3 member of the centrosomal proteins (however, in some cells, nocodazole-resistant centrosomal association has been observed; F. J. McNally, personal communication). Katanin exemplifies the very dynamic interaction between the pericentriolar material and the surrounding cytoplasm and also the experimental difficulties involved with assigning a protein to the centrosome and grasping the large number of activities localized to and present at centrosomes.

Recent investigations cast new light on the issues of centrosomal localization and MT-severing activity by Katanin (Hartman *et al.*, 1998). The 60-kDa subunit is a new member of the AAA family of ATPases and contains the ATP binding site, the ATPase activity, and the severing activity of Katanin. However, the severing and ATPase activity of the 60-kDa subunit were enhanced twofold when complexed to the 81-kDa subunit, and *in vivo* the 60-kDa subunit is exclusively found in a very tight complex with the 81-kDa subunit. Interestingly, rotary shadowing electron microscopy showed that the 60-kDa subunit and the complex between the 60- and 81-kDa subunits forms ring-like structures. Previous studies showed that the heterooligomeric complex can bind to a tubulin subunit in its ADP form. Moreover, the heterooligomeric complex binds to MTs but not to subtilisin-cleaved MTs (in which the C terminus of tubulin is missing; McNally and Vale, 1993. Microtubule binding characterization of the Baculovirus-expressed 60- and 81-kDa subunits was not performed by Hartman *et al.,* 1998.) Surprisingly, subtilisin-treated MTs still stimulated the ATPase activity without MT severing (McNally and Vale, 1993). Previous studies also indicated that Katanin removes single tubulin subunits from the wall of MTs rather than removing oligomeric tubulin complexes (McNally and Vale, 1993). In summary, it is known at present that the C terminus of tubulin, ATP hydrolysis, and binding to MTs are required for Katanin to sever MTs. More work is required to determine how MTs mechanistically are severed, but the ring-like structure of the complex may be a key.

The 81-kDa subunit contains a C-terminal domain that is required for interaction with the 60-kDa subunit. In addition, the N terminus of the 81-kDa subunit contains six "WD40" repeat motifs, and this domain was sufficient to localize a GFP fusion protein to the centrosome in a MT-independent way. This suggests that the WD40-containing domain targets Katanin to the centrosome. Since previous studies showed that MTs are required to target Katanin to the centrosome (McNally *et al.,* 1996), it is possible that the binding characteristics of the entire complex are somewhat different from those of the WD40–GFP fusion protein or that more proteins are involved.

The homology between the 220-kDa human and *Xenopus* **NuMa** is only 37% (Merdes *et al.,* 1996), but both proteins contain a central 150-kDa predicted coil–coil domain and are localized to the centrosome. NuMA is localized to the nucleus during interphase and in a MT-dependent way to the spindle pole/centrosome region during mitosis. Like Katanin, NuMA is not an integral member of the centrosome but a member of the outermost layer of the pericentriolar matrix. Merdes and Cleveland (1997) and Merdes *et al.* (1996) showed that NuMA forms a complex with the MT minus end-directed cytoplasmic dynein–dynactin motor complex and proposed that dynein drives NuMA to the minus end of MTs where it is involved in focalizing MTs by cross-linking. This localization mechanism is attractive and probably applies to more proteins that are localized to the MTOC/centrosome and that are not centrosomally associated (i.e., Katanin, XKLP2, and Eg5).

In addition to the data on CP190, recent data on **XKLP2** (Boleti *et al.,* 1996; Wittmann *et al.,* 1998) also give some insights on the requirements for localization of a protein to the centrosomal area. XKLP2 is a plus-end-directed motor but accumulates at the minus ends of MTs during mitosis. The minus-end-directed motor dynein is required for this localization. This bears resemblance with the localization mechanism of NuMA. Moreover, in the very C terminal of the motor (the tail) a leucine zipper is essential for localization, and a single point mutation in the leucine zipper is sufficient to prevent localization of the tail. The leucine zipper has been shown to interact with a 100-kDa MT-associated protein (TPX2). The dynein complex together with TPX2 somehow target the tail of XKLP2 to MT minus ends (Wittmann *et al.,* 1998). (It remains to be shown how the full length XKLP2 motor targets.)

IV. High-Resolution Analysis of Centrosome Structure and Assembly

In this section I discuss the basic principles that are at work during centrosome assembly. I divide the analysis in three parts. First, I discuss the role of centrioles for centrosome function. Then I examine how a centrosome may assemble "on top" of the centrioles. Lastly, I discuss cell cycle control of centriole and centrosome duplication.

A. Role of Centrioles for Centrosome Function

In the absence of centrioles, a structure that nucleates MTs is a MTOC (Pickett-Heaps, 1969; Porter, 1966). This definition of the centrosome simplifies the discussion on centrosome assembly since the complication arising from the absence of centrioles from many so-called centrosomes is eliminated.

The functional difference between a centrosome and a MTOC can be appreciated from two elegant studies. In the first, the centrosome was removed by microsurgery (Maniotis and Schliwa, 1991) from somatic vertebrate tissue culture cells. Cells without centrosomes continued to grow but failed to divide and the cell cycle was arrested (Sluder, 1992). However, a functional MTOC developed in the cells devoid of centrosomes. Thus, these cells appear to be dependent on centrosomes for cell division, indicating that the centrosome has functions besides being a MTOC. The second study was performed using sea urchin eggs (Picard *et al.,* 1987). Sea urchins oocytes are arrested in interphase. This is different from most other embryos, such as human or *Xenopus,* which arrest at the second meiotic metaphase (Schatten, 1994). Fertilization of *Xenopus* eggs results in a Ca^{2+} wave that causes cdc2 kinase inactivation and reinitiation of the cell cycle (Sagata *et al.,* 1989). In sea urchin a Ca^{2+} wave can, as in *Xenopus,* resume the cell cycle. Contrary to *Xenopus,* however, a centrosome is sufficient to start the cell cycle (in the absence of Ca^{2+}; Picard *et al.,* 1987). This study indicates that the centrosome in sea urchin either contains or assembles components that are involved in regulating the reinitiation of the cell cycle. The observations by Picard *et al.* complement the observations by Maniotis and Schliwa (1991) and suggest that the centrosome can be involved in the regulation of cell cycle progression and cell division. Other experiments from *Xenopus* show that the centrioles are not required for cell cycle initiation (a Ca^{2+} wave induced by pricking is sufficient) but are required for proper cleavage (Karsenti, 1991).

However, there are numerous examples of cells and eggs that have a cell cycle and that divide without centrosomes. Examples include the early stages of mouse oocyte cleavage, plant cell division, and a mutant *Drosophila* cell line (Debec *et al.,* 1995; Joshi and Palevitz, 1996; Lambert, 1993; Palacios *et al.,* 1993; Schatten, 1994). These exceptions probably reflect alternative ways of solving the "problem" of cell division: Cell division is such an essential process that different pathways must have evolved (Schatten, 1994; González *et al.,* 1998).

What is then the role of the centriole if it is not present in all systems? The short and simple answer is that in systems with centrioles, cell division seems to have become "addicted" to their presence. All vertebrate somatic cells, being rather young from an evolutionary perspective, contain centrioles. It is possible that the centriole developed as an extra regulatory mechanism for vertebrate somatic cell division (Grafen, 1988; Sluder, 1992). However, the function of centrioles and how they organize material for organization around them are unknown. In the words of Lange and Gull (1996, p. 351), "Comparison of the capacity of cells with and without centrioles can be informative, but really the question is: what is the function of centrioles in those cells that possess them?" However, currently this question cannot be answered.

B. Centrosome Assembly

Why is it that we know so little about centrosome assembly, when (perhaps) equally complicated assembly processes, such as ribosome assembly, are well characterized (Nomura and Held, 1974)? This is in part because of the low abundance of centrosomal proteins. One centrosome weighs approximately 10^{-14} g (Bornens *et al.,* 1987), and it is therefore difficult to obtain enough centrosomes to be able to characterize the attached components both morphologically and biochemically. In addition, it is difficult to assess centrosome function both *in vivo* and *in vitro.* For example, *in vitro* experiments show that the centrosome is capable of MT nucleation. However, as discussed in Section IV,A some cells do not divide without a centrosome, although a functional MTOC forms in the absence of centrioles. Centrosomes are clearly more than just MTOCs.

However, MTOCs that lack a centriole pair may provide insight into how centrosomes assemble since they are examples of self-organized MTOCs which in many ways behave like centrosomes. Such acentriolar centrosomes, MTOCs, are found in plants (Joshi and Palevitz, 1996; Lambert, 1993), the early mouse embryo (Palacios *et al.,* 1993), and an acentriolar *Drosophila* cell line (Debec *et al.,* 1995). In addition, they can "artificially" be induced by taxol or DMSO addition to *Xenopus* egg extracts and in other systems (De Brabander *et al.,* 1986; Sawin and Mitchison, 1994; Verde *et al.,* 1991).

I now discuss the formation of an "artificial" MTOC. The formation of MTOCs in DMSO-treated mitotic *Xenopus* egg extracts first involves *nucleation* of MTs followed by *stabilization* and *growth* by MT-associated proteins (MAPs). The MTs are then *sorted* by motors that *read the polarity* of MTs and MTOCs or asters form (Heald *et al.,* 1996; Heald *et al.,* 1997; Hyman and Karsenti, 1996; Karsenti, 1991; Verde *et al.,* 1991). Thus, these asters assemble by an interplay between motors and MTs. The end result is an aster of MTs with uniform polarity, in which the base of the aster morphologically and compositionally resembles a centrosome (Galio *et al.,* 1996; Heald *et al.,* 1997; Merdes *et al.,* 1996; Verde *et al.,* 1991). It is unknown whether these asters can actively nucleate MTs and thus whether they should be called MTOC. This issue, however, is of minor importance in this context because the motor-dependent MT sorting mechanism also is at work in the presence of a centrosome (Fig. 6) where MTs are attached to the centrosome with their minus end. Therefore, the term "artificial" MTOC is used for these structures. Thus, the motors that convey the sorting during artificial MTOC formation are also working when a centrosome is present. One possibility is therefore that duplication and *de novo* formation of centrioles/centrosomes ensues as a result of local high concentration of proteins obtained by transport on MTs to the center of the centrosome/MTOC (Verde *et al.,* 1991). (However, see the discussion on *de novo* centriole formation kinetics later in this section).

What makes a centrosomes different from a DMSO-induced MTOC is especially that it must have some kind of "protein glue" that tethers the MT nucleating material to the centriolar surface. For example, centrosomes can be purified with the MT nucleation material attached (Bornens *et al.,* 1987), whereas dissociation of DMSO-induced MTOCs is irreversible (i.e., Heald *et al.,* 1997). It is unresolved how a centrosome can maintain the MT nucleation material such as γ-tubulin/pericentrin on the centrosome, and currently there are no candidates for such centrosomal protein glue (Section III.A). Mazia and Schatten suggested hypothetical models for such glue. Thus, the template attached to the centriole that allows stable binding of pericentriolar matrix components is proposed to appear either as a sticky ribbon (for γ-tubulin) or as a crystalline mold into which the bulk of the centrosomal proteins fit (Mazia, 1984, 1987). There seems to be some experimental support for this idea (Dictenberg *et al.,* 1998). Schatten (1994) proposed that during the penetration of the egg the sperm could acquire some kind of a receptor system that would allow molecules such as γ-tubulin to stably attach to the centrioles and thus provide the starting point for a centrosome. Favorable to the latter idea is that the *Xenopus* egg cortex is full of γ-tubulin (Gard, 1994). However, since complementation directly in the extract is possible, there does not seem to be a need to invoke a cortex-associated receptor for molecules such as γ-tubulin (Félix *et al.,* 1994; Stearns and Kirschner, 1994).

In trying to imagine how a centrosome assembles on top of a centriole, it is useful to think about layers of proteins that attach to the surface of the centrioles in a defined order (Fig. 6). Central to the scheme is that the egg complements the sperm centrioles with proteinatious components.

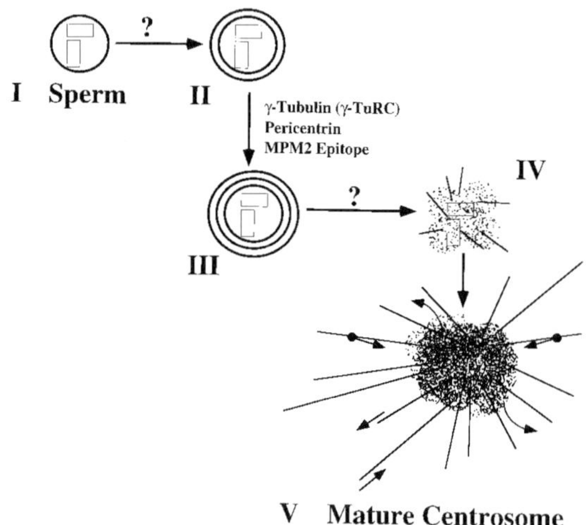

FIG. 6 Schematic representation of the pathyway leading to centrosome assembly in interphase or mitotic *Xenopus* egg extracts (Stearns and Kirschner, 1994; Félix *et al.,* 1994; Buendia *et al.,* 1992). The paternally delivered sperm-associated centrioles have some proteins attached (*I*). Upon entry into the egg some proteins probably immediately associate with the centriole, i.e., during penetration of the egg (transition I–II). During transition II–III, specific proteins attach to the centrosomal "glue" present on the centrioles. These components include the γ-TuRC, pericentrin, and the MPM2 phosphoepitope. At some point MTs start to nucleate on the nascent centrosome. This requires correct orientation and organization of the pericentriolar matrix components (transition III–IV). During transition IV–V, more MTs are nucleated and more pericentriolar material accumulate at the centrosome. Motor proteins drive components to the minus end of MTs localized in the pericentriolar matrix (circle with arrow), proteins have a dynamic attachment to the centrosomal area (curved arrows indicate a certain off-rate), and MTs growing from the centrosome are dynamic (arrows on MTs indicating growth and shortening events). Note: The activity of minus end-directed motors shown here seems to be considerably increased during mitosis and may not exist during interphase.

Imagine an assembly scheme as presented in Fig. 6: (I) Depict paternal proteins associated with the centrioles. At the time of delivery to the egg some proteins immediately attach (transition I–II). After and during transition II–III proteins required for MT nucleation are retrieved from the maternal pool. During transition III–IV the three-dimensional lattice required for MT nucleation is constructed (Dictenberg *et al.*, 1998). During transition IV–V, proteins are transported by MT-associated motors toward the centrosome (as in DMSO-induced MTOCs), terminating the assembly process. Arrows on MTs indicate movement by motors, and otherwise loss of proteins due to their intrinsic on/off rate.

The sequence of association of pericentrin and γ-tubulin has been characterized in detail and can be fitted to the scheme presented in Fig. 6 (Archer and Solomon, 1994; Dictenberg *et al.*, 1998; Doxsey *et al.*, 1994; Félix *et al.*, 1994; Stearns and Kirschner, 1994). Thus, MTs do not nucleate off *Xenopus* sperm nuclei *in vitro* prior to incubation in the *Xenopus* egg extract (Fig. 6, I). After incubation in the extract the sperm nuclei become competent for MT nucleation and this competence requires the recruitment of γ-tubulin (Archer and Solomon, 1994; Félix *et al.*, 1994; Stearns and Kirschner, 1994). The accumulation of γ-tubulin does not require MTs as shown by complementation in the absence of MTs. However, the recruitment of γ-tubulin requires ATP and occurs concomitantly with the recruitment of the phosphoepitope called MPM2 (Davis *et al.*, 1983). Whether the requirement for ATP is due to an energy-dependent step in the recruitment of γ-tubulin to the sperm or related to the simultaneous addition of the MPM2 epitope is unknown (Fig. 6, II and III). Interestingly, γ-tubulin is in a 25S complex (γ-TuRC) indicating that other proteins may be required for the attachment of γ-tubulin to the centrosome (Stearns and Kirschner, 1994; Zheng *et al.*, 1995; Section III.A).

In contrast to γ-tubulin, the proteins pericentrin, centrin, and α-tubulin are all present on the sperm prior to incubation in the extract. However, incubation of sperm heads in extracts containing anti-pericentrin antibodies causes an inhibition of MT nucleation (Doxsey *et al.*, 1994). Additional pericentrin is recruited in the extract that somehow is involved in forming a proper environment for MT nucleation. The recruitment of γ-tubulin to the sperm heads is not abolished by anti-pericentrin antibodies (it is unknown whether γ-tubulin is required to recruit pericentrin to the sperm head in the extract). This indicates that both γ-tubulin and pericentrin are required for MTs to be able to nucleate off centrosomes. Pericentrin may be required for the γ-TuRC to bind to the centrosome since purified γ-TuRC does not bind to salt-stripped (pericentrin-free) centrosomes (Dictenberg *et al.*, 1998; Moritz *et al.*, 1998; Section III.A).

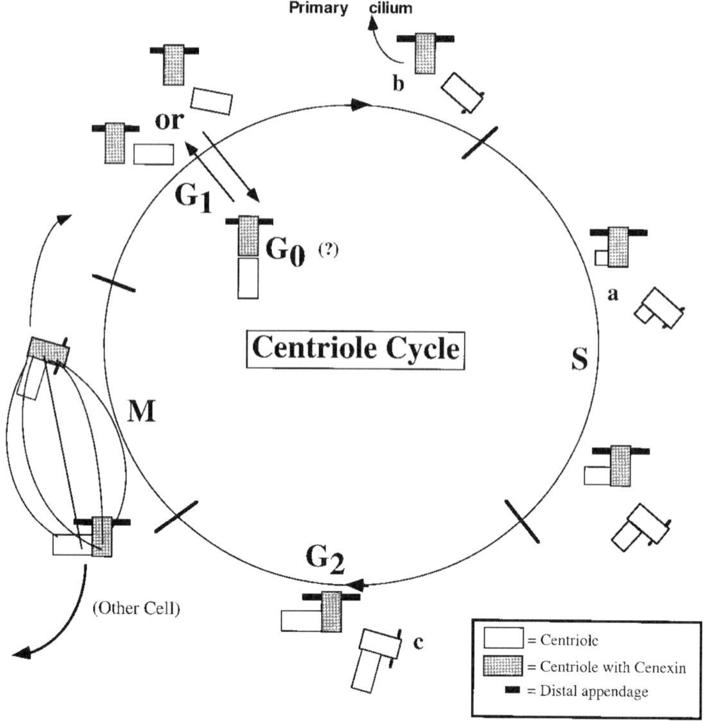

FIG. 8 Schematic representation of the centriole cycle. After cell division a pair of centrioles is delivered to the two daughter cells (G_1). The centrioles have an orthogonal or slightly tilted orientation with respect to each other (filamentous connections are not shown; see Fig. 3). The orthogonal configuration may be lost during M phase (Alvey, 1985). It has also been suggested that the orthogonal configuration is maintained during M phase. This configuration could presumably control and restrict centriole duplication to one duplication event per cell cycle (Tournier *et al.*, 1991b; Hinchcliffe *et al.*, 1998). It has been suggested that one of the morphological differences that follows, and perhaps regulates, entry into quiescent state (G_0) is a close linear conformation of the centriole pair (G_0) (Tournier *et al.*, 1991a, b). Note that one centriole is delivered with distal appendages and has Cenexin associated (mostly at the distal end; Lange and Gull, 1995). It is not clear when and how the distal appendages grow or are being initiated (see Chrétien *et al.*, 1997, for details on appendages and satellites). It is suggested here that the distal appendages are permanent structures that reach their final length sometime during G_1 (b). This centriole (b) corresponds to a "mature centriole" that is able to nucleate a primary cilium or a flagella during interphase (see Fig. 13). The other centriole is "immature" and does not have this capacity. The centriole pair completely loses the orthogonal connection during G_1, and seeding and duplication of the new daughter centrioles starts at the G_1–S transition. If we consider the seeding of the new centrioles as the starting point it can easily be appreciated that the time for formation of a mature centriole (from a to b) takes more than one full cell cycle. It is suggested that the distal appendages start to grow on the immature centriole in late G_1 and that growth continues until M phase. At the end of G_2 the new daughter centrioles have almost reached their final length. However, new data indicate that the daughter cells are significantly shorter than their parents at the

daughter centrioles grow perpendicularly on the parent centrioles during G_1–S and G_2 phases and the duplication process is semiconservative (meaning that the mother centriole is completely stable and the daughter centriole generated *de novo*; Kochanski and Borisy, 1990). It is unknown when the distal appendages start to grow. Full-size appendages are observed only on the mature mother centriole (Fig. 8, a). Due to experimental difficulties no studies have reported on the speed at which daughter centrioles grow, although it appears to be a continuous process. It is also unknown what determines the fixed length of the centrioles. When the parent centrioles have duplicated, the centrosomes divide at M phase and each centrosome harbors a centriole pair consisting of one parent and one daughter centriole. At the G_2–M transition the mother centriole achieves cenexin. Note that formation of a mature centriole takes almost two more than one cell cycle (Fig. 8, follow the new centriole from a to b).

Studies on the requirements for centriole assembly have shown that centriole duplication can occur in the absence of DNA replication, transcription, and translation (Balczon *et al.,* 1995; Debec *et al.,* 1996; Gard *et al.,* 1990; Kuriyama *et al.,* 1986; Rattner and Phillips, 1973; Sluder *et al.,* 1990). This is quite remarkable and shows that the centrosome cycle to some extent is independent of the nuclear cycle. However, in somatic cells there must be a link between the cell cycle regulating DNA replication and centrosome duplication because, like chromosomes, centrosomes duplicate only once during one cell cycle. There is no explanation for how this is possible. The regulatory machinery must to some extent be autonomous but also share components with the machinery that regulates the cell cycle (Debec *et al.,* 1996; Snaith *et al.,* 1996). Recent results indicate that blocking cyclin E/cdk2 activity in *Xenopus* and sea urchin eggs by injection of the p21 cdk inhibitor protein blocks the continued rounds of centrosome duplication, implicating cyclin E/cdk2 as the kinase driving centrosome duplication (Stearns and Winey, 1997). Moreover, Hinchcliffe *et al.* (1998) used fertilized sea urchin eggs (zygotes) to determine what regulates centrosome duplication. Continuous centrosome duplication was observed when S phase was prolonged by inhibition of DNA polymerase with aphidicolin. Duplication was not affected by Cdk1-B kinase activity (in this system it stably rises to supramitotic levels when DNA duplication is inhibited).

---

end of G_2, suggesting that elongation may continue into G_1 of the subsequent cell cycle (Chrétien *et al.,* 1997). At the G_2–M transition the centriole pairs separate to each end of the bipolar spindle (lines indicate MTs). At the same time the youngest of the parent centrioles acquire Cenexin. Note that MTs in the spindle do not nucleate from the centrioles but from the PCM (not shown).

Mitotic arrest (induced by injection of mRNA coding for nondegradable cyclin B) caused failure to duplicate the centrosome although anaphase still occurred. In accord with previous studies, multiple centrosome duplication occurred if protein translation was blocked at fertilization, a condition that also blocks Cdk1-B activation. Thus, regulation of centrosome duplication occurs independently of Cdk1-B activity and the proteolytic events during anaphase. Remarkably, if protein translation inhibitors were applied during prophase of the first cell cycle (rather than at fertilization), the zygotes arrested in the second cell cycle and subsequently duplicated their centrosome only once over an 8-h period (five cell cycles). Analysis of the arrests showed that zygotes treated with translation inhibitors at fertilization were arrested in active S phase (verified by refertilizing zygotes to monitor active DNA replication by BrdU incorporation), whereas no DNA replication occurred in those arrested during the second cell cycle (G_1 equivalent). Thus, genuine S phase arrest is apparently necessary and sufficient to support continuous centrosome duplication. Interestingly, during S phase arrest the time for centrosome reduplication was up to three times longer than one complete cell cycle (this also holds true in somatic cells; see Balczon *et al.,* 1995). A slow S phase-regulated clock may thus ensure that centrosomes duplicate only once during the faster cell cycle. Moreover, during a normal cell cycle the previous S phase may prime the centrioles/centrosome, "licensing" them to duplicate only once during the next cell cycle.

At the molecular level nothing is known about how the complicated 9 × 3 MT triplet structure assembles perpendicularly on top of the parent centriole. γ-Tubulin was found in the lumen of the centrioles and together with centrin proposed to be involved in centriole duplication (Fig. 9; Fuller *et al.,* 1995; Schiebel and Bornens, 1995). However, considering the 9 × 3 MT structure of the centriole, the assembly process of centriolar MTs most likely is complex and probably involves more than just γ-tubulin and pericentrin. Previous studies report how treatment of eggs with hypertonic solutions can induce the *de novo* formation of centrioles (Dirksen, 1991). Palazzo *et al.* (1992) described how centrioles could assemble *in vitro* within only 4 min in surf clam extracts. Considering these fast kinetics, it seems possible that the 9 × 3 MT structure may be initiated by a macromolecular proteinatious template containing γ-tubulin and present in the extract and in the pericentriolar matrix. It is possible that such a "centriolar assembly template" (CAT) to a large extent consists of γ-TuRC linked together at the base of the centriole in pairs of three. Although a CAT could explain seeding of the centriole, it is not easy to imagine how the number and alignment of the triplet MTs relative to one another changes from the proximal to the distal end of the centriole (Figs. 3–5).

The components and structure of this hypothetical CAT are unknown. Considering the low abundance of centrioles in a cell, a genetic approach

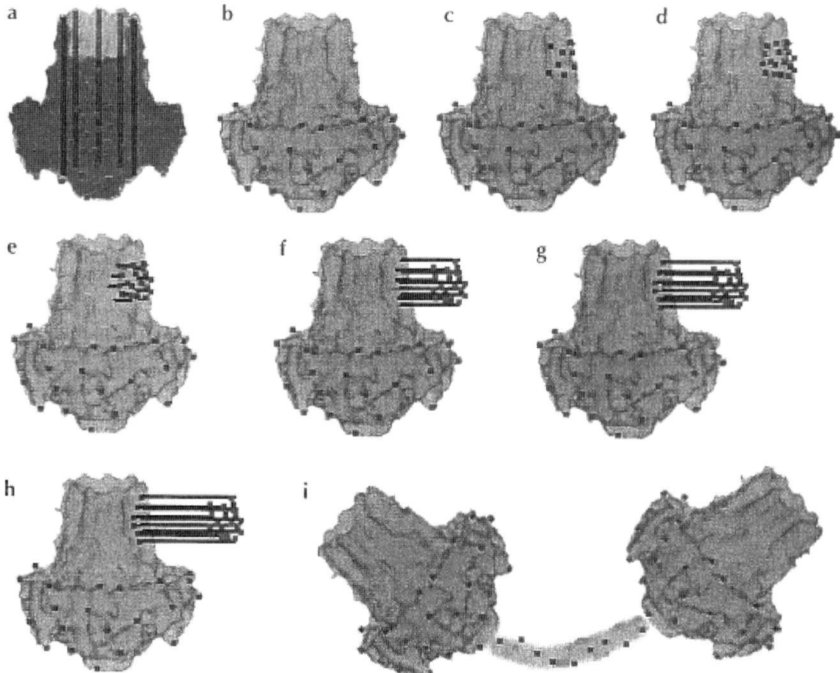

FIG. 9 Centriolar association of γ-tubulin and hypothetical function for centriole duplication. The suggested distribution of γ-tubulin and other components during centriole duplication is shown using a low-resolution centriole reconstruction to indicate positions; γ-tubulin is show in black. The maternal centriole has γ-tubulin (a) in its core as well as (b) on the periphery of the pericentriolar material. The MTs of the centriolar barrel are made up of α- and β-tubulin and are represented as straight black lines. (c, d) Duplication begins with the association of γ-tubulin and other components of the centriolar assembly template (CAT; see text) at a proximal site on the side of the barrel. (e–g) The components of the CAT form a template from which the MTs of the daughter centriole grow. (h, i) Once the daughter's barrel has grown to full length, the centrioles separate but remain linked by a structure that contains γ-tubulin (see also Fig. 3) (partially reproduced from Fuller *et al.*, 1995, with permission from Current Biology Ltd.).

(as opposed to a biochemical) will probably be required to define its components. The most promising organism for this purpose seems to be the biflagellated green algae *Chlamydomonas reinhardtii* (Dirksen, 1991; Dutcher, 1989). Thus, a study on *Chlamydomonas* from Ehler *et al.* (1995) gives hope of identifying molecules involved in centriole assembly and function. Based on a study by Goodenough and St. Clair (1975), Ehler *et al.,* showed that the *bld2–1* mutation in *Chlamydomonas* did not assemble basal bodies, and that the genetic locus did not match with actin, α-, β-, or γ-tubulin, or centrin. However, it was not possible to identify additional alleles with similar phenotypes and the gene(s) has not been identified (Dutcher and Trabuco, 1998). Interestingly, the cells with the *bld2-1* mutation are viable and form functional spindles, although spindles are mispositioned compared to wild-type strains. Ehler *et al.* (1995, p. 959) conclude:

> It is clear from this study that centrioles/basal bodies are not necessary for the nucleation of cytoplasmic, rootlet/astral, or spindle microtubules during the mitotic cell cycle in *Chlamydomonas.* Rather, it appears that the role of the centriole in some cell types may be to provide an organized cytoskeletal superstructure that is critical for the correct placement of the spindle and the cleavage furrow during cell division.

To understand centriolar function and assembly, one could take a genetic approach using *C. reinhardtii* and *Drosophila* to determine components and functions, and then use biochemical systems, such as *Xenopus* or surf clam, to determine a biochemical pathway of assembly.

The next level one must understand is centrosome separation or division as a process—not the regulation of the process. Thus, how does the cell at prophase manage to separate the pericentriolar matrix surrounding the "double centrosome" into two equally big pools? Two mechanisms have been proposed for how the centrosomes are separated. The first proposes that minus-end-directed motors attaching to astral MTs emanating from the centrosome pull centrosomes apart and also help in the migration of centrosomes during prophase (i.e., Waters *et al.,* 1993). The other model suggests that plus-end-directed motors push centrosomes apart. By sitting on MTs emanating from one centrosome and moving toward the plus end of MTs emanating from the other centrosome, separation would occur (i.e., Boleti *et al.,* 1996, Fig. 10).

In addition to an active motor-driven movement, it seems reasonable to invoke a mechanism that weakens the adhesiveness of the pool of pericentriolar matrix to itself to separate centrosomes at prophase. A recent study on Nek2 supports this idea (Fig. 10). It is proposed that phosphorylation of the pericentriolar matrix by Nek2 leads to a weakening of the interaction/adhesiveness of the pericentriolar matrix. Through the subsequent movement by motors the centrosomes are separated. In the model the plus-end-directed motor Eg5 is proposed to push the centrosomes apart. As indicated in the speculative Fig. 10, recruitment of the motor Eg5 is a step that occurs after Nek2 has loosened the pericentriolar matrix in preparation for centrosome separation. This is based on observations showing that phosphorylation of a single cdc2 site is required to localize Eg5 to the centrosome during mitosis and that this site can be phosphorylated by cdc2 kinase *in vitro* (Blangy *et al.,* 1995; Sawin and Mitchison, 1995). Interestingly, there is evidence that Eg5 associates with p150 (glued), which is a member

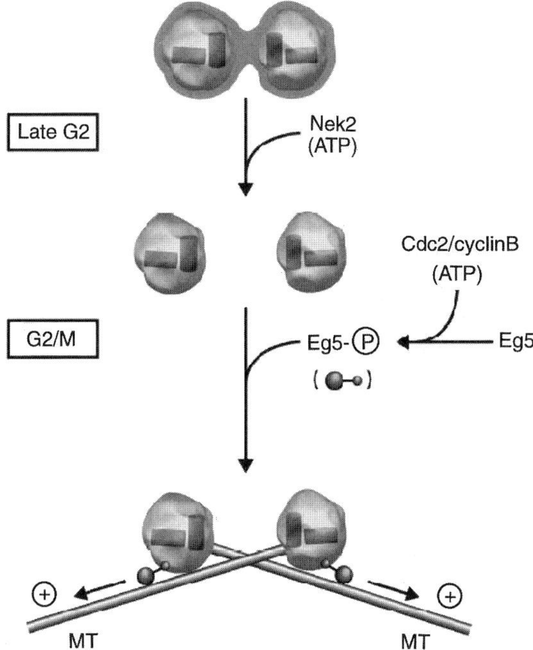

FIG. 10 A two-step model for centrosome separation. In this model, the process of centrosome separation is described in terms of two independent steps which must occur at the onset of mitosis. The first step involves the severing of a connection (e.g., the dissolution of a hypothetical cage) which holds the centrosomes in close proximity throughout interphase. Under physiological conditions, this step is proposed to be triggered in late G_2 by the activation of Nek2 (and/or the inactivation of a counteracting phosphatase). Under the experimental conditions described here, premature separation (i.e., splitting) of centrosomes results from ectopic expression of active Nek2. In both cases, phosphorylation of as yet unidentified centrosomal "glue" proteins may cause their depolymerization or target them for proteolytic degradation. The second step involves several MT-based motor proteins, only one of which is depicted here. This kinesin-related motor, Eg5, is recruited to the centrosome by the dynein complex and in response to phosphorylation by Cdc2/cyclin B, and its activity contributes to the separation of centrosomes along MTs. Note that objects are not drawn to scale, and the precise mode of action of Eg5 remains to be determined (a similar function has been proposed for the motor XKLP2, see text) (reproduced from Fry *et al.,* 1998, with permission from Oxford University Press).

of the dynactin complex that interacts with the minus-end-directed motor dynein (Blangy *et al.,* 1997). These findings indicate that it is possible that localization of a plus-end-directed motor (i.e., Eg5) may be regulated by a minus-end-directed motor (i.e., dynein), and vice versa. This is interesting because it strongly suggests that motors localized by such means do not simply transport cargo around but serve other functions, i.e., push centrosomes apart (Fig. 10). (A similar function and localization mechanism has been proposed for XKLP2.)

An interesting addition to the regulatory hierarchy of mechanisms that control the frequency of centrosome duplication came with the discovery that the tumor suppressor p53 can influence centrosome duplication (Fukasawa *et al.,* 1996). In a mouse cell line lacking p53, multiple copies of functionally competent centrosomes were generated during a single cell cycle. p53 is a tetrameric transcription factor that can activate or inhibit the expression of groups of genes and the effect on centrosome duplication may therefore be rather indirect. It is unclear how p53 affects centrosome duplication, but it most likely involves a change in the signaling cascades regulating the cell cycle. Since multiple centrosomes lead to missegregation of chromosomes, this may explain how lack of p53 ultimately can lead to development of cancer (Agarwal *et al.,* 1998; Winey, 1996).

V. Microtubule Nucleation

The principal function of the centrosome is to be a site of MT nucleation. Since nucleation is the first step in MT assembly it is a process that is important to understand. The aim of this section is to analyze and compare different mechanisms of MT nucleation. I will distinguish between three types of MT nucleation: self-nucleation, catalyzed nucleation, and centrosome-mediated nucleation (Fig. 11).

A. Bulk Microtubule Assembly: A Thermodynamic Problem

Only the tubulin dimer and GTP are needed to form a MT. A MT is an assembly of approximately 13 protofilaments, consisting of $\alpha\beta$-tubulin heterodimers, closed into a hollow filament 25 nm in diameter. Centrosomally nucleated MTs always contain 13 protofilaments, whereas self-nucleated MTs may have a variable number of protofilaments (12–16). Both α- and β-tubulin can bind GTP but only GTP bound to the β-tubulin subunit can be hydrolyzed. Since the building block is the $\alpha\beta$-tubulin heterodimer, MTs have an intrinsic polarity: a fast growing plus end with β-tubulin at the end and a slow growing minus end with α-tubulin at the end (localized at the centrosome). In the cell we mostly see MTs growing off centrosomes because they ease the nucleation step. However, centrosomes are not strictly required for nucleation and MTs can self-nucleate in the absence of centrosomes. However, self-nucleation of MTs require a much higher tubulin concentration than the concentration required to grow MTs off

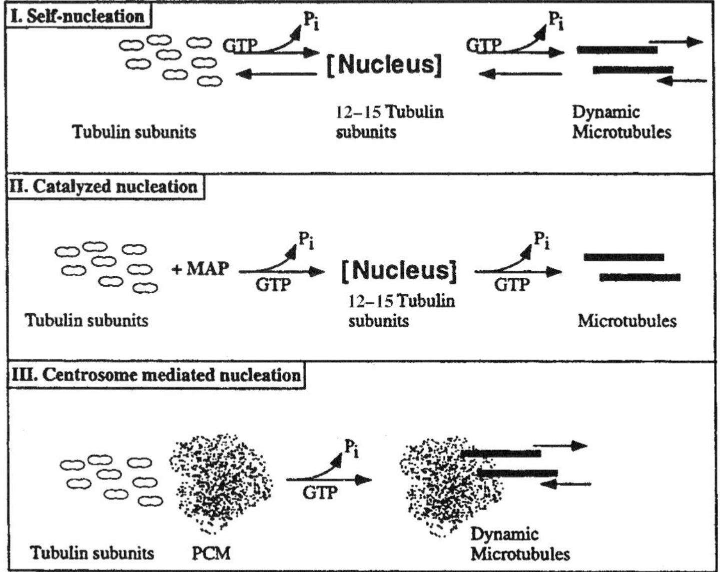

FIG. 11 Microtubule nucleation. (I) Pure tubulin can spontaneously assemble into MTs. This requires tubulin concentrations that are much higher than those *in vivo* and occur via a nucleation core of twelve to fifteen tubulin subunits. Concomitantly with MT assembly, the GTP bound to tubulin hydrolyzes into GDP and P_i. The exact timing of polymerization and phosphate release is unclear (Erickson and O'Brien, 1992). (II) In the presence of MAPs the mechanism of nucleation is the same as that for pure tubulin. However, the MAPs manage (see text) to stabilize the critical nucleation core consisting of five to seven subunits. MAPs that enhance nucleation (i.e., τ) also dampen MT dynamics, and stable MTs are the end result. (III) The pericentriolar matrix contains the components required for nucleation off centrosomes (i.e., γ-TuRC and pericentrin). The γ-TuRC helps in the formation of the first nucleation core, probably in a manner comparable with a template for MT assembly (Pereira and Schiebel, 1997).

centrosomes. The requirement for a higher tubulin concentration reflects the transition energy that has to be surmounted to form the first part of a MT. It is thought that formation of a core of approximately twelve to fifteen tubulin heterodimers is required to form the critical starting nucleus for MT assembly (Carlier *et al.*, 1997; Carlier and Pantaloni, 1978, 1981). The formation of this nucleation core is the rate-limiting step for MTs assembled in solution (Fig. 11, I). Using the second law of thermodynamics, it can be appreciated why formation of the nucleation core is the rate-limiting step (the second law of thermodynamics states that $\Sigma\Delta G = \Sigma\Delta H - T\Sigma\Delta S$. For a process to occur, ΔG must be negative). The entropy (S) of the system that is going to form a MT is high before nucleation because there is a high probability of finding the freely diffusing dimers in a given position at a given time. In contrast, the probability of simultaneous association of five to seven dimers is low and this corresponds to a decrease in entropy. Thus, the sign of $\Sigma\Delta S$ is negative during nucleation. (Note, however, that if the dimers have water molecules associated that dissociate upon interaction with other tubulin dimers, this will increase the entropy.) The entropy change during MT nucleation and MT growth is therefore most likely unfavorable for the formation of MTs (Inoué and Salmon, 1995). It therefore seems that the driving force must come from a decrease in internal energy of the tubulin dimer during MT polymerization. Where does the energy come from? The β-subunit hydrolyzes GTP to GDP upon MT polymerization. The hydrolysis of GTP results in the release of energy that perhaps could be used for nucleation and polymerization. However, studies from a slowly hydrolyzable GTP analog (GMPCPP) show that MTs can nucleate much better in the absence of GTP hydrolysis (Carlier *et al.,* 1997; Carlier and Pantaloni, 1978; Hyman *et al.,* 1992). Moreover, calculations have shown that the energy of GTP hydrolysis is stored in the MT lattice (Caplow *et al.,* 1994). Thus, GTP hydrolysis cannot be used to fuel the nucleation process. On the contrary, GTP hydrolysis destabilizes the interaction between tubulin subunits and is required for the process of MT dynamics (Chrétien *et al.,* 1995; Erickson and O'Brien, 1992; Hyman *et al.,* 1992). It is possible that tubulin in the GTP state has a conformation favorable for nucleation and that rapid GTP hydrolysis evolved to control the nucleation step. We must conclude that the transition energy for formation of the nucleus cannot be very high, and it remains unclear exactly how the first nucleation core forms during MT self-nucleation. With the structure of tubulin now solved (Nogales *et al.,* 1998) we may soon have answers to these questions. For example, how does the structure of tubulin change upon GTP hydrolysis?

Although the centrosome is the major site of MT nucleation *in vivo,* it has recently been shown that MTs can assemble spontaneously in the cytoplasm *in vivo* (Keating *et al.,* 1997; Vorobjev *et al.,* 1997). However, using very pure tubulin it has been observed that nucleation of MTs in solution does not occur *in vitro* even at tubulin concentrations that are five- to sevenfold higher than those *in vivo.* This suggests that nucleation of cytoplasmic MTs *in vivo* must involve some catalysts. *In vivo* there are two types of catalysts: MAPs and the centrosome (Fig. 11, II–III). In the cytoplasm MAPs most likely are involved in regulating the spontaneous assembly of MTs (Andersen, 1998). For example, the MAP τ is an example of a catalyst that very potently promotes MT polymerization in solution (Brandt and Lee, 1993) and suppresses MT dynamics (in particular by suppressing the catastrophe frequency). How can τ promote MT nucleation? Recent studies have shown that MT binding of τ occurs through two types of sites (Preuss *et al.,* 1997). One site targets the MAP to the

MT or to the MT nucleation unit, consisting of twelve to fifteen tubulin subunits. This site is called a "jaw" since it attaches to the nascent MT. The other site is the repeat domains that are required for inducing MT nucleation and polymerization. It is suggested that MAPs such as τ can cross-link protofilaments and in this way facilitate nucleation. Whether τ or other MAPs influence the hydrolysis rate of GTP is unknown. Most likely τ promotes nucleation simply by stabilizing interaction between tubulin dimers in the nucleation center. As already described and further discussed in Section V,B, centrosome nucleation is distinct from self-nucleation and MAP-catalyzed nucleation. Centrosome-mediated nucleation utilizes the γ-TuRC and always results in MTs with 13 protofilaments.

B. Microtubule Nucleation by Centrosomes

In vitro the centrosome nucleates MTs at very low tubulin concentrations (around 5 μM; Andersen *et al.,* 1994; Andersen and Karsenti, 1997; Bré and Karsenti, 1990; Hill and Kirschner, 1983). As described in Section V,A, it has been known for more than 20 years that MAPs can lower the critical concentration for MT nucleation. However, a search for MAPs that could explain the extraordinary nucleation capacity of centrosomes has been unsuccessful (but see Domínguez *et al.,* 1994). It appears that γ-tubulin in the form of the γ-TuRC is the unit that is responsible for all MT nucleation by the centrosome (Section III.A). Most likely, γ-tubulin requires some additional proteins to be able to induce MT nucleation, some of which may be members of the γ-TuRC and pericentrin. Thus, the γ-TuRC has been purified and shown to induce MT nucleation *in vitro* (Zheng *et al.,* 1995; Moritz *et al.,* 1998). There is no direct knowledge about the physical chemistry of the facilitation of nucleation but a recent review examines two models for how the γ-TuRC serves as a template for MT nucleation at the centrosome (Pereira and Schiebel, 1997).

The next level of complexity for centrosomal nucleation involves regulation of the nucleation capacity. I split the discussion of regulation of the nucleation capacity into two components. First, I shed some light on the biophysical aspects of MT nucleation by the centrosome, and then I discuss cell cycle regulation in Section V.C.

Microtubules are nucleated very closely to one another at the centrosome and a study reports on the possibility of local depletion of tubulin dimers around the centrosome during massive nucleation of MTs off the centrosome (Dogterom *et al.,* 1995). Assuming that the centrosome is a smooth sphere and by changing the number of nucleation sites on the surface, Dogterom *et al.* calculate that diffusion may effectively limit the number of MTs growing off a centrosome. How would this be possible?

The dynamics of MTs is very dependent on the tubulin concentration and by local depletion MTs could become so dynamic that they would not at all polymerize (Fygenson *et al.*, 1994). The result of Dogterom *et al.* (1995) is interesting since it could indicate that it is not the number of nucleation sites that regulates how many MTs can grow off the centrosome but perhaps the diffusion of tubulin to the nucleation sites. However, Méda *et al.* (1997) show that purified centrosomes have a capacity to bind 25,000 tubulin dimers with a K_D of 5 μM and with a half-saturation time of 3 min. Interestingly, this K_D closely matches the critical concentration for MT growth off centrosomes (Andersen *et al.*, 1994). Moreover, if such low-affinity binding sites exist the diffusion limit (Dogterom *et al.*, 1995) for the total number of MTs growing off a centrosome would be annulled since they could fuel the real nucleation sites with tubulin. The composition of the low-affinity binding sites is unknown, but the data by Méda *et al.* (1997) indicate that this binding is not to the γ-TuRC.

The study by Méda *et al.* (1997) is also interesting because it brings a new dimension to centrosome-mediated MT nucleation by suggesting that the nucleation step is a two-step process. The first step consists of local increased concentration at the centrosome of tubulin dimers by an unknown mechanism. The next step consists of nucleation of MTs facilitated by the γ-TuRC. Méda *et al.* make clear that two-step processes are common in biological systems in which a few high-affinity structures (in this case the γ-TuRC) are bound to insoluble structures (in this case the centrosome) since ligand (in this case tubulin) targeting is greatly facilitated under such conditions (e.g., DNA-binding proteins, phages binding to cell membranes, and actin nucleation).

It has been observed that soluble MAPs affect the nucleation capacity of centrosomes *in vitro,* which shows that soluble factors not associated with the centrosome can also influence nucleation (Andersen *et al.*, 1994; Bré and Karsenti, 1990). However, these soluble factors do not directly enhance nucleation of MTs on the centrosome but simply allow more growth off the centrosome through a regulation of MT dynamic instability. This mechanism is thus an indirect way of regulating the nucleating capacity of the centrosome. This type of modulation may be very important *in vivo* in which both soluble and insoluble factors are subject to cell cycle regulation.

C. Regulation of Centrosomal Microtubule Nucleation Activity during the Cell Cycle and Differentiation

Why is it important that the cell regulates the MT nucleation capacity of the centrosome? There are several answers to this question. Here I will approach this question by contemplating the properties of MTs from a biophysical point of view. A MT is a very stiff polymer. Its resistance toward bending is expressed as "flexural rigidity" and has been measured to be in the range of 2–40 10^{-24} Nm^2, comparable with the stiffness of plast polymers (Feigner *et al.,* 1996; Kurachi *et al.,* 1995; Mickey and Howard, 1995). Thus, such a polymer can be used to generate shape in three-dimensions, like poles holding a tent. In a way, MTs are part of the mechanism responsible for transforming the linear two-dimensional DNA code into the three-dimensional realm of a cell. Because MTs have this function for morphogenesis (Hyman and Karsenti, 1996; Kirschner and Mitchison, 1986), it is important for the cell to regulate the length, the number, and the orientation of MTs. Moreover, MTs are involved in directed vesicular transport, organelle movement, and cell division. Therefore, it is important to regulate MT assembly both spatially and temporally.

An important parameter in this regulation consists of regulation of the nucleation of MTs in general and by centrosomes in particular. The following two examples will illustrate why it is important to regulate the nucleation capacity of the centrosome.

The first example concerns morphogenesis of neurons. Extending axons build an elaborate cytoskeleton mainly composed of MTs (Matus, 1994). There is lively debate about the origin of these MTs (Baas and Brown, 1997; Bray, 1997; Hirokawa, 1997; Hirokawa *et al.,* 1997). One model proposes that tubulin subunits are transported to the tip of the axon where they are polymerized into MTs (Hirokawa, 1997; Hirokawa *et al.,* 1997). The other model suggests that MTs are nucleated by the centrosome in the cell body, released from the centrosome, and then transported out to the tip of the axon (Baas and Brown, 1997; Bray, 1997). In the latter model the centrosome serves as an engine for the generation of MTs. Thus, the ability of the centrosome to nucleate MTs indirectly becomes essential for axon outgrowth. One way to resolve the controversy would be to remove the centrosome by microsurgery (Maniotis and Schliwa, 1991). It remains unresolved which model most faithfully reflects reality.

The second example relates to regulation of MT assembly during the cell cycle. In all cells there is a dramatic rearrangement of the cytoskeletal network from interphase to mitosis. A schematic representation of the transition from interphase to mitosis is shown in Fig. 12. Note the following: (i) In interphase many MTs are not only found associated with the centrosome but also free in the cytoplasm and (ii) in mitosis, all MTs seem to emanate from the centrosome and are almost only observed in the direction of the chromosomes (Fig. 12). How is this rearrangement brought about? What is the origin of the MTs? Are they derived from the centrosomes or spontaneously nucleated in the cytoplasm?

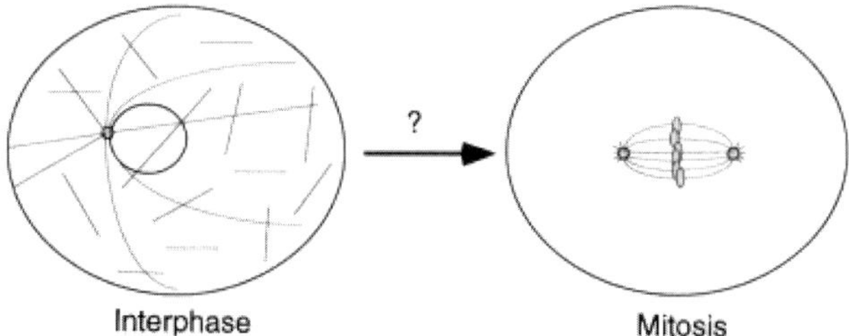

FIG. 12 Schematic representation of the rearrangement of MTs between interphase and mitosis. Interphase MTs (black lines) are long and stable and are found associated with the centrosomes (circle) or assemble freely in the cytoplasm (Vorobjev *et al.,* 1997; Andersen, 1999). In mitosis all free cytoplasmic MTs depolymerize and MTs only polymerize off centrosomes and in the vicinity of the chromosomes (ellipses), forming the elliptical bipolar spindle.

Recent studies have shown that cytoplasmic MTs during interphase are derived both by ejection from the centrosome and from spontaneous assembly in the cytoplasm (Keating *et al.,* 1997; Vorobjev *et al.,* 1997; Andersen, 1998). In mitosis MTs are also ejected from centrosomes, as observed in mitotic *Xenopus* egg extracts, but no spontaneous assembly occurs. Thus, in both interphase and mitosis, MTs are ejected from the centrosome into cytoplasm. Why then are no cytoplasmic MTs observed during mitosis (Fig. 12)? Research during the past decade has shown that MTs become more dynamic during mitosis, primarily due to a dramatic increase in the catastrophe frequency, which subsequently leads to a 10-fold increase of MT turnover from interphase to mitosis (Belmont *et al.,* 1990; Tournebize *et al.,* 1997; Verde *et al.,* 1992). Thus, the increase in MT dynamics during mitosis explains why the cytoplasmic MTs disappear from interphase to mitosis (Fig. 12).

How do the MTs manage to preferentially grow in the area around chromosomes during mitosis if cytoplasmic MTs are so dynamic that they depolymerize? Research during the past 20 years has shown that the mitotic chromosomes change the environment for MT polymerization (Andersen, 1999; Andersen *et al.,* 1997; Heald *et al.,* 1996; Karsenti, 1991; Karsenti *et al.,* 1984; Marek, 1978; Nicklas, 1988; Zhang and Nicklas, 1995). Thus, the mitotic chromosomes somehow stimulate MT polymerization in an area around them, leading to the preferential distribution of MTs around the mitotic chromosomes. These findings may also explain how spindle assembly is possible in some systems devoid of centrosomes (i.e., plants and eggs) because spindle MT assembly may be largely chromosome driven in such systems (Andersen, 1999; Andersen *et al.,* 1997; González *et al.,* 1998).

Regarding specifically the nucleation capacity of centrosomes, it is seen from this analysis that in mitosing vertebrate cells the only source of MTs is the centrosome. All MTs must be generated by the centrosome. Therefore, the centrosomal ability to generate MTs becomes essential for cell division. In order to safely and efficiently build the mitotic spindle, the centrosome as "microtubule source" must work at a maximum rate, i.e., a priori have a higher MT nucleation capacity during mitosis than during interphase. The benefit of an augmented capacity to generate MTs during mitosis is an increased chance of proper chromosome attachment to the MTs.

How does the cell manage to change the nucleation capacity of the centrosome? The first important discovery was that it is the pericentriolar matrix that nucleates the MTs and not the centrioles (Gould and Borisy, 1977). Several studies showed that there is more pericentriolar matrix around the centrioles during mitosis than during interphase, indicating that this could explain the increased MT nucleation activity observed at mitosis (Dictenberg *et al.,* 1998; Rieder and Borisy, 1982; Snyder and McIntosh, 1975). Concurrent with the increase in volume of pericentriolar matrix the MPM2 antibody that recognizes a phosphoepitope shows increased staining of the centrosome (Davis *et al.,* 1983; Vandre *et al.,* 1986). Subsequent studies showed that the increased MT nucleation capacity during mitosis is dependent on the MPM2 epitope and thus phosphorylation events (Centonze and Borisy, 1990). Additional *in vitro* experiments showed that treatment of centrosomes with cyclin A-dependent cdc2 kinase caused a specific increase in MT nucleation. In contrast, cyclin B-dependent cdc2 kinase which had no effect (Buendia *et al.,* 1992; Buendia and Karsenti, 1995). Thus, through a cascade of phosphorylation events, more MTs are nucleated by centrosomes during mitosis than during interphase.

How is this increased nucleation capacity brought about by phosphorylation of centrosomal proteins? There are several possibilities: (i) phosphorylation-dependent removal of inhibitors of MT nucleation; (ii) phosphorylation-dependent binding of activators of MT nucleation; (iii) phosphorylation-dependent reorganization of the pericentriolar matrix, opening up for more nucleation sites; and (iv) regulation of soluble factors involved in MT stabilization. The first possibility seems the least likely since *in vitro* phosphorylation of purified centrosome results in increased nucleation capacity (but see Balczon *et al.,* 1994). Moreover, this first possibility may be thought of as a variation of the third possibility.

The latter three possibilities are probably all at work. Thus, the increase in the amount of pericentriolar material probably to some extent reflects recruitment of material required for MT nucleation. Moreover, it seems likely that reorganization of the pericentriolar matrix can open up otherwise hidden nucleation sites. With the discovery of the γ-TuRC, and its direct localization to the base of centrosome nucleated MTs, this possibility seems even more attractive (Dictenberg et al., 1998; Moritz *et al.,* 1995a, b; Vogel *et al.,* 1997; Zheng *et al.,* 1995). The proteins associated with γ-tubulin in the γ-TuRC are beginning to become characterized and may be the target of the cyclin A-dependent phosphorylation events leading to increased nucleation capacity (Section III.A). For example, by tilting the γ-TuRC attached to the pericentrin rings (Fig. 7) it may be possible to regulate the number of accessible nucleation sites at the centrosome. Much more work is required to clarify these issues.

As a last example of the regulation of MT nucleation by centrosomes, I discuss the change in MT nucleation pattern during the differentiation of cells into epithelial cell layers. The MT rearrangement shown in Fig. 12 is primarily found in fibroblast cells or cells in suspension. If one examines an epithelial cell layer, the behavior of the centrosome is markedly different (Fig. 13, see color plate). How the cell regulates this change in location and activity of the centrosome during the cell cycle is unknown. It is notably interesting in this example that during interphase the epithelial cells may be said to be devoid of a centrosome but apparently contain numerous apically localized MTOCs.

Where do all these noncentrosomal apical MTs come from? One possibility is that the material mediating MT nucleation (i.e., the γ-TuRC and pericentrin) is still localized to the periphery of the centrioles and that this area serves as a MT factory at the apical side. This would work much like the proposed centrosome-driven MT engine for axon extension. Another possibility is that MTs may nucleate throughout the entire apical side through dispersed γ-TuRCs and pericentrin. Alternatively, MTs may be spontaneously nucleated independently from γ-TuRC and pericentrin through the activity of MAPs and subsequently sorted by motors with their minus end localized to the apical surface (Vorobjev *et al.,* 1997; Andersen, 1998). Which of these possibilities is most likely to work?

Although a study reports localization of dispersed γ-TuRC and pericentrin in some epithelial cells at the apical surface (Meads and Schroer, 1995), others have obtained negative results on this issue (Meads and Schroer, 1995; Mogensen *et al.,* 1997; Reinsch and Karsenti, 1994; Vogl *et al.,* 1995). In the latter three studies, γ-TuRC and pericentrin were only found localized to the material surrounding the apically located centrioles. This could indeed suggest that MTs become generated close to the centrioles through nucleation by γ-TuRC and pericentrin and subsequently transported to their destination at the apical membrane. Figure 14 summarizes how Mogensen *et al.* (1997) imagine that such a MT factory could work. Central to the model is that MTs are nucleated in a γ-tubulin-dependent way (Fig. 14A). The minus end is then stabilized through binding of proteins (Figs.

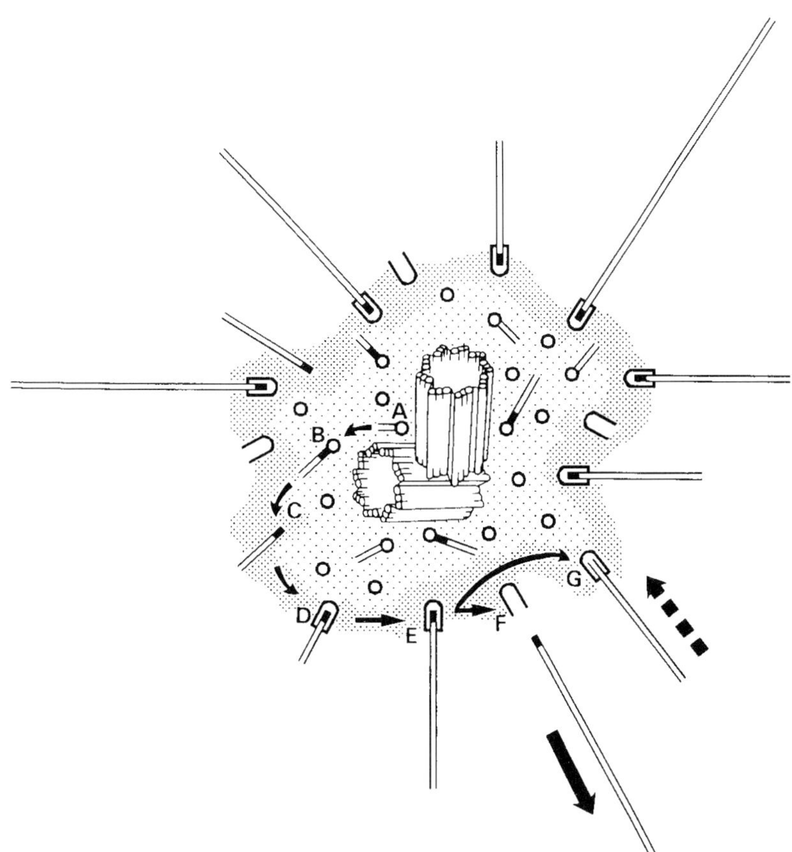

FIG. 14 Schematic diagram of a typical animal cell centrosome outlining the main features of the docking element hypothesis. The pericentriolar matrix is differentiated into a mainly central nucleating domain (lightly stippled) containing nucleating elements (○) and a more peripheral docking domain (densly stippled) with docking elements (U). (A) Nucleation: Microtubule assembly is initiated by a nucleating element. (B) Capping: Addition of a minus-end cap (black stain on MTs). (C) Translocation: Release from a nucleating element is followed by translocation. (D) Docking: A docking element changes its configuration when it effects anchorage of a capped minus end (schematically represented by closure lines at the previously open end of the docking element). (E) Elongation: The proximal minus end remains anchored to the centrosome via a docking element as the MT continues to elongate by addition of tubulin to its distal plus end. (F) Escape: The minus end retains its protective cap as it is released by a docking element. The MT escapes from the centrosome and migrates (arrow) to a new cytoplasmic location. This transport is mediated by motors and has been directly observed *in vivo* (Keating *et al.*, 1997; Vorobjev *et al.*, 1997). (G) Minus-end dynamics: Anchorage by a docking element is maintained after cap loss, which permits minus-end loss of tubulin that can contribute to a flux of tubulin or MT shortening (dashed arrow) (reproduced from Centrosomal deployment of γ-tubulin and pericentrin: Evidence for a microtubule-nucleating domain and a minus-end docking domain in certain mouse epithelial cells, Mogensen, M. M., Mackie, J. B., Doxsey, S. J., Stearns, T., and Tucker, J. B., *Cell Motil. Cytoskel.,* copyright 1997 Wiley-Liss, Inc. Reprinted by permission from Wiley-Liss, Inc., a subsidiary of John Wiley & Sons, Inc).

14B and 14C) which allows the MT to be inserted in so-called docking sites either at the centrosome (Fig. 14D) or elsewhere at the apical surface after transportation (Fig. 14F). In this model the area around the centrioles functions like a MT "factory" and MTs are then transported to other locations (Keating *et al.,* 1997; Rodionov and Borisy, 1997).

VI. Concluding Remarks

The challenges for the future lie in integrating the vast amount of molecular information about the centrosome into a coherent story that mechanistically answers the questions addressed in the Introduction. With regard to the questions surrounding centriole assembly and duplication, we are still in very uncharted territory. We need to understand how centrioles can affect the cell cycle and cell division in some cells. We also need to determine more thoroughly how nucleation is facilitated by the centrosome and how the nucleation capacity of the centrosome is regulated during the life cycle of a cell. It will also be interesting to contrast and compare differences and similarities of MTOCs from different organisms. The following are the major questions in the field: What is the precise function of the centrioles and how do they assemble? The major goals within the centrosome field will be to describe the time and place of the association of centrosomal proteins to the centrioles and to build a dynamic three-dimensional model of the pathway leading to a mature centrosome.

Acknowledgments

I am grateful to M. Bornens, D. Chrétien, S. Doxsey, S. Fuller, E. Nigg, and S. Reinsch for permission to reproduce figures. Thanks to M. Rose for providing facilities for the production of the review. Many thanks to K. Ganguly, W. Khalfan, and Academic Press for proofreading the manuscript. This work was partially supported by EMBO long-term fellowship No. ALTF97-721.

References

Agarwal, M. L., Taylor, W. R., Chernov, M. V., Chernova, O. B., and Stark, G. P. (1998). The p53 network. *J. Biol. Chem.* **273,** 1–4.

Alvey, P. L. (1985). An investigation of the centriole cycle using 3T3 and CHO cells. *J. Cell Sci.* **78,** 147–162.

Anderson, R. G. W., and Brenner, R. M. (1971). The formation of basal bodies (centrioles) in the rhesus monkey oviduct. *J. Cell Biol.* **50,** 10–34.

Andersen, S. S. L. (1998). *Xenopus* interphase and mitotic microtubule-associated proteins differentially suppress microtubule dynamics *in vitro*. *Cell Motil. Cytoskel.* **41,** 202–213.

Andersen, S. S. L. (1999). Balanced regulation of microtubule dynamics during the cell cycle: A contemporary view. *BioEssays* **21,** in press.

Andersen, S. S. L., and Karsenti, E. (1997). XMAP310: A Xenopus rescue promoting factor localized to the mitotic spindle. *J. Cell Biol.* **139,** 975–983.

Andersen, S. S. L., Buendia, B., Domínguez, J. E., Sawyer, A., and Karsenti, E. (1994). Effect on microtubule dynamics of XMAP230, a microtubule-associated protein present in *Xenopus laevis* eggs and dividing cells. *J. Cell Biol.* **127,** 1289–1299.

Andersen, S. S. L., Ashford, A. J., Tournebize, R., Gavet, O., Sobel, A., Hyman, A. A., and Karsenti, E. (1997). Mitotic chromatin regulates phosphorylation of Stathmin/Op18. *Nature* **389,** 640–643.

Archer, J., and Solomon, F. (1994). Deconstructing the microtubule-organizing center. *Cell* **76,** 589–591.

Baas, P. W., and Brown, A. (1997). Slow axonal transport: The polymer transport model. *Trends Cell Biol.* **7,** 380–384.

Bailly, E., Doree, M., Nurse, P., and Bornens, M. (1989). p34cdc2 is located in both nucleus and cytoplasm; Part is centrosomally associated at G2/M and enters vesicles at anaphase. *EMBO J.* **8,** 3985–3995.

Bailly, E., Bordes, N., Bornens, M., and Klotz, C. (1992a). A high molecular weight centrosomal protein of mammalian cells is antigenically related to myosin II. *Cell Motil. Cytoskel.* **23,** 122–132.

Bailly, E., Pines, J., Hunter, T., and Bornens, M. (1992b). Cytoplasmic accumulation of cyclin B1 in human cells: Association with a detergent-resistant compartment and with the centrosome. *J. Cell Sci.* **101,** 529–545.

Balczon, R. (1996). The centrosome in animal cells and its functional homologs in plant and yeast cells. *Int. Rev. Cytol.* **169,** 25–82.

Balczon, R., and West, K. (1991). The identification of mammalian centrosomal antigens using human autoimmune anticentrosome antisera. *Cell Motil. Cytoskel.* **20,** 121–135.

Balczon, R., Bao, L., and Zimmer, W. E. (1994). PCM-1, a 228-kD centrosome autoantigen with a distinct cell cycle distribution. *J. Cell Biol.* **124,** 783–793.

Balczon, R., Bao, L., Zimmer, W. E., Brown, K., Zinkowski, R. P., and Brinkley, B. R. (1995). Dissociation of centrosome replication events from cycles of DNA synthesis and mitotic division in hydroxyurea-arrested Chinese hamster ovary cells. *J. Cell Biol.* **130,** 105–115.

Baron, A. T., and Salisbury, J. L. (1992). Role of centrin in spindle pole dynamics. *In* "The Centrosome" (V. I. Kalnins, ed.), pp. 167–187. Academic Press, San Diego.

Belmont, L. D., Hyman, A. A., Sawin, K. E., and Mitchison, T. J. (1990). Real-time visualization of cell cycle-dependent changes in microtubule dynamics in cytoplasmic extracts. *Cell* **62,** 579–589.

Blangy, A., Lane, H. A., d'Herin, P., Harper, M., Kress, M., and Nigg, E. A. (1995). Phosphorylation by p34cdc2 regulates spindle association of human Eg5, a kinesin-related motor essential for bipolar spindle formation in vivo. *Cell* **83,** 1159–1169.

Blangy, A., Arnaud, L., and Nigg, E. A. (1997). Phosphorylation by p34cdc2 protein kinase regulates binding of the kinesin-related motor HsEg5 to the dynactin subunit p150. *J. Biol. Chem.* **272,** 19418–19424.

Boleti, H., Karsenti, E., and Vernos, I. (1996). Xklp2, a novel Xenopus centrosomal kinesin-like protein required for centrosome separation during mitosis. *Cell* **84,** 49–59.

Bornens, M. (1992). Structure and function of isolated centrosomes. *In* "The Centrosome" (V. Kalnins, ed.), pp. 1–43. Academic Press, San Diego.

Bornens, M., Paintrand, M., Berges, J., Marty, M. C., and Karsenti, E. (1987). Structural and chemical characterization of isolated centrosomes. *Cell Motil. Cytoskel.* **8,** 238–249.

Bouckson Castaing, V., Moudjou, M., Ferguson, D. J. P., Mucklow, S., Belkaid, Y., Milon, G., and Crocker, P. R. (1996). Molecular characterisation of ninein, a new coiled-coil protein of the centrosome. *J. Cell Sci.* **109,** 179–190.

Brandt, R., and Lee, G. (1993). Functional organization of microtubule-associated protein tau. Identification of regions which affect microtubule growth, nucleation, and bundle formation in vitro. *J. Biol. Chem.* **15,** 3414–3419.

Bray, D. (1997). The riddle of slow transport—An introduction. *Trends Cell Biol.* **7,** 379.

Bré, M. H., and Karsenti, E. (1990). Effects of brain microtubule-associated proteins on microtubule dynamics and the nucleating activity of centrosomes. *Cell Motil, Cytoskel.* **15,** 88–98.

Brewis, N. D., Street, A. J., Prescott, A. R., and Cohen, P. T. (1993). PPX, a novel protein serine/threonine phosphatase localized to centrosomes. *EMBO J.* **12,** 987–996.

Brinkley, B. R. (1985). Microtubule organizing centers. *Annu. Rev. Cell Biol.* **1,** 145–172.

Brown, C. R., Doxsey, S. J., Hong Brown, L. Q., Martin, R. L., and Welch, W. J. (1996a). Molecular chaperones and the centrosome. A role for TCP-1 in microtubule nucleation. *J. Biol. Chem.* **271,** 824–832.

Brown, C. R., Hong Brown, L. Q., Doxsey, S. J., and Welch, W. J. (1996b). Molecular chaperones and the centrosome. A role for HSP 73 in centrosomal repair following heat shock treatment. *J. Biol. Chem.* **271,** 833–840.

Buendia, B., and Karsenti, E. (1995). Regulation of centrosome function during mitosis. *In* "Advances in Molecular Cell Biology Vol. 13" (E. E. Bittar, ed.) pp. 44–67. JAI Press, London.

Buendia, B., Draetta, G., and Karsenti, E. (1992). Regulation of the microtubule nucleating activity of centrosomes in Xenopus egg extracts: Role of cyclin A-associated protein kinase. *J. Cell Biol.* **116,** 1431–1442.

Caplow, M., Ruhlen, R. L., and Shanks, J. (1994). The free energy for hydrolysis of a microtubule-bound nucleotide triphosphate is near zero: All of the free energy for hydrolysis is stored in the microtubule lattice. *J. Cell Biol.* **127,** 779–788.

Carlier, M. F., and Pantaloni, D. (1978). Kinetic analysis of cooperativity in tubulin polymerization in the presence of guanosine di- or triphosphate nucleotides. *Biochemistry* **17,** 1908–1915.

Carlier, M. F., and Pantaloni, D. (1981). Kinetic analysis of guanosine 5-triphosphate hydrolysis associated with tubulin polymerization. *Biochemistry* **20,** 1918–1924.

Carlier, M. F., Didry, D., and Pantaloni, D. (1997). Hydrolysis of GTP associated with the formation of tubulin oligomers is involved in microtubule nucleation. *Biophys J.* **73,** 418–427.

Centonze, V. E., and Borisy, G. G. (1990). Nucleation of microtubules from mitotic centrosomes is modulated by a phosphorylated epitope. *J. Cell Sci.* **95,** 405–411.

Chen, R., Perrone, C. A., Amos, L. A., and Linck, R. W. (1993). Tektin B1 from ciliary microtubules: Primary structure as deduced from the cDNA sequence and comparison with tektin A1. *J. Cell Sci.* **106,** 909–918.

Chrétien, D., Fuller, S. D., and Karsenti, E. (1995). Structure of growing microtubule ends: Two-dimensional sheets close into tubes at variable rates. *J. Cell Biol.* **129,** 1311–1328.

Chrétien, D., Buendia, B., Fuller, S. D., and Karsenti, E. (1997). Reconstruction of the centrosome cycle from cryoelectron micrographs. *J. Struct. Biol.* **120,** 117–133.

Clark, S. W., and Meyer, D. I. (1992). Centractin is an actin homologue associated with the centrosome [see comments]. *Nature* **359,** 246–250.

Davis, F. M., Tsao, T. Y., Fowler, S. K., and Rao, P. N. (1983). Monoclonal antibodies to mitotic cells. *Proc. Natl. Acad. Sci. USA* **80,** 2926–2930.

Debec, A., Detraves, C., Montmory, C., Geraud, G., and Wright, M. (1995). Polar organization of gamma-tubulin in acentriolar mitotic spindles of Drosophila melanogaster cells. *J. Cell Sci.* **108,** 2645–2653.

Debec, A., Kalpin, R. F., Daily, D. R., McCallum, P. D., Rothwell, W. F., and Sullivan, W. (1996). Live analysis of free centrosomes in normal and aphidicolin-treated Drosophila embryos. *J. Cell Biol.* **134,** 103–115.

De Brabander, M., Guens, G., Nuydens, R., Willebrords, R., Aerts, F., and De Mey, J. (1986). Microtubule dynamics during the cell cycle: The effects of taxol and nocodazole on the microtubule system of PtK2 cells at different stages of the mitotic cycle. *Int. Rev. Cytol.* **101,** 215–274.

Dictenberg, J. B., Zimmerman, W., Sparks, C. A., Young, A., Vidair, C., Zheng, Y., Carrington, W., Fay, F. S., and Doxsey, S. J. (1998). Pericentrin and γ-tubulin form a protein complex and are organized into a novel lattice at the centrosome. *J. Cell Biol.* **141,** 163–174.

Dirksen, E. R. (1991). Centriole and basal body formation during ciliogenesis revisited. *Biol. Cell* **72,** 31–38.

Dogterom, M., Maggs, A. C., and Leibler, S. (1995). Diffusion and formation of microtubule asters: Physical processes versus biochemical regulation. *Proc. Natl. Acad. Sci. USA* **92,** 6683–6688.

Domínguez, J. E., Buendia, B., Lopez Otin, C., Antony, C., Karsenti, E., and Avila, J. (1994). A protein related to brain microtubule-associated protein MAP1B is a component of the mammalian centrosome. *J. Cell Sci.* **107,** 601–611.

Doxsey, S. J., Stein, P., Evans, L., Calarco, P. D., and Kirschner, M. (1994). Pericentrin, a highly conserved centrosome protein involved in microtubule organization [see comments]. *Cell* **76,** 639–650.

Dutcher, S. K. (1989). Genetic analysis of microtubule organizing centers. *In* "Cell Movement: Kinesins," Vol. 2, pp. 83–112. A. R. Liss, New York.

Dutcher, S. K., and Trabuco, E. C. (1998). The UNI3 gene is required for assembly of basal bodies of *Chlamydomonas* and encodes δ-tubulin, a new member of the tubulin superfamily. *Mol. Biol. Cell* **9,** 1293–1308.

Ehler, L. L., Holmes, J. A., and Dutcher, S. K. (1995). Loss of spatial control of the mitotic spindle apparatus in a *Chlamydomonas reinhardtii* mutant strain lacking basal bodies. *Genetics* **141,** 945–960.

Erickson, H. P., and O'Brien, E. T. (1992). Microtubule dynamic instability and GTP hydrolysis. *Annu. Rev. Biophys. Biomol. Struct.* **21,** 145–166.

Errabolu, R., Sanders, M. A., and Salisbury, J. L. (1994). Cloning of a cDNA encoding human centrin, an EF-hand protein of centrosomes and mitotic spindle poles. *J. Cell Sci.* **107,** 9–16.

Feigner, H., Frank, R., and Schliwa, M. (1996). Flexural rigidity of microtubules measured with the use of optical tweezers. *J. Cell Sci.* **109,** 509–516.

Félix, M. A., Antony, C., Wright, M., and Maro, B. (1994). Centrosome assembly in vitro: Role of gamma-tubulin recruitment in *Xenopus* sperm aster formation. *J. Cell Biol.* **124,** 19–31.

Friedman, D. B., Sundberg, H. A., Huang, E. Y., and Davis, T. N. (1996). The 110-kD spindle pole body component of Saccharomyces cerevisiae is a phosphoprotein that is modified in a cell cycle-dependent manner. *J. Cell Biol.* **132,** 903–914.

Fry, A. M., and Nigg, E. A. (1995). The NIMA kinase joins forces with Cdc2. *Curr. Biol.* **5,** 1122–1125.

Fry, A. M., Meraldi, P., and Nigg, E. A. (1998). A centrosomal function for the human Nek2 protein kinase, a member of the NIMA-family of cell cycle regulators. *EMBO J.* **17,** 470–481.

Frydman, J., and Hartl, F. U. (1996). Principles of chaperone-assisted protein folding: Differences between in vitro and in vivo mechanisms [see comments]. *Science* **272,** 1497–1502.

Fukasawa, K., Choi, T., Kuriyama, R., Rulong, S., and Vande Voude, G. F. (1996). Abnormal centrosome amplification in the absence of p53. *Science* **271,** 1744–1747.

Fuller, S. D., Gowen, B. E., Reinsch, S., Sawyer, A., Buendia, B., Wepf, R., and Karsenti, E. (1995). The core of the mammalian centriole contains gamma-tubulin. *Curr. Biol.* **5,** 1384–1393.

Fulton, C. (1971). Centrioles. *In* "Origin and Continuity of Cell Organelles" (J. Reinert and H. Ursprung, eds.), pp. 170–221. Springer-Verlag, Berlin.

Fygenson, D. K., Braun, E., and Libchaber, A. (1994). Phase-diagram of microtubules. *Phys. Rev. E* **50,** 1579–1588.

Gaglio, T., Saredi, A., Bingham, J. B., Hasbani, M. J., Gill, S. R., Schroer, T. A., and Compton, D. A. (1996). Opposing motor activities are required for the organization of the mammalian mitotic spindle pole. *J. Cell Biol.* **135,** 399–414.

Gard, D. L. (1994). Gamma-tubulin is asymmetrically distributed in the cortex of Xenopus oocytes. *Dev. Biol.* **161,** 131–140.

Gard, D. L., Hafezi, S., Zhang, T., and Doxsey, S. J. (1990). Centrosome duplication continues in cycloheximide-treated Xenopus blastulae in the absence of a detectable cell cycle. *J. Cell Biol.* **110,** 2033–2042.

Geissler, S., Pereira, G., Spang, A., Knop, M., Souès, S., Kilmartin, J., and Schiebel, E. (1996). The spindle pole body component Spc98p interacts with the γ-tubulin-like Tub4p of *Saccharomyces cerevisiae* at the sites of microtubule attachment. *EMBO J.* **15,** 3899–3911.

Gibbons, I. R., and Grimstone, A. V. (1960). On flagellar structure in certain flagellates. *J. Biophys. Biochem.* **7,** 697–716.

González, C., Tavosanis, G., and Mollinari, C. (1998). Centrosomes and microtubule organization during *Drosophila* development. *J. Cell Sci.* **111,** 2697–2706.

Goodenough, U. W., and St. Clair, H. S. (1975). Bald-2: A mutation affecting the formation of doublet and triplet sets of microtubules in *Chlamydomonas reinhardtii. J. Cell Biol.* **66,** 480–491.

Gould, R. R., and Borisy, G. G. (1977). The pericentriolar material in Chinese hamster ovary cells nucleates microtubule formation. *J. Cell Biol.* **73,** 601–615.

Grafen, A. (1988). A centrosomal theory of the short term evolutionary maintenance of sexual reproduction. *J. Theor. Biol.* **131,** 163–173.

Hagan, I., and Yanagida, M. (1995). The product of the spindle formation gene sad1+ associates with the fission yeast spindle pole body and is essential for viability. *J. Cell Biol.* **129,** 1033–1047.

Hall, J. L., Ramanis, Z., and Luck, D. J. (1989). Basal body/centriolar DNA: Molecular genetics studies in *Chlamydomonas. Cell* **59,** 121–132.

Hartman, J. J., Mahr, J., McNally, K., Okawa, K., Iwamatsu, A., Thomas, S., Cheesman, S., Heuser, J., Vale, R. D., and McNally, F. J. (1998). Katanin, a microtubule-severing protein, is a novel AAA ATPase that targets to the centrosome using a WD40-containing subunit. *Cell* **93,** 277–287.

Heald, R., Tournebize, R., Blank, T., Sandaltzopoulos, R., Becker, P., Hyman, A., and Karsenti, E. (1996). Self-organization of microtubules into bipolar spindles around chromatin beads in *Xenopus* egg extracts. *Nature* **382,** 420–425.

Heald, R., Tournebize, R., Habermann, A., Karsenti, E., and Hyman, A. (1997). Spindle assembly in Xenopus egg extracts: Respective roles of centrosomes and microtubule self-organization. *J. Cell Biol.* **138,** 615–628.

Heidemann, S. R., Sander, G., and Kirschner, M. W. (1977). Evidence for a functional role of RNA in centrioles. *Cell* **10,** 337–350.

Hill, T. L., and Kirschner, M. W. (1983). Regulation of microtubule and actin filament assembly–disassembly by associated small and large molecules. *Int. Rev. Cytol* **84,** 185–234.

Hinchcliffe, E. H., Cassels, G. O., Rieder, C. L., and Sluder, G. (1998). The coordination of centrosome reproduction with nuclear events of the cell cycle in the sea urchin zygote. *J. Cell Biol.* **140,** 1417–1426.

Hirokawa, N. (1997). The mechanism of fast and slow transport in neurons: Identification and characterization of the new kinesin superfamily motors. *Curr. Opin. Cell Biol.* **7,** 605–614.

Hirokawa, N., Terada, S., Funakoshi, T., and Takeda, S. (1997). Slow axonal transport: The subunit transport model. *Trends Cell Biol.* **7,** 384–388.

Huang, B. (1990). Genetics and biochemistry of centrosomes and spindle poles. *Curr. Opin. Cell Biol.* **2,** 28–32.

Hyman, A. A., and Karsenti, E. (1996). Morphogenetic properties of microtubules and mitotic spindle assembly. *Cell* **84,** 401–411.

Hyman, A. A., Salser, S., Drechsel, D. N., Unwin, N., and Mitchison, T. J. (1992). Role of GTP hydrolysis in microtubule dynamics: Information from a slowly hydrolyzable analogue, GMPCPP. *Mol. Biol. Cell* **3,** 1155–1167.

Inoué, S., and Salmon, E. D. (1995). Force generation by microtubule assembly/disassembly in mitosis and related movements. *Mol. Biol. Cell* **6,** 1619–1640.

Johnson, K. A., and Rosenbaum, J. L. (1990). The basal bodies of *Chlamydomonas reinhardtii* do not contain immunologically detectable DNA. *Cell* **62,** 615–619.

Joshi, H. C., and Palevitz, B. A. (1996). Gamma-tubulin and microtubule organization in plants. *Trends Cell Biol.* **6,** 41–44.

Joshi, H. C., Palacios, M. J., McNamara, L., and Cleveland, D. W. (1992). Gamma-tubulin is a centrosomal protein required for cell cycle-dependent microtubule nucleation. *Nature* **356,** 80–83.

Kalnins, V. I. (1992). "The Centrosome" (D. E. Buetow, I. L. Cameron, G. M. Padilla, and A. M. Zimmerman, eds.) p. 368. Academic Press, San Diego.

Kalt, A., and Schliwa, M. (1996). A novel structural component of the Dictyostelium centrosome. *J. Cell Sci.* **109,** 3103–3112.

Kapeller, R., Toker, A., Cantley, L. C., and Carpenter, C. L. (1995). Phosphoinositide 3-kinase binds constitutively to alpha/beta-tubulin and binds to gamma-tubulin in response to insulin. *J. Biol. Chem.* **270,** 25985–2591.

Karsenti, E. (1991). Mitotic spindle morphogenesis in animal cells. *Sem. Cell Biol.* **2,** 251–260.

Karsenti, E. (1993). Severing microtubules in mitosis. *Curr. Biol.* **3,** 208–210.

Karsenti, E., Newport, J, and Kirschner, M. (1984). Respective roles of centrosomes and chromatin in the conversion of microtubule arrays from interphase to metaphase. *J. Cell Biol.* **99,** 47s–57s.

Keating, T. J., Peloquin, J. G., Rodionov, V. I., Momcilovic, D., and Borisy, G. G. (1997). Microtubule release from the centrosome. *Proc. Natl. Acad. Sci. USA* **94,** 5078–5083.

Kellogg, D. R., Field, C. M., and Alberts, B. M. (1989). Identification of microtubule-associated proteins in the centrosome, spindle, and kinetochore of the early Drosophila embryo. *J. Cell Biol.* **109,** 2977–2991.

Kellogg, D. R., Moritz, M., and Alberts, B. M. (1994). The centrosome and cellular organization. *Annu. Rev. Biochem.* **63,** 639–674.

Kellogg, D. R., Oegema, K., Raff, J., Schneider, K., and Alberts, B. M. (1995). CP60: A microtubule-associated protein that is localized to the centrosome in a cell cycle-specific manner. *Mol. Biol. Cell* **6,** 1673–1684.

Kenny, J., Karsenti, E., Gowen, B., and Fuller, S. D. (1997). Three-dimensional reconstruction of the mammalian centriole from cryoelectron micrographs: The use of common lines for orientation and lignment. *J. Struct. Biol.* **120,** 320–328.

Keryer, G., Rios, R. M., Landmark, B. F., Skalhegg, B., Lohmann, S. M., and Bornens, M. (1993). A high-affinity binding protein for the regulatory subunit of cAMP-dependent protein kinase II in the centrosome of human cells. *Exp. Cell Res.* **204,** 230–240.

Keryer, G., Celati, C., and Klotz, C. (1995). In isolated human centrosomes, the associated kinases phosphorylate a specific subset of centrosomal proteins. *Biol. Cell* **84,** 155–165.

Kidd, D., and Raff, J. W. (1997). LK6, a short lived protein kinase in *Drosophila* that can associate with microtubules and centrosomes. *J. Cell Sci.* **110,** 209–219.

Kirschner, M., and Mitchison, T. (1986). Beyond self-assembly: From microtubules to morphogenesis. *Cell* **45,** 329–342.

Knop, M., Pereira, G., Geissler, S., Grein, K., and Schiebel, E. (1997). The spindle pole body component Spc97p interacts with the γ-tubulin of Saccharomyces cerevisiae and functions in microtubule organization and spindle pole body duplication. *EMBO J.* **16,** 1550–1564.

Kochanski, R. S., and Borisy, G. G. (1990). Mode of centriole duplication and distribution. *J. Cell Biol.* **110,** 1599–1605.

Kurachi, M., Hoshi, M., and Tashiro, H. (1995). Buckling of a single microtubule by optical trapping forces: Direct measurement of microtubule rigidity. *Cell Motil. Cytoskel.* **30,** 221–228.

Kuriyama, R., and Borisy, G. G. (1981). Centriole cycle in Chinese hamster ovary cells as determined by whole-mount electron microscopy. *J. Cell Biol.* **91,** 814–821.

Kuriyama, R., Dasgupta, S., and Borisy, G. G. (1986). Independence of centriole formation and initiation of DNA synthesis in Chinese hamster ovary cells. *Cell Motil. Cytoskel.* **6,** 355–362.

Lambert, A.-M. (1993). Microtubule-organizing centers in higher plants. *Curr. Opin. Cell Biol.* **5,** 116–122.

Lange, B. M., and Gull, K. (1995). A molecular marker for centriole maturation in the mammalian cell cycle. *J. Cell Biol.* **130,** 919–927.

Lange, B. M. H., and Gull, K. (1996). Structure and function of the centriole in animal cells: Progress and questions. *Trends Cell Biol.* **6,** 348–352.

Levy, Y. Y., Lai, E. Y., Remillard, S. P., Heintzelman, M. B., and Fulton, C. (1996). Centrin is a conserved protein that forms diverse associations with centrioles and MTOCs in Naegleria and other organisms. *Cell Motil. Cytoskel.* **33,** 298–323.

Li, Q., and Joshi, H. C. (1995). Gamma-tubulin is a minus end-specific microtubule binding protein. *J. Cell Biol.* **131,** 207–214.

Linck, R. W., and Stephens, R. E. (1987). Biochemical characterization of tektins from sperm flagellar doublet microtubules. *J. Cell Biol.* **104,** 1069–1075.

Maldonado Codina, G., and Glover, D. M. (1992). Cyclins A and B associate with chromatin and the polar regions of spindles, respectively, and do not undergo complete degradation at anaphase in syncytial Drosophila embryos. *J. Cell Biol.* **116,** 967–976.

Maniotis, A., and Schliwa, M. (1991). Microsurgical removal of centrosomes blocks cell reproduction and centriole generation in BSC-1 cells. *Cell* **67,** 495–504.

Marek, L. F. (1978). Control of spindle form and function in Grasshopper spermatocytes. *Chromosoma* **68,** 367–398.

Margolis, R. L., and Wilson, L. (1998). Microtubule treadmilling: What goes around comes around. *BioEssays* **20,** 830–836.

Marschall, L. G., and Stearns, T. (1997). Cytoskeleton: Anatomy of an organizing center. *Curr. Biol.* **7,** R754–R756.

Marschall, L. G., Jeng, R. L., Mulholland, J., and Stearns, T. (1996). Analysis of Tub4p, a yeast γ-tubulin-like protein: Implications for microtubule-organizing center function. *J. Cell Biol.* **134,** 443–454.

Martin, O. C., Gunawardane, R. N., Iwamatsu, A., and Zheng, Y. (1998). Xgrip109: A γ-tubulin-associated protein with an essential role in γ tubulin ring complex (γTuRC) assembly and centrosome function. *J. Cell Biol.* **141,** 675–687.

Matus, A. (1994). Stiff microtubules and neuronal morphology. *Trends Neurosci.* **17,** 19–22.

Mazia, D. (1984). Centrosomes and mitotic poles. *Exp. Cell Res.* **153,** 1–15.

Mazia, D. (1987). The chromosome cycle and the centrosome cycle in the mitotic cycle. *Int. Rev. Cytol.* **100,** 49–92.

McIntosh, J. R. (1983). The centrosome as an organizer of the cytoskeleton. *Modern Cell Biol.* **2,** 115–142.

McIntosh, J. R. (1984). Microtubule catastrophe. *Nature* **312,** 196–197.

McNally, F. J., and Vale, R. D. (1993). Identification of katanin, an ATPase that severs and disassembles stable microtubules. *Cell* **75,** 419–429.

McNally, F. J., Okawa, K., Iwamatsu, A., and Vale, R. D. (1996). Katanin, the microtubule-severing ATPase, is concentrated at centrosomes. *J. Cell Sci.* **109,** 561–567.

Meads, T., and Schroer, T. A. (1995). Polarity and nucleation of microtubules in polarized epithelial cells. *Cell Motil. Cytoskel.* **32,** 273–288.

Méda, P., Chevrier, V., Eddé, B., and Job, D. (1997). Demonstration and analysis of tubulin binding sites on centrosomes. *Biochemistry* **36,** 2550–2558.

Merdes, A., and Cleveland, D. W. (1997). Pathways of spindle formation: Different mechanisms, conserved components. *J. Cell Biol.* **138,** 953–956.

Merdes, A., Ramyar, K., Vechio, J. D., and Cleveland, D. W. (1996). A complex of NuMA and cytoplasmic dynein is essential for mitotic spindle assembly. *Cell* **87,** 447–458.

Mickey, B., and Howard, J. (1995). Rigidity of microtubules is increased by stabilizing agents. *J. Cell Biol.* **130,** 909–917.

Mogensen, M. M., Mackie, J. B., Doxsey, S. J., Stearns, T., and Tucker, J. B. (1997). Centrosomal deployment of γ-tubulin and pericentrin: Evidence for a microtubule-nucleating domain and a minus-end docking domain in certain mouse epithelial cells. *Cell Motil. Cytoskel.* **36,** 276–290.

Moritz, M., Zheng, Y., Alberts, B. M., and Oegema, K. (1998). Recruitment of the γ-tubulin ring complex to *Drosophila* salt-stripped centrosome scaffolds. *J. Cell Biol.* **142,** 775–786.

Moritz, M., Braunfeld, M. B., Fung, J. C., Sedat, J. W., Alberts, B. M., and Agard, D. A. (1995a). Three-dimensional structural characterization of centrosomes from early Drosophila embryos. *J. Cell Biol.* **130,** 1149–1159.

Moritz, M., Braunfeld, M. B., Sedat, J. W., Alberts, B., and Agard, D. A. (1995b). Microtubule nucleation by gamma-tubulin-containing rings in the centrosome. *Nature* **378,** 638–640.

Moudjou, M., Bordes, N., Paintrand, M., and Bornens, M. (1996). Gamma-tubulin in mammalian cells: The centrosomal and the cytosolic forms. *J. Cell Sci.* **109,** 875–887.

Murphy, S. M., Urbani, L., and Stearns, T. (1998). The mammalian γ-tubulin complex contains homologues of the yeast spindle pole body components Spc97p and Spc98p. *J. Cell Biol.* **141,** 663–674.

Nelson, R. J., and Craig, E. A. (1992). TCP1-molecular chaperonin of the cytoplasm? *Curr. Biol.* **2,** 487–489.

Nicklas, R. B. (1988). Chromosomes and kinetochores do more in mitosis than previously thought. *In* "Chromosome Structure and Function: The Impact of New Concepts" (J. P. Gustafson and R. Appels, eds.), pp. 53–74. Plenum, New York.

Nogales, E., Wolf, S. G., and Downing, K. H. (1998). Structure of the alphabeta tubulin dimer by electron crystallography. *Nature* **391,** 199–203.

Nomura, M., and Held, W. A. (1974). Reconstitution of ribosomes: Studies of ribosome structure, function and assembly. *In* "Ribosomes" (M. Nomura, A. Tissières, and P. Lengyel, eds.), pp. 193–223. Cold Spring Harbor Laboratory Press, Cold Spring Harbor, NY.

Oakley, C. E., and Oakley, B. R. (1989). Identification of γ-tubulin, a new member of the tubulin superfamily encoded by *mipA* gene of *Aspergillus nidulans. Nature* **338,** 662–664.

Oegema, K., Whitfield, W. G., and Alberts, B. (1995). The cell cycle-dependent localization of the CP190 centrosomal protein is determined by the coordinate action of two separable domains. *J. Cell Biol.* **131,** 1261–1273.

Oegema, K., Marshall, W. F., Sedat, J. W., and Alberts, B. M. (1997). Two proteins that cycle asynchronously between centrosomes and nuclear structures: *Drosophila* CP60 and CP190. *J. Cell Sci.* **110,** 1573–1583.

Osborne, M. A., Schlenstedt, G., Jinks, T., and Silver, P. A. (1994). Nuf2, a spindle pole body-associated protein required for nuclear division in yeast. *J. Cell Biol.* **125,** 853–866.

Paintrand, M., Moudjou, M., Delacroix, H., and Bornens, M. (1992). Centrosome organization and centriole architecture: Their sensitivity to divalent cations. *J. Struct. Biol.* **108,** 107–128.

Palacios, M. J., Joshi, H. C., Simerly, C., and Schatten, G., (1993). Gamma-tubulin reorganization during mouse fertilization and early development. *J. Cell Sci.* **104,** 383–389.

Palazzo, R. E., Vaisberg, E., Cole, R. W., and Rieder, C. L. (1992). Centriole duplication in lysates of Spisula solidissima oocytes (published erratum appears in *Science* **256**(5065), 1746 1992). *Science* **256,** 219–221.

Paoletti, A., Moudjou, M., Paintrand, M., Salisbury, J. L., and Bornens, M. (1996). Most of centrin in animal cells is not centrosome-associated and centrosomal centrin is confined to the distal lumen of centrioles. *J. Cell Sci.* **109,** 3089–3102.

Pereira, G., and Schiebel, E. (1997). Centrosome–microtubule nucleation. *J. Cell Sci.* **110,** 295–300.

Picard, A., Karsenti, E., Dabauvalle, M. C., and Doree, M. (1987). Release of mature starfish oocytes from interphase arrest by microinjection of human centrosomes. *Nature* **327,** 170–172.

Pickett-Heaps, J. D. (1969). The evolution of the mitotic apparatus: An attempt at comparative ultrastructural cytology in dividing plant cells. *Cytobios* **3,** 257–280.

Pockwinse, S. M., Krockmalnic, G., Doxsey, S. J., Nickerson, J., Lian J. B., van Wijnen, A. J., Stein, J. L., Stein, G. S., and Penman, S. (1997). Cell cycle independent interaction of CDC2 with the centrosome which is associated with the nuclear matrix-intermediate filament scaffold. *Proc. Natl. Acad. Sci. USA* **94,** 3022–3027.

Porter, K. R. (1966). Cytoplasmic microtubules and their functions. *In* "Principles of Biomolecular Organization" (G. E. W. Wolstenholme and M. O'Connor, eds.), pp. 308–357 Churchill, London.

Preuss, U., Biernat, J., Mandelkow, E.-M., and Mandelkow, E. (1997). The "jaws" model of tau-microtubule interaction examined in CHO cells. *J. Cell Sci.* **110,** 789–800.

Raff, J. W. (1996). Centrosomes and microtubules: Wedded with a ring. *Trends Cell Biol.* **6,** 248–251.

Raff, J. W., Kellogg, D. R., and Alberts, B. M. (1993). Drosophila gamma-tubulin is part of a complex containing two previously identified MAPs. *J. Cell Biol.* **121,** 823–835.

Rattner, J. B., and Phillips, S. G. (1973). Independence of centriole formation and DNA synthesis. *J. Cell Biol.* **57,** 359–372.

Reinsch, S., and Karsenti, E. (1994). Orientation of spindle axis and distribution of plasma membrane proteins during cell division in polarized MDCKII cells. *J. Cell Biol.* **126,** 1509–1526.

Rieder, C. L., and Borisy, G. G. (1982). The centrosome cycle in PTK2 cells: Asymetric distribution and structural changes in the pericentriolar material. *Biol. Cell* **44,** 117–132.

Rodionov, V. I., and Borisy, G. G. 1997. Microtubule treadmilling *in vivo. Science* **275,** 215–218.

Roghi, C., Giet, R., Uzbekov, R., Morin, N., Chartrain, I., Le Guellec, R., Couturier, A., Dorée, M., Philippe, M., and Prigent, C. The Xenopus protein kinase pEg2 associates with the centrosome in a cell cycle-dependent manner, binds to the spindle microtubules, and is involved in bipolar mitotic spindle assembly. *J. Cell Sci.* **111,** 557–572.

Rose, M. D., Biggins, S., and Satterwhite, L. L. (1993). Unravelling the tangled web at the microtubule-organizing center. *Curr. Opin. Cell Biol.* **5,** 105–115.

Sagan, L. (1967). On the origin of mitosing cells. *J. Theor. Biol.* **14,** 225–274.

Sagata, N., Watanabe, N., Vande Woude, G. F., and Ikawa, Y. (1989). The c-*mos* proto-oncogene is a cytostatic factor (CSF) responsible for meiotic arrest in vertebrate eggs. *Nature* **342,** 512–518.

Salisbury, J. L. (1995). Centrin, centrosomes, and mitotic spindle poles. *Curr. Opin. Cell Biol.* **7,** 39–45.

Sanders, M. A., and Salisbury, J. L. (1994). Centrin plays an essential role in microtubule severing during flagellar excision in *Chlamydomans reinhardtii. J. Cell Biol.* **124,** 795–805.

Sawin, K. E., and Mitchison, T. J. (1994). Microtubule flux in mitosis is independent of chromosomes, centrosomes, and antiparallel microtubules. *Mol. Biol. Cell* **5,** 217–226.

Sawin, K. E., and Mitchison, T. J. (1995). Mutations in the kinesin-like protein Eg5 disrupting localization to the mitotic spindle. *Proc. Natl. Acad. Sci. USA* **92,** 4289–4293.

Schatten, G. (1994). The centrosome and its mode of inheritance: The reduction of the centrosome during gametogenesis and its restoration during fertilization. *Dev. Biol.* **165,** 299–335.

Schiebel, E., and Bornens, M. (1995). In search of a function for centrins. *Trends Cell Biol.* **5,** 197–201.

Schliwa, M., and Höner, B. (1993). Microtubules, centrosomes and intermediate filaments in directed cell movement. *Trends Cell Biol.* **3,** 377–380.

Shu, H. B., and Joshi, H. C. (1995). Gamma-tubulin can both nucleate microtubule assembly and self-assemble into novel tubular structures in mammalian cells. *J. Cell Biol.* **130,** 1137–1147.

Simerly, C., Wu, G.-J., Zoran, S., Ord, T., Rawlins, R., Jones, J., Navara, C., Gerrity, M., Rinehart, J., Binor, Z., Asch, R., and Schatten, G. (1995). The paternal inheritance of the centrosome, the cells microtubule organizing center, in humans, and the implications for infertility. *Nat. Med.* **1,** 47–52.

Sluder, G. (1992). Double or nothing. *Curr. Biol.* **2,** 243–245.

Sluder, G., Miller, F. J., Lewis, K., Davison, E. D., and Rieder, C. L. (1989). Centrosome inheritance in starfish zygotes: Selective loss of the maternal centrosome after fertilization. *Dev. Biol.* **131,** 567–579.

Sluder, G., Miller, F. J., Cole, R., and Rieder, C. L. (1990). Protein synthesis and the cell cycle: Centrosome reproduction in sea urchin eggs is not under translational control. *J. Cell Biol.* **110,** 2025–2032.

Sluder, G., Miller, F. J., and Lewis, K. (1993). Centrosome inheritance in starfish zygotes. II: Selective suppression of the maternal centrosome during meiosis. *Dev. Biol.* **155,** 58–67.

Smirnova, E. A., and Bajer, A. S. (1992). Spindle poles in higher plant mitosis. *Cell Motil. Cytoskel.* **23,** 1–7.

Snaith, H. A., Armstrong, C. G., Guo, Y., Kaiser, K., and Cohen, P. T. W. (1996). Deficiency of protein phosphatase 2A uncouples the nuclear and centrosome cycles and prevents attachment of microtubules to the kinetochore in *Drosophilia microtubule star* (mts) embryos. *J. Cell Sci.* **109,** 3001–3012.

Snyder, J. A., and McIntosh, J. R. (1975). Initiation and growth of microtubules from mitotic centers in lysed mammalian cells. *J. Cell Biol.* **67,** 744–760.

Sobel, S. G., and Snyder, M. (1995). A highly divergent γ-tubulin gene is essential for cell growth and proper microtubule organization in *Saccharomyces cerevisiae. J. Cell Biol.* **131,** 1775–1788.

Stearns, T., and Kirschner, M. (1994). In vitro reconstitution of centrosome assembly and function: The central role of gamma-tubulin [see comments]. *Cell* **76,** 623–637.

Stearns, T., and Winey, M. (1997). The cell center at 100. *Cell* **91,** 303–309.

Steffen, W., and Linck, R. W. (1988). Evidence for tektins in centrioles and axonemal microtubules. *Proc. Natl. Acad. Sci. USA* **85,** 2643–2647.

Steffen, W., Fajer, E. A., and Linck, R. W. (1994). Centrosomal components immunologically related to tektins from ciliary and flagellar microtubules. *J. Cell Sci.* **107,** 2095–2105.

Tash, J. S., Means, A. R., Brinkley, B. R., Dedman, J. R., and Cox, S. M. (1980). Cyclic nucleotide and Ca^{2+} regulation of microtubule initiation and elongation. *In* "Microtubules and Microtubule Inhibitors" (M. De Brabander and J. DeMey, eds.), pp. 269–279. Elsevier/North-Holland, Amsterdam.

Tassin, A.-M., Celati, C., Moudjou, M., and Bornens, M. (1998). Characterization of the human homologue of the yeast Spc98p and its association with γ-tubulin. *J. Cell Biol.* **141,** 689–701.

Tian, G., Huang, Y., Rommelaere, H., Vandekerckhove, J., Ampe, C., and Cowan, N. J. (1996). Pathway leading to correctly folded beta-tubulin. *Cell* **86,** 287–296.

Tilney, L. G., Bryan, J., Bush, D. J., Fujiwara, K., Mooseker, M. S., Murphy, D. B., and Snyder, D. H. (1973). Microtubules: Evidence for 13 protofilaments. *J. Cell Biol.* **59,** 267–275.

Tournebize, R., Andersen, S. S. L., Verde, F., Dorée, M., Karsenti, E., and Hyman, A. A. (1997). Distinct roles of PP1 and PP2A-like phosphatases in control of microtubule dynamics during mitosis. *EMBO J.* **16,** 5537–5549.

Tournier, F., Komesli, S., Paintrand, M., Job, D., and Bornens, M. (1991a). The intercentriolar linkage is critical for the ability of heterologous centrosomes to induce parthenogenesis in *Xenopus. J. Cell Biol.* **113,** 1361–1369.

Tournier, F., Cyrklaff, M., Karsenti, E., and Bornens, M. (1991b). Centrosomes competent for parthenogenesis in *Xenopus* eggs support procentriole budding in cell-free extracts. *Proc. Natl. Acad. Sci. USA* **88,** 9929–9933.

Vandre, D. D., Davis, F. M., Rao, P. N., and Borisy, G. G. (1986). Distribution of cytoskeletal proteins sharing a conserved phosphorylated epitope. *Eur. J. Cell Biol.* **41,** 72–81.

Verde, F., Berrez, J.-M., Antony, C., and Karsenti, E. (1991). Taxol-induced microtubule asters in mitotic extracts of *Xenopus* eggs: Requirement for phosphorylated factors and cytoplasmic dynein. *J. Cell Biol.* **112,** 1177–1187.

Verde, F., Dogterom, M., Stelzer, E., Karsenti, E., and Leibler, S. (1992). Control of microtubule dynamics and length by cyclin A- and cyclin B-dependent kinases in Xenopus egg extracts. *J. Cell Biol.* **118,** 1097–1108.

Vogel, J. M., Stearns, T., Rieder, C. L., and Palazzo, R. E. (1997). Centrosomes isolated from *Spisula solidissima* oocytes contain rings and an unusual stoichiometric ratio of α/β tubulin. *J. Cell Biol.* **137,** 193–202.

Vogl, A. W., Weis, M., and Pfeiffer, D. C. (1995). The perinuclear centriole-containing centrosome is not the major microtubule organizing center in Sertoli cells. *Eur. J. Cell Biol.* **66,** 165–179.

Vorobjev, I. A., Svitkina, T. M., and Borisy, G. G. (1997). Cytoplasmic assembly of microtubules in cultured cells. *J. Cell Sci.* **110,** 2635–2645.

Walker, R. A., O'Brien, E. T., Pryer, N. K., Sobeiro, M. F., Voter, W. A., Erickson, H. P., and Salmon, E. D. (1988). Dynamic instability of individual microtubules analysed by video light microscopy: Rate constants and transition frequencies. *J. Cell Biol.* **107,** 1437.

Waters, J. C., Cole, R. W., and Rieder, C. L. (1993). The force-producing mechanism for centrosome separation during spindle formation in vertebrates is intrinsic to each aster. *J. Cell Biol.* **122,** 361–372.

Waters, J. C., Mitchison, T. J., Rieder, C. L., and Salmon, E. D. (1996). The kinetochore microtubule minus-end disassembly associated with poleward flux produces a force that can do work. *Mol. Biol. Cell* **7,** 1547–1558.

Whitfield, W. G., Millar, S. E., Saumweber, H., Frasch, M., and Glover, D. M. (1988). Cloning of a gene encoding an antigen associated with the centrosome in Drosophila. *J. Cell Sci.* **89,** 467–480.

Whitfield, W. G., Chaplin, M. A., Oegema, K., Parry, H., and Glover, D. M. (1995). The 190 kDa centrosome-associated protein of Drosophila melanogaster contains four zinc finger motifs and binds to specific sites on polytene chromosomes. *J. Cell Sci.* **108,** 3377–3387.

Wigge, P. A., Jensen, O. N., Holmes, S., Soues, S., Mann, M., and Kilmartin, J. V. (1998). Analysis of the *Saccharomyces* spindle pole by matrix-assisted laser desorption/ionization (MALDI) mass spectroscopy. *J. Cell Biol.* **141,** 967–977.

Wilsman, N. J., and Farnum, C. E. (1983). Arrangement of C tubule protofilaments in mammalian basal bodies. *J. Ultstruct. Res.* **84,** 205–212.

Wilson, E. B. (1896). "The Cell in Development and Inheritance." Macmillan, New York.

Wilson, E. B. (1925). Some problems of cell organization. *In* "The Cell in Development and Heredity," pp. 670–700. Macmillan, New York.

Winey, M. (1996). Genome stability: Keeping the centrosome cycle on track. *Curr. Biol.* **6,** 962–964.

Wittmann, T., Boleti, H., Antony, C., Karsenti, E., and Vernos, I. (1998). Localization of the kinesin-like protein Xklp2 to spindle poles requires a leucine zipper, a microtubule-associated protein, and dynein. *J. Cell Biol.* **143,** 673–685.

Yang, Z.-H., Gallicanom, G. I., Yu, Q.-C., and Fuchs, E. (1997). An unexpected localization of basonuclin in the centrosome, mitochondria, and acrosome of developing spermatids. *J. Cell Biol.* **137,** 657–669.

Zhang, D., and Nicklas, R. B. (1995). The impact of chromosomes and centrosomes on spindle assembly as observed in living cells. *J. Cell Biol.* **129,** 1287–1300.

Zheng, Y., Jung, M. K., and Oakley, B. R. (1991). Gamma-tubulin is present in Drosophila melanogaster and Homo sapiens and is associated with the centrosome. *Cell* **65,** 817–823.

Zheng, Y., Wong, M. L., Alberts, B., and Mitchison, T. (1995). Nucleation of microtubule assembly by a γ-tubulin-containing ring complex. *Nature* **378,** 578–583.

Zhu, J., Bloom, S. E., Lazarides, E., and Woods, C. (1995). Identification of a novel Ca(2+)-regulated protein that is associated with the marginal band and centrosomes of chicken erythrocytes. *J. Cell Sci.* **108,** 685–698.

The Pecten Oculi of the Chicken: A Model System for Vascular Differentiation and Barrier Maturation

Hartwig Wolburg,* Stefan Liebner,* Andreas Reichenbach,† and Holger Gerhardt*
*Institute of Pathology, University of Tübingen, D-72076 Tübingen, Germany; and †Paul Flechsig Institute for Brain Research, University of Leipzig, D-04109 Leipzig, Germany

The pecten oculi is a convolute of blood vessels in the vitreous body of the avian eye. This structure is well known for more than a century, but its functions are still a matter of controversies. One of these functions must be the formation of a blood–retina barrier because there is no diffusion barrier for blood-borne compounds available between the pecten and the retina. Surprisingly, the blood–retina barrier characteristics of this organ have not been studied so far, although the pecten oculi may constitute a fascinating model of vascular differentiation and barrier maturation: Pectinate endothelial cells grow by angiogenesis from the ophthalmotemporal artery into the pecten primordium and consecutively gain barrier properties. The pectinate pigmented cells arise during development from retinal pigment epithelial cells and subsequently lose barrier properties. These inverse transdifferentiation processes may be triggered by the peculiar microenvironment in the vitreous body. In addition, the question is discussed whether the avascularity of the avian retina may be due to the specific metabolic activity of the pecten.

KEY WORDS: Pecten oculi, Blood–brain barrier, Blood–retina barrier, Retina, Tight junctions, Angiogenesis, Vascularization, Glucose metabolism.

I. Introduction

The blood–brain barrier (BBB) is destined to protect the central nervous system (CNS) against neurotoxic compounds and variations of the composition of the blood. It is established in the endothelium of brain capillaries

0074-7696/99 $30.00

(Reese and Karnovsky, 1967; Brightman and Reese, 1969), the arachnoid layer as the outer site of the blood–cerebrospinal fluid barrier (Nabeshima *et al.,* 1975; Rascher and Wolburg, 1997), and both the tanycytes and the choroid plexus as the inner site of the blood–cerebrospinal fluid barrier (Van Deurs and Koehler, 1979; Mollgard and Saunders, 1986). A special part of the BBB is the blood–retina barrier, which is established mainly by the pigment epithelium (Vinores, 1995). The paracellular route in all these barriers is obstructed by tight junctions (TJs; Kniesel and Wolburg, 1999). This obstruction requires the expression of many specific carrier and transport systems that allow molecular access to the brain (Dermietzel and Krause, 1991; Wolburg and Risau, 1995). Extensive investigations of endothelial cells including *in vitro* models have been undertaken to achieve a better understanding of the complex processes of formation and maintenance of these BBB characteristics (Rubin, 1991; Rubin *et al.,* 1991; Abbott *et al.,* 1992; Joo, 1993; Wolburg *et al.,* 1994; Biegel *et al.,* 1995; Lane *et al.,* 1995; El Hafny *et al.,* 1996; Muruganandam *et al.,* 1997; Stanness *et al.,* 1997). Astrocytes are believed to play a role in both the underlying induction mechanism and the further stabilization of acquired endothelial barrier (Beck *et al.,* 1986; Janzer and Raff, 1987; Tao-Cheng *et al.,* 1987; Tontsch and Bauer, 1991). However, this view was challenged by Holash *et al.* (1993), who emphasized that there is no conclusive evidence for a significant role of astrocytes in the initial expression of the BBB.

In a more general approach, Engelhardt and Risau (1995) introduced the hypothesis that the initial commitment of endothelial cells to express BBB properties requires the interaction of these cells with neuroectodermal cells which do not necessarily have to be astrocytes. The idea of a primary commitment by neuroectodermal cells, and a secondary stabilization or maintenance by astrocytes, is supported by observations showing that invading vascular cells acquire barrier properties before astrocytes are differentiated and (even later) intimately associated with endothelial cells. Many experiments have shown that cultured astrocytes are able to induce BBB properties in cocultured endothelial cells (Joo, 1993; Wolburg and Risau, 1995). However, these *in vitro* models suffer from a lack of the peculiar spatial topology and polarization of cells participating in the formation of the BBB complex *in vivo.* Thus, these model systems are incomplete and are insufficient in the induction and maintenance of the real barrier function requiring TJs with high electrical resistance and low permeability. Therefore, we have inaugurated the pecten oculi of the avian eye as a new useful model of the BBB (Gerhardt *et al.,* 1996; Liebner *et al.,* 1997). This article provides a comprehensive presentation of this model, which we hope will provide a better understanding of the formation of barrier properties in the CNS *in vivo.*

The pecten oculi is a convolute of blood vessels emerging from the optic disc into the vitreous body. The morphology of the pecten, including the microvasculature of the avian eye, has been described in detail by using light and electron microscopy including freeze fracturing and scanning electron microscopy (Mann, 1924; Rohen, 1955; Romanoff, 1960; O'Rahilly and Meyer, 1961; Wingstrand and Munk, 1965; Raviola and Raviola, 1967; Welsch, 1972; Dieterich *et al.,* 1973; Bawa and YashRoy, 1974; Braekevelt, 1984; Hossler and Olson, 1984; Uehara *et al.,* 1990; Kiama *et al.,* 1994; Smith *et al.,* 1996; Figs. 1 and 2). In the adult chicken, the pecten oculi mainly consists of two cell types: endothelial cells lining blood vessels, and pigmented cells intimately associated with the former and filling spaces between blood vessels. There are only a few pericytes on mature blood vessels. The abluminal membrane of endothelial cells is separated from the vitreous only by a double basal lamina and a thick sheath of collagenous extracellular matrix. If this unusual situation is compared with other parts of the CNS, two suggestions can be made: (i) Because every neural tissue in the CNS is protected by a diffusion barrier, there should also be a corresponding blood–retina barrier, and (ii) because there is no cellular element between

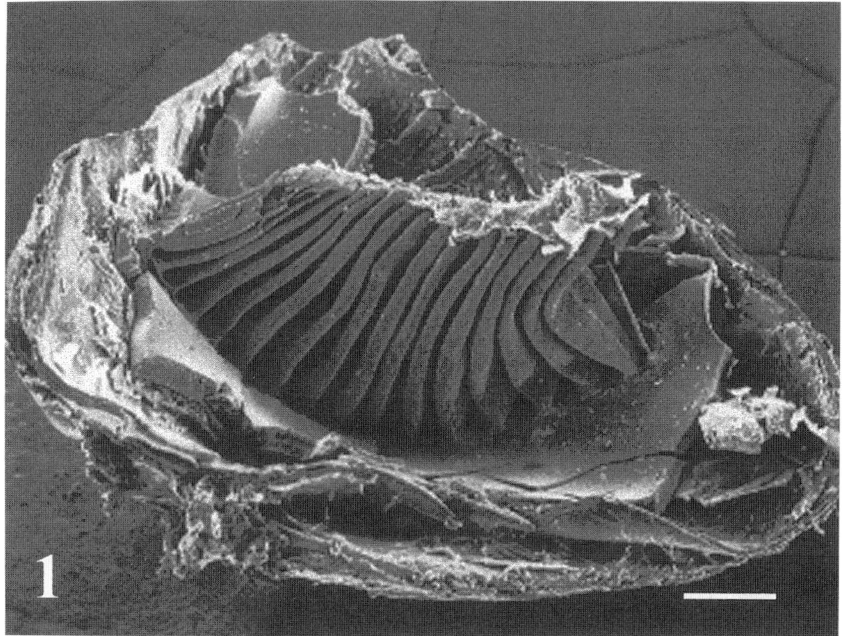

FIG. 1 Scanning electron micrograph of the pecten oculi of the chicken. Pleats arise from the optic disc. At the top of the pleats, the bridge is visible. Scale bar = 500 μm.

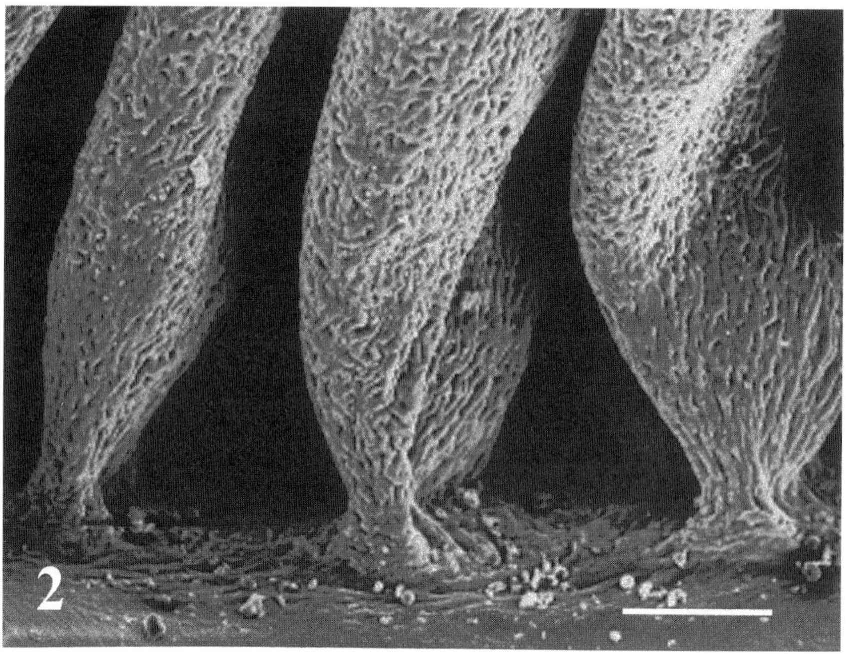

FIG. 2 Higher magnification from Fig. 1. The large number of blood vessels appear as a dense network on the surface of the pecten pleats. Scale bar = 100 μm (reproduced with permission from Gerhardt *et al.*, 1996, Fig. 1, copyright 1996 by Springer-Verlag).

the endothelial cells and the retina, this barrier must be composed of endothelial cells.

During ontogenetic development, pigmented cells migrate from the retinal pigment epithelium to the growing pecten which is raised from an accumulation of undifferentiated glial cells at the papilla nervi optici, called pecten primordium. At the time when the migrating retinal pigment epithelial cells gain access to the pecten, they (as well as their "sister cells" remaining at the site of origin, between the retina and the choroid) already express barrier properties, such as complex TJs, the glucose transporter isoform Glut1, and the barrier-specific antigen HT7. After integration into the pecten, however, pigmented cells gradually lose these barrier properties. Primarily, pigmented cells form a border between vitreous and blood vessels; later on, they retract and move between blood vessels which, as a result of these changes, achieve direct contact to the vitreous.

By contrast, in inverse relation to pigmented cells, endothelial cells of the pecten acquire BBB properties during embryonic development. Because there are no other cell types in this organ (with the exception of a few pericytes and peripectinate cells), pigmented (glial) cells are good candidates for inductors of barrier properties in endothelial cells. Thus, due to its relative simplicity, pecten may be an advantageous model to study the significance of neuroectodermal cells for the induction and maintenance of an endothelial blood–brain or blood–retina barrier.

II. Morphology of the Mature and Developing Blood–Brain Barrier

In this review, it is not possible to consider all aspects of structure and formation of the BBB. Therefore, we restrict our description to those topics regarding pecten oculi of the avian eye.

The morphological properties of mature BBB capillaries in the avian and mammalian brain include the small height of endothelial cells, the presence of complex TJs between the processes of the same cell (Nagy *et al.*, 1984), the small number of caveolae at the luminal surface of the cell (Coomber and Stewart, 1986; Peters *et al.*, 1991), the high number of mitochondria in the endothelial cells (Stewart and Coomber, 1986), and the regular presence of subendothelial pericytes which are completely surrounded by a basal lamina and by perivascular astrocytic processes or lamellae. The extracellular space between the pericyte/endothelial complex and astrocytes is extremely narrow (Dermietzel and Krause, 1991), and the membrane of astrocytes directly adjacent to the blood vessel is crowded with a high density of so-called orthogonal arrays of particles (Wolburg, 1995). These intramembranous arrays are probably formed by transport proteins; recently it was found that the water channel protein, aquaporin-4, is a principal constitutent of these arrays (Verbavatz *et al.*, 1997). Furthermore, aquaporin-4 has been demonstrated to be confined to astrocytes and to be expressed predominantly in perivascular and subpial membrane domains (Nielsen *et al.*, 1997). The functional significance of all these glial–vascular interrelationships is corroborated by the observation that they arise along with the step by step maturation of the BBB.

The vasculature of the CNS develops by angiogenesis which is characterized by ingrowth of mesodermal vascular sprouts from predifferentiated vascular cells into the neural tube. This ingrowth depends on the delivery of growth factors and expression of appropriate receptors (Moses *et al.*, 1995), among which the vascular endothelial growth factor (VEGF) and its receptor flk-1, as well as the receptor tyrosine kinases tie-1 and tie-2, appear to be of particular importance (Mustonen and Alitalo, 1995; Breier and Risau, 1996; Flamme *et al.*, 1997). For example, a dominant-negative tie-2 mouse mutant revealed an inability to grow blood vessels into the embryonic cerebral wall (Sato *et al.,* 1995). When the BBB in cerebral vessels begins to become tight, there are neither differentiated astrocytes nor astrocytic precursors in close association to the early vessels, and the basal lamina is still lacking or ill-defined (Caley and Maxwell, 1970; Bauer *et al.,* 1993; Stewart and Hayakawa, 1994). During embryonic and postnatal development, endothelial cells become flattened, the astrocytic ensheathment around cerebral capillaries is gradually completed, and the subendothelial extracellular space is greatly reduced (Caley and Maxwell, 1970; Hannah and Nathaniel, 1974; Bertossi *et al.,* 1989, 1993).

Of particular interest is the embryonic development of interendothelial junctions in brain microvessels. In the murine and rat brain, endothelial cells invade the neural parenchyma between the embryonic days (E) 10 or 11 and E13. Whereas the early endothelium displays fenestrations as the most prominent landmark of high permeability, these fenestrations disappear after invasion of the CNS tissue, and TJs begin to appear (Yoshida *et al.,* 1988; Bauer *et al.,* 1993; Stewart and Hayakawa, 1994). However, the maturation of barrier features in the brain does not develop synchronously. By means of horseradish peroxidase (HRP) injection into the allantoic vein of chick embryos (Wakai and Hirokawa, 1978) it has been shown that between E12 and E15 there occurs a stepwise progression of the endothelial barrier against HRP from the superficial to the medullary region of both the cerebellum and the spinal cord. In a combined morphometric and tracer study in the developing mouse, Stewart and Hayakawa (1987) demonstrated a gradual decline of the "permeability index," which was defined as the ratio of brain/plasma HRP activity divided by the vessel density. This decline was similar to that of the "interendothelial cleft index," which is defined as the proportion of the junctional profile that is composed of junctional clefts. In order to define these clefts more closely, Schulze and Firth (1992) characterized the maturation of the BBB in the rat brain as an increase of the ratio of "narrow zones" to "wide zones" in the interendothelial clefts. Schulze and Firth (1993) also identified the molecular composition of interendothelial junction(s). Throughout the entire length of the junction, including narrow and wide zones, adherens junction components such as α-actinin, vinculin, and cadherins were expressed. This finding is of particular importance since in epithelial cells, tight and adherens junctions are spatially separated (Woods and Bryant, 1993). As we will discuss later, the close intercalation of tight and adherens junction components appears to be a significant property of BBB endothelial cells.

Despite considerable progress in understanding molecular constituents of TJs, the freeze-fracture technique is still one of the most powerful methods for the investigation of TJs. According to results of studies on epithelia, it is commonly believed that TJ particles of partner strands are separated by the cleaving process involved in this method. In this way, the strands are almost completely associated with the protoplasmatic fracture face (P-face). It is not clear whether or not the grooves on the external fracture face (E-face), which are mainly free of particles, represent imprints caused by the P-face-associated strands. In pigment epithelial cells of the ciliary body, Noske and Hirsch (1986) have shown that E-face grooves were occupied by particles at those places where the complementary P-face-associated strands were interrupted. On the other hand, discontinuous P-face ridges frequently do correspond to continuous E-face grooves, suggesting that each E-face groove may not necessarily reflect the equivalent of a P-face ridge. Therefore, it has been proposed that TJ-specific membrane grooves may represent cytoskeleton-dependent membrane deformations rather than imprints of opposite membrane particles (Kniesel *et al.,* 1996). It is evident in this respect that if TJ particles occur at the E-face they are arranged as chains or as single globular particles, whereas TJs at the P-face frequently occur as cylindrical profiles. This may suggest that the extracellular surface of TJs is continuous and smooth, and the cytoplasmic surface is discontinuous and irregular due to linkage sites to the cytoskeleton (Suzuki and Nagano, 1991; Lane *et al.,* 1992). In rat brain endothelium *in situ,* the P-face association (PFA; i.e., the percentage of particles associated with the P-face) is approximately 60%, which is significantly less than that in epithelia and more than that in non-BBB endothelia. The E-face association (EFA) of rat brain endothelium is on the order of 50% (Kniesel *et al.,* 1996). Again, TJ particles are frequently formed as cylindrical profiles at the P-face but as chains of round particles at the E-face. In submammalian species, however, BBB endothelial TJs are almost completely associated with the P-face and are therefore reminiscent of the epithelial type of TJs (Shivers, 1979; Nico *et al.,* 1992; Gerhardt *et al.,* 1996; Liebner *et al.,* 1997).

The molecular composition of tight junctions has recently been investigated by many groups. Generally, there are two main hypotheses concerning the significance of lipids and proteins for establishing barrier properties. The lipid model proposes inverse lipidic micelles to represent TJ strands (Kachar and Reese, 1982; Pinto da Silva and Kachar, 1982; Grebenkämper and Galla, 1994); this model implies continuity of external leaflets of two adjacent cells. The protein model suggests insertion of a protein(s) into the membrane occluding the intercellular cleft. A candidate molecule for this protein has been described—occludin (Furuse *et al.,* 1993, 1994; Ando-Akatsuka *et al.,* 1996; McCarthy *et al.,* 1996; Tsukita *et al.,* 1996). This protein has four transmembrane domains, and it was confirmed to be a constituent of TJ strands by means of the label-fracture technique (Hirase *et al.,* 1997; Saitou *et al.,* 1997). In addition, several TJ-associated proteins have been identified in the past few years, including ZO-1, ZO-2, p130, 7H6, and cingulin (Anderson and Van Itallie, 1995). All these proteins are

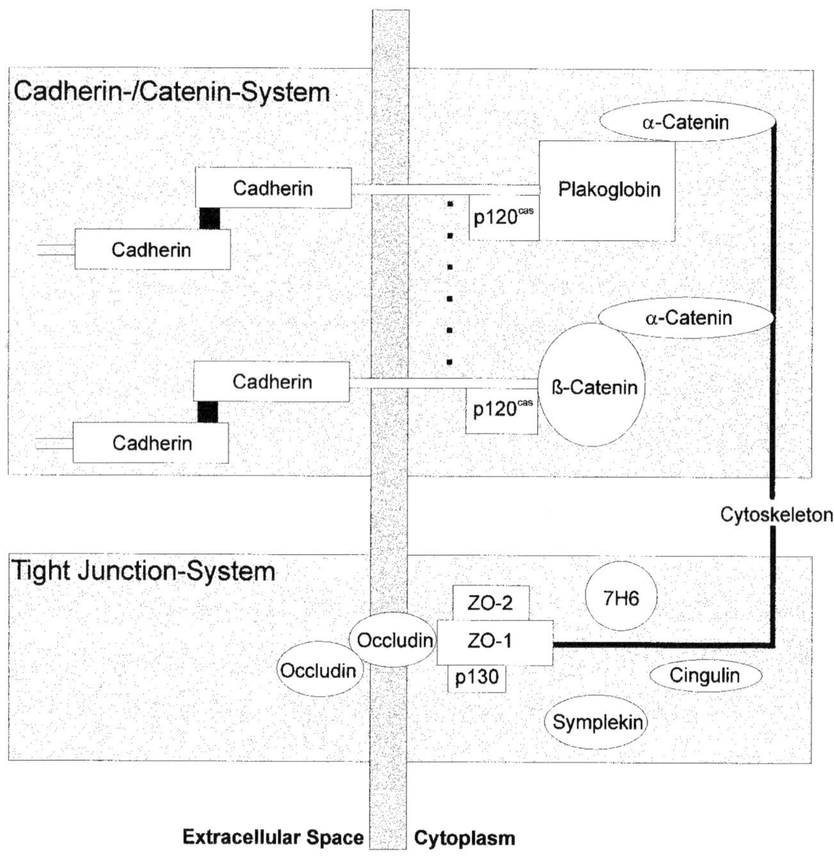

FIG. 3 Model of protein interactions at the tight junction (bottom) and the adherens junction (top). For simplicity, only the membrane of one partner cell is shown (bar in the middle). Components of both junctional systems are connected by the cytoskeleton. Only for completion, symplekin as a member of the tight junction associated proteins was included into the scheme, although it is not found in endothelial tight junctions.

under the control of different kinases and their phosphorylation states correlate with barrier function (Kniesel and Wolburg, 1999; Fig. 3).

III. The Blood–Brain Barrier *in Vitro*

In order to gain facilitated access to the brain's vascular system for studies of its barrier characteristics, many groups have established *in vitro* systems of BBB endothelial cells (Méresse *et al.,* 1989; Rubin, 1991, 1992; Rubin *et al.,* 1991; Abbott *et al.,* 1992; Raub *et al.,* 1992; Joo, 1993; Boado *et al.,* 1994; Biegel *et al.,* 1995; Ceballos and Rubio, 1995; Hayashi *et al.,* 1997; Muruganandam *et al.,* 1997; Spatz *et al.,* 1997; Stanness *et al.,* 1997). Whereas endothelial cells lost their barrier properties *in vitro,* some BBB properties were shown to be reestablished when endothelial cells were cultured together with astrocytes or glioma cells. These included activity of the γ-glutamyl transpeptidase (De Bault and Cancilla, 1980; Maxwell *et al.,* 1987; Dehouck *et al.,* 1990; Tontsch and Bauer, 1991), transport of neutral amino acids (Cancilla and De Bault, 1983), polarity and activity of Na^+-K^+-ATPase (Beck *et al.,* 1986; Tontsch and Bauer, 1991), glucose uptake (Maxwell *et al.,* 1989), and expression of the barrier-specific antigen HT7 or neurothelin (Lobrinus *et al.,* 1992). Moreover, astrocytes and C6 glioma cells have been shown to induce reorganization of microvascular endothelial cells into capillary-like structures (Laterra *et al.,* 1990; Laterra and Goldstein, 1991). The question as to whether or not the expression of the glucose transporter isoform Glut1, high electrical resistance, and formation of complex TJs are among the properties of endothelial cells that can be reinduced by astrocytes may be of particular interest.

The 55-kDa form of the glucose transporter isoform Glut1 is highly restricted to microvascular cells in the brain (Pardridge *et al.,* 1990; Maher *et al.,* 1994). Its gradual restriction appears to reflect the onset of BBB function during development (Dermietzel *et al.,* 1992; Bauer *et al.,* 1995). Cultured endothelial cells undergo a marked loss of Glut1 expression. Brain-derived factors have been shown to enhance the transcription rate of the Glut1 gene (Boado *et al.,* 1994) by increasing Glut1 transcript stability (Boado, 1996).

There are two ways astrocytic cells may induce formation of TJs by endothelial cells. Arthur *et al.* (1987) and Shivers *et al.* (1988) provided evidence for humoral factors released by astrocytes to be responsible for TJ formation in endothelial cells. In contrast, Tao-Cheng and Brightman (1987) found that a direct contact between astrocytes and endothelial cells is required for TJ induction. Frequently, the mere existence of TJs in cultured endothelial cells is considered as evidence for the validity of culture systems (Joo, 1993). However, even under the influences of cocultivation or conditioning with astroglial or glioma cells or other factors, the physiological parameters of endothelial cells, such as high electrical resistance and transendothelial impermeability, are much less developed than in BBB microvessels *in vivo* (Butt *et al.,* 1990; Rubin *et al.,* 1991; Raub *et al.,* 1992; Wolburg *et al.,* 1994; Guerin and Bobilya, 1997; Muruganandam *et al.,* 1997; Schulze *et al.,* 1997). Also, the fine structure of TJs is altered *in vitro*; cultured BBB endothelial cells show a dramatic decrease of PFA and an increase of EFA which only partially can be compensated by cocultured astrocytes (Wolburg *et al.,* 1994). This finding corroborates the suggestion that outside the intact brain, endothelial TJs alter their molecular configuration within the membrane, with the result being decreased transendothelial resistance and increased permeability. However, it should be stressed that alteration of TJ morphology after cultivation of endothelial cells appears to be confined to mammalian species. As will be described later, endothelial TJs in the submammalian BBB differ from those in mammals by a PFA that is approximately 100%. This complete P-face association suggests that intercellular adhesion of TJ strands or particles is weaker than their attachment to the protoplasmic leaflet of the membrane.

IV. The Pecten as a Reliable *in Vivo* Model of the Blood–Retina Barrier

A. General Considerations

The pecten oculi of the avian eye was originally described in 1675 by N. Stensen (N. Steno) [cited in Seaman and Storm (1963) and Wingstrand and Munk (1965)] and in 1676 by C. Perrault [cited in O'Rahilly and Meyer (1961) and Wingstrand and Munk (1965)]. Despite many detailed morphological, developmental, and biochemical investigations, the function of the pecten remains unclear. It is generally believed to improve the nutrition of the inner parts of the avascular retina (Mann, 1924; Wingstrand and Munk, 1965; Pettigrew *et al.,* 1990), but other functions have also been proposed: (i) pH regulation [because of the decrease of intraocular pH from 7.56 in control chicken eyes to 7.35 after intraocular ablation of the pecten by electrocautery (Brach, 1975) due to the localization of carboanhydrase in pecten cells (Leiner, 1951; Kauth and Sommer, 1953; Eichhorn and Flügel, 1988]; (ii) eye pressure regulation (Seaman and Storm, 1963); (iii) thermic regulation (Bachsich and Gellért, 1935); (iv) stabilization of the vitreous (Tucker, 1975); (v) direct [Franz (1910) as cited in Blochmann and von Husen (1911) and Wingstrand and Munk (1965)]; and (vi) supportive sensory actions (Crozier and Wolf, 1944a,b). A more complete list of assumed functions of the pecten is found in Wingstrand and Munk (1965).

Pettigrew *et al.* (1990) found an extravasation of fluorescent dyes out of the pecten vessels across the retinal surface, synchronous with saccadic eye movements. This observation strengthens the hypothesis of nutritional support of the retina by the pecten but also evokes the question of whether the pecten selectively delivers nutrients to the retina (but restrains noxious compounds). However, except for a description of TJs between pecten endothelial cells (Dieterich *et al.,* 1973), information on the barrier characteristics of this tissue is limited or even misleading. Abelsdorff and Wessely (1909; cited in Bawa and YashRoy (1974) stressed its high permeability and diffusion rate, whereas Rodriguez-Peralta (1975) first mentioned the pecten as a part of hematic and fluid barriers of the retina and the vitreous, and Latker and Beebe (1984) described a continuous impermeability to HRP of the pecten blood vessels throughout ontogeny.

Since there is no metabolic barrier between the vitreous body (with the pecten oculi) and the retina, pectinate blood vessels must establish an endothelial blood–retina barrier in the same way as is done by intraretinal blood vessels in other species. In contrast to intraretinal or intracerebral blood vessels, however, pectinate vessels are devoid of the complex perivascular neural environment. Besides a few pericytes and peripectinate cells, only pigmented cells are located around endothelial cells. These pigmented cells are glial cells which share a common origin with other neuroectodermal cells, including astrocytes and retinal pigment epithelial cells. Thus, the pecten oculi is a simple system composed by only a few cell types, allowing study of the ill-understood mechanisms of how and by which cells or factors the BBB is induced. Indeed, this system is as simple as a cell culture but has the valuable advantage of being an *in vivo* system.

B. The Structure of the Pecten Oculi in the Avian Eye

In semithin sections, pigmented cells of the pecten were observed to be continuous with pigmented cells in the optic disc, which, in turn, seemed to be oriented toward the region of transition between the glia limitans of optic nerve head and the pigment epithelium/Bruch's membrane underneath the retina. This observation was consistent with earlier developmental investigations (Yew, 1978) describing pigmented cells of the pecten as originating in the retinal pigment epithelium. The pecten itself is composed of pigmented cells and endothelial cells (Figs. 4 and 5). The circumference of the large vessels is generally lined by more than six elongated endothelial cells, showing smooth luminal and abluminal membranes. In contrast, those of capillaries are enlarged by microfolds leading to an epithelioid appearance. However, even capillaries of very small diameter generally reveal two to four or more cells per circumference rather than only one, as usually observed in brain microvasculature.

1. The Endothelial Cells of the Pecten Oculi

At the electron-microscopical level, endothelial cells of small vessels reveal lateral interdigitations, equipped with extended junctional domains (Fig. 5). The cytoplasm contains stacks of the Golgi apparatus, rough endoplasmic

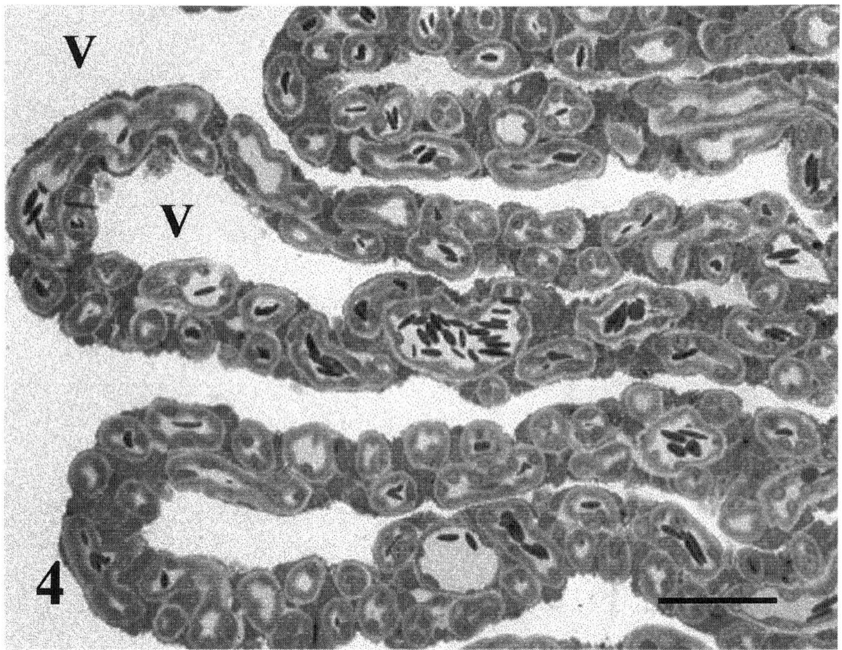

FIG. 4 Semithin section through the pleated adult pecten oculi of the chicken. In contrast to brain microvessels, the circumference of pecten vessels is formed by more than one endothelial cell. Pigmented cells are located in between the vessels. V, vitreous. Scale bar = 25 μm.

cisterns, many mitochondria, and bundles of intermediate filaments. Both apical and basal surfaces are usually enlarged by numerous microfolds and a basal labyrinth, respectively. There is a subendothelial space filled with extracellular matrix (collagen fibrils and amorphous substances) between the basal lamina and pigmented cells. At some places, cellular debris lies within the matrix. Remnants of apoptotic pericytes are the most probable candidates to explain this debris.

Freeze-fracture replicas of the pecten oculi reveal differences in the membrane architecture between pigmented and endothelial cells, in the early posthatching as well as the adult chicken. Endothelial cells show a

---

FIG. 5 Electron micrograph of a traverse section through an adult pecten oculi of the chicken. In the middle, the lumen of a vessel is obstructed by a leukocyte. The endothelial cells enlarge their surfaces by the formation of a basal labyrinth and apical microfolds. Note the multiple layers of extracellular matrix between the endothelial and the pigmented cells. V, vitreous; PC, pigmented cell; EC, endothelial cell. Scale bar = 0.25 μm.

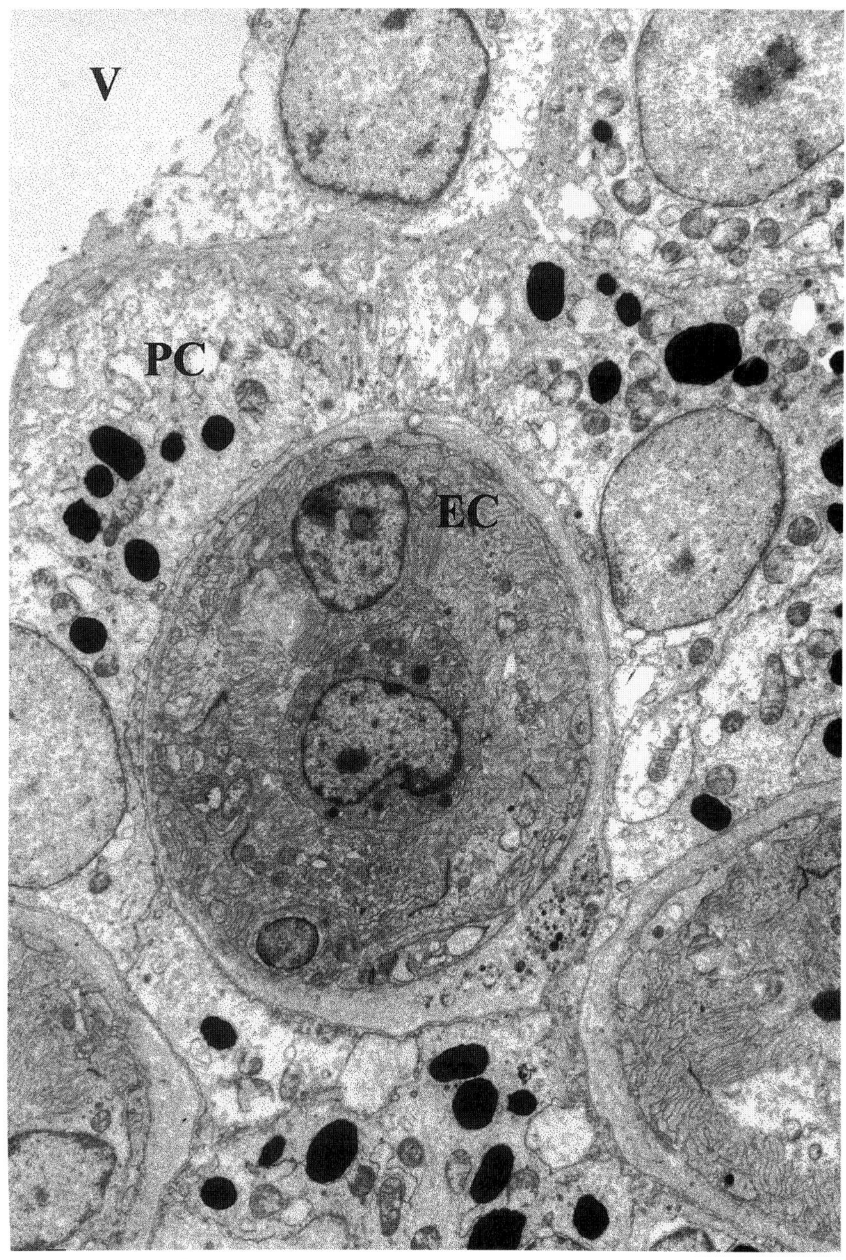
V
PC
EC

relatively high density of intramembranous particles. The cells are connected laterally by a complex network of TJ strands that are almost completely associated with the P-face. Grooves at the E-face only rarely show single particles; these may correspond to gaps in the strands at the P-face (Fig. 6). It should be pointed out that this TJ structure is also found in brain microvessels of the chicken (Nico *et al.*, 1992) and reptiles (Shivers, 1979) but differs from that in mammalian BBB endothelia.

2. The Glial Nature of Pigmented Cells of the Pecten Oculi

Pigmented cells, also called stroma cells (Wingstrand and Munk, 1965), consistently separate blood vessels from the vitreous during development. Posthatch, they lose their epithelial arrangement and no longer form a continuous sheath around the vessels. As a result, endothelial cells locally gain access to the vitreous and are covered by both the endothelial and the pigment cell-derived basal lamina. The latter is continuous with the inner limiting membrane of the retina and envelops the entire pecten oculi

FIG. 6 Freeze-fracture replica of endothelial cell tight junctions in the pecten on P8. The strands are associated mainly to the P-face (P), leaving empty grooves on the E-face (E). MF, microfolds. Scale bar = 0.1 μm.

toward the vitreous. Pigment granules are present almost exclusively in the perikarya of the pigmented cells. Only a few intermediate junctions occur between pigmented cells; gap and tight junctions are not present at the posthatch stages. Pigmented cells are generally poor in cellular organelles. Because pigmented cells of the pecten are clearly of neuroectodermal origin, they are defined as glial cells. Their glial nature, which was stressed by authors such as Wingstrand and Munk (1965), is underlined by the specific expression of glutamine synthetase (Fig. 7) and vimentin. Whereas vimentin is not expressed in glial cells exclusively, glutamine synthetase is restricted to glial cells in the mature nervous system of both vertebrates and invertebrates (Wiesinger, 1995; Linser *et al.,* 1997). Recent studies indicate that this enzyme serves an important and conserved function in providing glutamine for transmitter synthesis and in removing excess (i.e., toxic) amounts of glutamate and ammonia in the nervous system, including the retina (Linser and Moscona, 1984; Derouiche and Frotscher, 1991; Derouiche and Rauen, 1995; Gorovits *et al.,* 1997). Thus, the abundant presence of

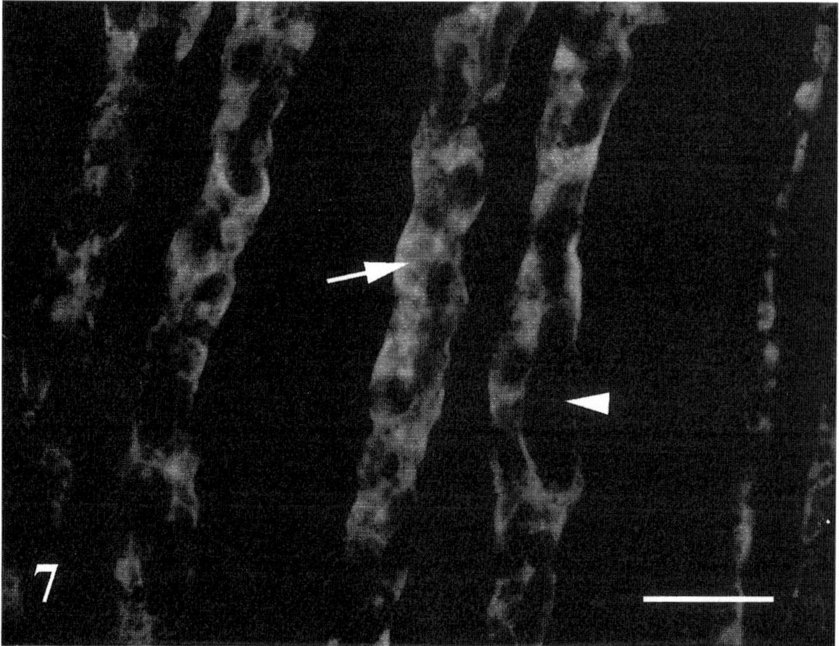

FIG. 7 Immunofluorescence labeling of glutamine synthetase in the pecten oculi of the chicken on P1. The pigmented cells are highly positive for this enzyme. The arrow points to pigmented cell; the arrowhead points to an endothelial cell. Scale bar = 50 μm.

glutamine synthetase not only stresses the glial nature of the pigmented cells but also sheds new light onto the functional importance of the pecten oculi.

If pigmented cells of the pecten oculi are compared to other glial cells, such as Müller (radial glial) cells or astrocytes, it becomes apparent that they share a peculiar pattern of antigen expression with Müller cells but not with astrocytes. The presence of vimentin and glutamine synthetase, and a lack of glial fibrillary acidic protein (GFAP), are characteristic features of mature retinal Müller cells, whereas the expression of GFAP is restricted to differentiated astrocytes in the optic nerve head. Interestingly, pigmented cells of the pecten show the same set of markers as Müller cells, although they derive and additionally express features of the retinal pigment epithelium. In the avian retina, Müller cells display ample contacts to neurons but none to blood vessels. Pigmented cells of the pecten contact blood vessels but have lost contact to neurons. During their migration into the pecten, they transiently contact neurons and concomitantly express GFAP. Future investigations should consider whether or not spatial relationships may trigger the expression pattern of glial intermediate filaments; GFAP is only expressed by astrocytes of the optic nerve head which are in direct contact to both neuronal elements and blood vessels.

C. The Development of the Pecten Oculi

1. Origin and Differentiation of Pectinate Pigmented Cells

The development of the pecten oculi is closely related to the closure of the choroid fissure and the degeneration of the arteria cupulae opticae (Romanoff, 1960; O'Rahilly and Meyer, 1961; Uehara *et al.,* 1990). The arteria cupulae opticae, the earliest blood vessel supplying the primordial tissue of the eye, enters the eye cup near the lens and runs through the vitreous in a loop-like manner. The environment of the vessel is occupied by a spongy network of mesenchymal cells that arise from the perineural plexus. At the electron microscopic level, the arteria cupulae opticae consists of several elongated endothelial cells. They are interconnected by poorly developed TJs. At these early stages (E4 or E5), the margins of the eye cup are composed of undifferentiated cells that connect the inner and outer sheet of the prospective retina. As a consequence, the inner limiting membrane continues into Bruch's membrane. On the sixth day of incubation (E6), the optic fissure starts to close bidirectionally from its intermediate portion in both proximal and distal directions. Distally, the tissue differentiates into retinal cells. Proximally, an intensive proliferation of the prospective pigment epithelium contributes to the formation of the pecten primordium (Romanoff, 1960). This primordium then gradually overgrows and suppresses the arteria cupulae opticae and the mesenchymal network (Fig. 8). As indicated by the integrity of Bruch's membrane, the arteria cupulae opticae remains outside the neuroectoderm at the base of the developing pecten; hence, the pecten primordium is completely devoid of blood vessels. Neuroectodermal (glial) cells forming the pecten primordium derive from the retinal pigment epithelium (Gerhardt *et al.,* 1996). Nevertheless, on E6 the pecten is still nonpigmented (whereas there is already some pigment in the pigment epithelium). At the surface of the pecten primordium toward the vitreous, the glial cells are densely interconnected, thereby forming an epithelial layer. Freeze-fracture replicas of the pecten primordium on E6 and E7 indicate the presence of TJs in the epithelial layer at the interface to the vitreous humor (Liebner *et al.,* 1997). At this time, blood vessels from the perineural vasculature start to invade into the basal parts of the glial tissue. Elongated endothelial cells exhibit poorly developed contact zones lining the circumference of these early blood

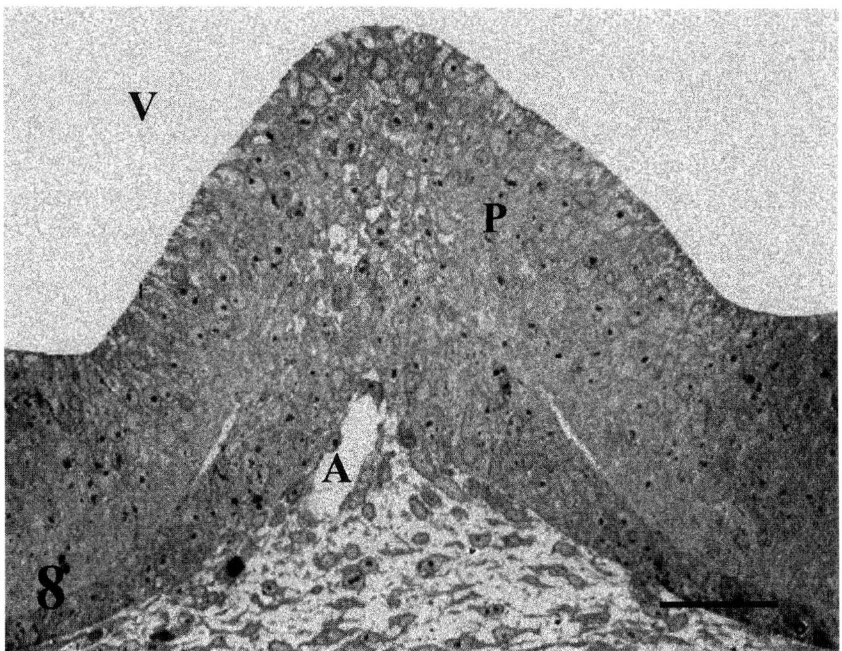

FIG. 8 Semithin section through the pecten primordium on E6. A, arteria cupulae opticae; P, pecten primordium; V, vitreous. Scale bar = 25 μm. (Reprinted from *Dev. Brain Res.* **100,** S. Liebner, H. Gerhardt, and H. Wolburg, Maturation of the blood–retina barrier in the developing pecten oculi of the chicken, pp. 205–219, 1997 with kind permission of Elsevier Science - NL, Sara Burgerhartstraat 25, 1055 KV Amsterdam, The Netherlands).

vessels. An endothelial basal lamina is not yet present. From E11 on, TJs are formed in the pectinate endothelial cells. During further development, their P-face association and complexity increase step by step, reaching mature values posthatch (Figs. 6 and 12).

On E11, the pecten appears as a faintly pigmented, pleated structure. The apex of the pecten seems not to be involved in pleat formation but remains as a smooth ridge of glial cells called the bridge, which connects the pecten pleats. Although vascularization is no longer restricted to the basal portion of the pecten, light microscopy reveals a compact glial tissue, occasionally interrupted by blood vessels. At this stage, large numbers of peculiar macrophage-like cells, called "peripectinate cells," appear on the surface of the pecten (Fischlschweiger and O'Rahilly, 1966; Uehara *et al.*, 1990). Concomitant with pleat formation and the beginning pigmentation, the pecten exhibits a more differentiated ultrastructure of pigmented and endothelial cells. Pigmented cells comprising the border of the vitreous humor form a compact layer, although, in contrast to earlier stages, prominent junctions were no longer found.

2. The Development of Barrier Properties in the Endothelial Cells of the Pecten

Previously, it was demonstrated that the outer glial cells lose their TJs between E7 and E11. In contrast, the endothelial cells gain them during the identical time period. Notably, on E11 these endothelia acquire a basal lamina surrounding pericytes endowed with large nuclei. As mentioned earlier, cellular debris can be found in the subendothelial space of the mature pecten. The observation that pericytes are seen regularly during embryonic development but only rarely in the mature stage may suggest that pericytes undergo apoptotic cell death. Frank *et al.* (1987) and Stewart and Tuor (1994) investigated the endothelial blood–retina barrier in the rat and compared it with the vasculature in the brain. The latter authors observed a fourfold higher density of pericytes in retinal vessels than in cerebral vessels. They suggested that pericytes may constitute a "second line of defense" to compensate for a more permeable barrier of retinal capillaries. The decline of pectinate pericytes during development and their absence in the mature system seem to indicate a certain disposal of this second line of defense; probably, the maintenance of barrier properties in the mature pecten is largely independent of pericytes.

Now, we examine these junctions and compare their development with that of other (molecular) markers or events during barrier genesis. First, the decrease of transendothelial permeability will be described because it is of particular importance for the formation of the barrier. Then, we will demonstrate the expression patterns of two barrier-associated antigens, the glucose transporter isoform Glut1 and the HT7 antigen. Finally, the development of TJs in pectinate endothelial cells will be described in detail and compared with recent results on the alterations in the expression of adhesion molecules on pectinate cells.

a. Lanthanum Permeability during Development Perfusion experiments with the low-molecular-weight tracer lanthanum nitrate demonstrated the decrease of paracellular and transcellular permeability during development. On E7, early blood vessels at the base of the primordial pecten exhibited a high paracellular permeability for the tracer which was widely distributed in the extracellular space (Fig. 9). Thus, the poorly developed tight junctions between endothelial cells did not occlude the intercellular space. On E8, the pecten vasculature was still leaky for lanthanum nitrate. Three days later, no extravasation of lanthanum was observed, a finding which indicates

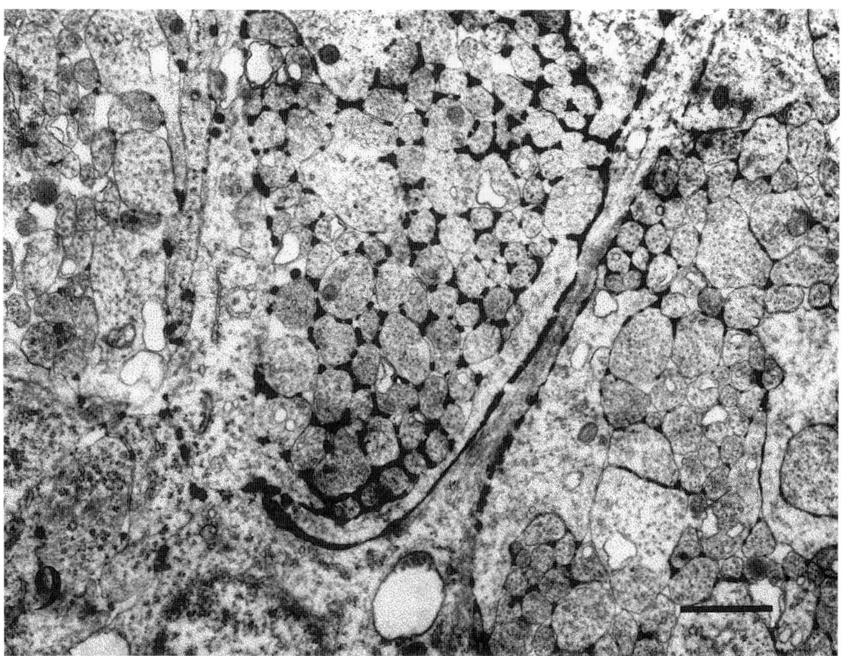

FIG. 9 Perfusion of the pecten oculi with lanthanum nitrate on E7. The tracer was distributed in the interstitium surrounding axons of retinal ganglion cells in the basis of the pecten primordium. Scale bar = 1 μm (Reprinted from *Dev. Brain Res.* **100,** S. Liebner, H. Gerhardt, and H. Wolburg, Maturation of the blood–retina barrier in the developing pecten oculi of the chicken, pp. 205–219, 1997 with kind permission of Elsevier Science - NL, Sara Burgerhartstraat 25, 1055 KV Amsterdam, The Netherlands).

a great improvement of the junctional seal (Fig. 10). Although still immature TJs were able to prevent paracellular leakage of the tracer, the blood–retina barrier was not fully developed at this stage; pecten endothelial cells exhibited an extensive endocytosis of lanthanum nitrate, indicating a physiological immaturity of the blood–retina barrier. On E16, no extravasation of the tracer could be detected, and there was a remarkable decrease in endocytosis. At all posthatching stages, no paracellular leakage was revealed. Vesicular endocytosis was successively decreased and not substantially detectable after P6.

Our observations of a gradual development and maturation of BBB properties in the endothelial cells of the pecten oculi conflict with those of Latker and Beebe (1984), who demonstrated an impermeability of pecten blood vessels for horseradish peroxidase from the time of first appearance. These authors concluded that the barrier was already mature. Because they

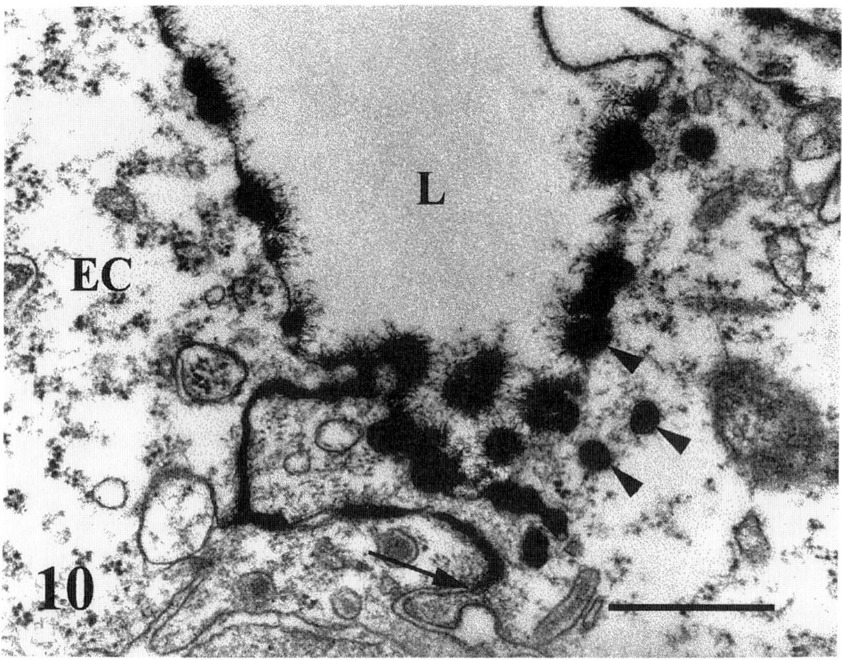

FIG. 10 Perfusion of the pecten oculi with lanthanum nitrate on E11. In contrast to E7, the blood vessels in the basal parts of the pecten showed no paracellular extravasation of the tracer (arrow points to the occluding junction) but a remarkable lumenal endocytosis. L, lumen; EC, endothelial cell; arrow points to membrane kisses; arrowheads point to vesicles. Scale bar = 0.5 μm (reprinted from *Dev. Brain Res.* **100,** S. Liebner, H. Gerhardt, and H. Wolburg, Maturation of the blood–retina barrier in the developing pecten oculi of the chicken, pp. 205–219, 1997 with kind permission of Elsevier Science - NL, Sara Burgerhartstraat 25, 1055 KV Amsterdam, The Netherlands).

used a tracer with a substantially higher molecular weight than lanthanum nitrate, a methodological difference might be the reason for the inconsistent results.

b. Expression of Glut1 and HT7 during Development The importance of the glucose transport for the energy supply of the brain has been stressed many times (Forsyth, 1996; Tsacopoulos and Magistretti, 1996). Although several isoforms of the glucose transporter are expressed in the brain, the 55-kDa form of the Glut1 isoform is specifically restricted to microvascular cells (Maher *et al.,* 1994). The time course of restriction of the Glut1 expression correlates well with barrier maturation (Dermietzel *et al.,* 1992; Bauer *et al.,* 1995). The expression and asymmetrical distribution of this molecule in abluminal versus luminal endothelial membranes are hallmarks of barrier integrity and maturity (Farrell and Pardridge, 1991; Bolz *et al.,* 1996).

Another, albeit less well understood, marker for barrier-forming endothelial cells in the brain is the HT7 molecule, which is a cell surface immunoglobulin-like glykoprotein (Seulberger *et al.,* 1990). It is expressed in endothelial cells of the chicken brain but also in epithelial cells of the kidney tubules, the retinal pigment epithelium, and the choroid plexus (Risau *et al.,* 1986; Finnemann *et al.,* 1997). In the chicken, it is also called neurothelin (Schlosshauer and Herzog, 1990) or 5A11 (Fadool and Linser, 1993); the murine molecules basigin and gp42 show 45% homology to chicken HT7, and the sequence identity of CE9/OX-47 in the rat and basigin/gp42 in the mouse is 94% (Seulberger *et al.,* 1992). The attributive of the HT7 antigen as a reliable BBB marker is confirmed by the observation that it is not expressed in fenestrated vessels of circumventricular organs (Albrecht *et al.,* 1990). Interestingly, OX-47 is not downregulated in rat brain endothelial cells *in vitro* (Risau, 1991). However, the function of this molecule in the maintenance or regulation of the BBB is unknown.

If the pecten oculi is a model of the BBB, at least the glucose transporter Glut1 and the HT7 antigen should be expressed specifically in the endothelial cells. In early developmental stages (E4 and E5), the retina, the lens, and the optic stalk were highly immunopositive for Glut1 and the HT7 antigen (Liebner *et al.,* 1997). The central parts of the prospective retinal pigment epithelium showed the most intense labeling. In contrast, the perineural plexus only faintly expressed the antigen. From E6 to E13, the entire pecten primordium, including invading endothelial cells, uniformly expressed Glut1 and the HT7 antigen. On E16, Glut1 and HT7 labeling revealed an altered expression in the pecten. However, at this stage the distribution of the two antigens was slightly different: Glut1 was found to be almost completely restricted to endothelial cells, whereas considerable amounts of the HT7 antigen were still found on pigmented cells. Immunofluorescence and immunogold labeling showed a further decrease in HT7, and Glut1 immunoreactivity on pigmented cells and a concommitant increase of both antigens on endothelial cells. On the first day after hatching (P1), pigmented cells of the pecten proper were devoid of specific labeling. In contrast, pigmented cells of the bridge (showing an epithelial arrangement) were still highly immunopositive for Glut1 and HT7. This pattern was observed in all posthatching stages investigated. In the pecten proper, HT7 became an exclusive marker of endothelial cells, being localized in the membranes of basal and apical microfolds (Gerhardt *et al.*, 1996; Albrecht *et al.*, 1990). Double-labeling experiments revealed a partial colocalization of Glut1 and HT7 (Fig. 11).

Regarding the distribution of anti-Glut1 immunogold particles in the pecten endothelial cells, we were unable to quantify reliably their density per standard membrane area. This was due to the strong folding of both basal and apical membranes. Therefore, we could not confirm or deny

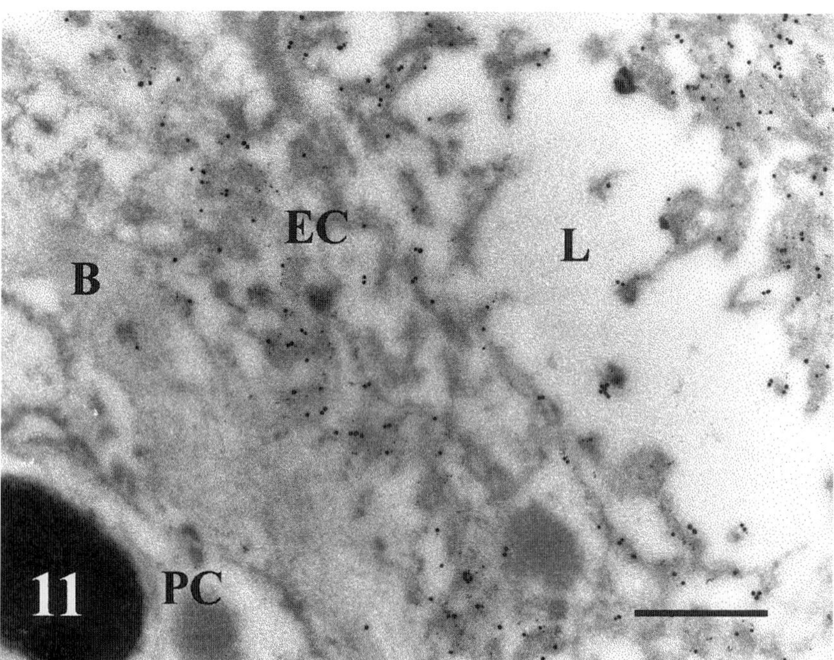

FIG. 11 Immunogold double labeling of HT7 (15 nm gold) and Glut1 (6 nm gold) of an endothelial cell of the chicken pecten oculi on P8. On the left, a pigmented cell is seen which is devoid of any gold particles. Some of the differently sized particles are directly colocalized suggesting that HT7 and Glut1 are partly codistributed. L, lumen; PC, pigmented cell; EC, endothelial cell; B, basal lamina. Scale bar = 0.5 μm.

the asymmetry of Glut1 localization in the luminal versus the abluminal membrane of the pecten endothelial cells. Nevertheless, first attempts at quantification (by counting particles in rectangles superimposed over luminal and abluminal membrane compartments, using the assumption of an almost equal depth of membrane folding) revealed no significant asymmetry. Thus, a four times higher expression of Glut1 in ablumenal versus lumenal membranes as described in rat brain endothelia (Farrell and Pardridge, 1991; Bolz *et al.,* 1996) was certainly not present in endothelial cells of the pecten oculi. In contrast, this asymmetrical distribution was found in BBB endothelia of the optic nerve in 6-day-old chicken. Therefore, we consider the described expression pattern as another unique feature of pecten endothelial cells.

Interestingly, the Müller cells, which span the entire retina and form the inner retinal surface, were also highly immunopositive for Glut1 (H. Gerhardt, unpublished results). Therefore, we suggest that glucose is transported via Glut1 through pectinal endothelial cells into the vitreous, is dispersed over the retinal surface by mechanical agitation of the pecten during saccadic eye movements and/or diffusion (Pettigrew *et al.,* 1990), and is transported into the retina via Glut1 of the Müller cells. Likewise, Glut1 is responsible for glucose supply of the outer retina because it is expressed by the retinal pigment epithelium. The question of glucose uptake into the neurons has been addressed in the human retina, in which ganglion cells and photoreceptors show Glut1 immunoreactivity (Kumagai *et al.,* 1994).

c. Formation of TJs in the Pecten during Development Freeze-fracture investigation of the pecten primordium on E7 revealed TJ strands on the P-face exhibiting a very high PFA. Although the strands on the P-face were not completely continuous, grooves on the E-face did not show any discontinuity. Interestingly, TJs were often single stranded, only occasionally displaying a more complex pattern (Fig. 12). Notably, no blind-ending strands were observed at this stage. At bifurcation points of the strands, small clusters of gap junctions were intercalated which consisted of only four to six connexons. Taking into account that only the most basal part of the pecten primordium was vascularized on E7, TJs found in more apical parts had to be attributed to epithelial cells. On E8, TJ morphology did not substantially differ from that on E7.

On E11, TJs exhibited a completely different morphology which is presumably due to expression of another type of TJ (Liebner *et al.,* 1997). The P-face strands were more fragmentary on E11 than on E7; now, they presented only a poor copy of the complete network. Additionally, blind-ending P-face strands were frequently observed. The E-face exhibited a

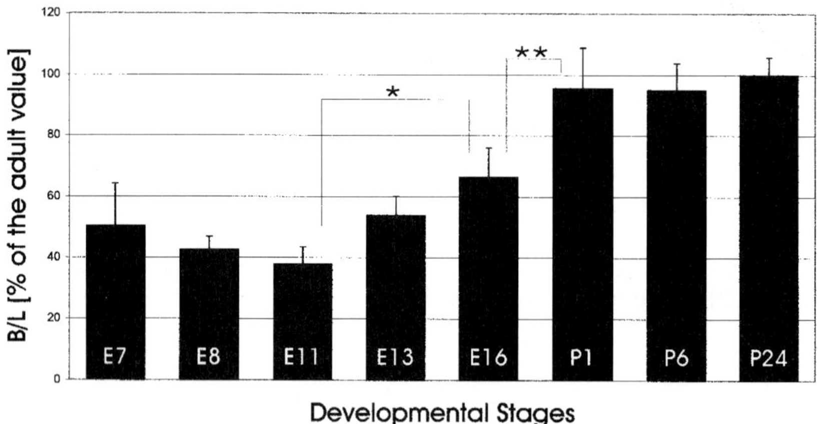

FIG. 12 Complexity index of the endothelial tight junctional strands as evaluated in freeze-fracture replicas for different stages of the pecten oculi of the chicken. The complexity index was determined by counting the branchpoints of tight junctional strands per micrometer strand length. On E11, another type of tight junctions appears in comparison to the earlier tight junctions (see text). These tight junctions show the lowest complexity seen during the period of observation. From E11 to P1, there is a continuous increase in tight junctional complexity. Posthatching, there is no further increase. As shown by the Welch test there is a significant increase in tight junction complexity between E11 and E16 ($*p \leq 0.05$) and between E16 and P1 ($**p \leq 0.01$).

more elaborate network of TJ grooves. E-face grooves were only sparsely occupied by particles which did not systematically correspond to gaps on the P-face. At this stage, intercalated gap junctions were not found.

On E13, TJs were slightly more mature (Fig. 12), showing a decrease of fragmentation and blind-ending P-face strands. On E16, TJs showed a complex network of almost continuous P-face strands, whereas E-face grooves were almost devoid of any particles. Fragmented strands, mostly oriented parallel to the basal–apical axis, were only rarely observed. Intercalated gap junctions were occasionally detected. In contrast to the P-face, the E-face acquired few new characteristics. Similar to the vessel ultrastructure noted previously, on P1 endothelial cells were easily identified by their lumenal and ablumenal microfolds. The morphology of TJs was very similar to the adult stage. The P-face strands were completely continuous; hence, the PFA was more than 90%. E-face grooves were completely devoid of particles. Interestingly, small clusters of gap junctions were often found at this stage. Pecten endothelial TJs acquired no other characteristics during further development (Fig. 12).

TJ particles in endothelial cells of the pecten are predominantly associated with the P-face, thus resembling epithelial TJs. This situation is similar

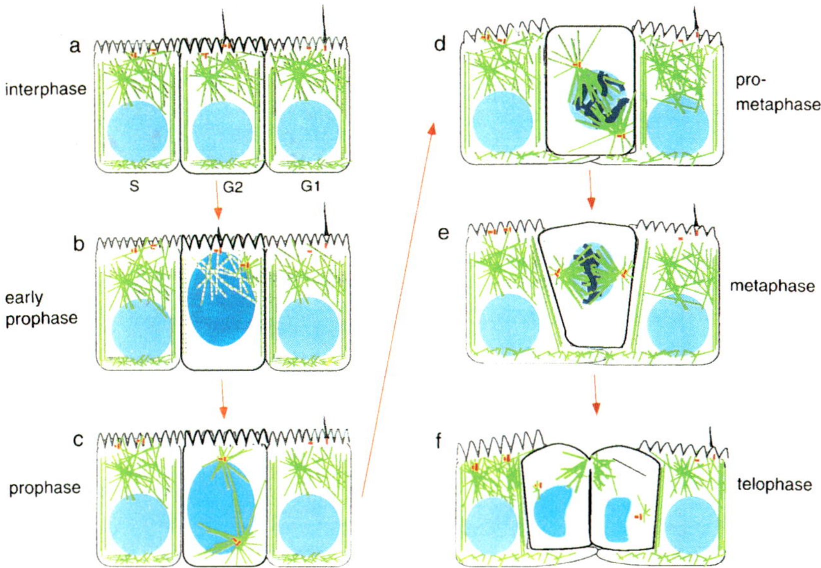

CH. 2, FIG. 13 Schematic representation of the changes in the organization of MTs (green) and the localization of centrioles (red) and the nuclei (blue) as observed in MDCKII epithelial cells. In interphase the mature centriole is at the base of the primary cilium (a; see also Fig. 8). Microtubules are very concentrated toward the apical surface and many minus ends are found in this location (see also Fig. 14). At the onset of prophase the centrioles start to migrate toward the nucleus to act as poles for the mitotic spindle (b, c). The spindle orients in a way so that cleavage primarily results in expansion of the epithelium, as opposed to the loss of a cell out of the epithelium during division (d, e). The centrosome plays a critical role in the correct positioning of the spindle. After telophase and cytokinesis (f) the cell returns to the interphase situation (a) (reproduced from Reinsch and Karsenti, *The Journal of Cell Biology*, 1994, **126,** 1509–1526 by copyright permission of The Rockefeller University Press).

to that found in endothelial cells in the brain of the chicken (Nico *et al.*, 1992) and reptiles (Shivers, 1979). However, these TJs differ considerably from those in bovine brain endothelial cells, which are associated with the P-face by only 50–60% (Wolburg *et al.*, 1994; Kniesel *et al.*, 1996). Furthermore, pecten endothelial cells, when cultured *in vitro,* do not lose their high PFA as do their mammalian counterparts (S. Liebner, unpublished observation). This difference is possibly related to an enhanced dependence of barrier formation on the brain microenvironment in mammals compared to submammalian animals.

d. Expression Pattern of the TJ Protein Occludin in the Pecten Endothelial Cells As mentioned previously, occludin is the first known transmembrane protein associated with TJs. To monitore the development of TJs and the BRB in pecten endothelial cells we investigated the pecten oculi in terms of occludin antigen expression (data not shown). The first blood vessels invading the pecten primordium on E7 were already positive for occludin but immunofluorescence showed a very diffuse labeling pattern, indicating that occludin was not yet restricted to junctional membrane areas. It remains to be determined whether the diffuse labeling was due to an intracellular labeling or to a broad distribution of occludin in the cell membrane. During further development, occludin labeling became more restricted to the inter-endothelial cell–cell junctions. This developmental scheme became most apparent in posthatching stages in which occludin was almost completely restricted to the cell–cell borders and only spot-like intracellular labeling was observed (Fig. 13, see color plate). How occludin is inserted into the junctional structure and the impact of this protein on TJ function and/or structure are not clear. Nevertheless, the labeling of occludin differed between barrier and nonbarrier endothelia in respect to both labeling intensity and pattern of intercellular junctions. Nonbarrier endothelia choroid vessels showed a weaker staining and a more fragile network of labeled cell–cell borders than barrier endothelia of the pecten (data not shown).

e. The Differentiation of Cellular Surface Properties in the Pecten As mentioned previously, differentiation of endothelial cells into the BBB phenotype seems to require specific induction signals. In this context, surface properties of endothelial cells and their direct neighbors are of special interest. Numerous cell culture experiments on barrier formation in epithelial and endothelial cells have been reported; the results of these experiments support the idea that a family of cell adhesion receptors, the cadherins, may play an important role in this process. For example, impairing cadherin function by antibodies and calcium depletion causes both loss of barrier function and TJ remodeling in Madin-Darby canine kidney

(MDCK) cells (Gumbiner *et al.*, 1988; Rajasekaran *et al.*, 1996). Cytoplasmic components of the adherens junctions, β-catenin and plakoglobin, bind directly to cadherins. Both molecules are members of the armadillo family (Huber *et al.*, 1996). These proteins have been shown to provide an important link between cell adhesion and transcriptional gene regulation in morphogenetic processes (Molenaar *et al.*, 1996; Orsulic and Peifer, 1996; Nusse, 1997). Depending on the degree of phosphorylation, β-catenin is either active in contact with cadherins or translocated into the nucleus where it interacts with Lef-1 as a transcription factor (Behrens *et al.*, 1996). Furthermore, as shown by its interaction with the adematous polyposis coli protein, β-catenin also plays a role in the regulation of tumor suppressor and oncogenic proteins (Hülsken *et al.*, 1994). How these various functions are related to the relevance of the cadherin/catenin system for BBB formation is unclear (Lampugnani and Dejana, 1997). It is noteworthy that during TJ remodeling after calcium depletion, β-catenin can be coimmunoprecipitated with the TJ component ZO-1 (Rajasekaran *et al.*, 1996). Lampugnani *et al.* (1997) demonstrated that the distribution and phosphorylation of β-catenin, plakoglobin, and ZO-1 are dependent on the degree of confluency of cultured human endothelial cells. Accordingly, low confluency and weak adhesion correspond to predominant β-catenin/VE-cadherin interaction, whereas long-lasting confluent cultures show mostly plakoglobin/VE-cadherin complexes. To date, there is only little data on the distribution, interaction, and developmental regulation of adherens and tight junction components *in vivo.*

During the development of the pecten, N-cadherin is expressed by the first endothelial cells invading the pecten primordium from the perineural plexus, which itself does not contain N-cadherin (data not shown). Interestingly, the endothelial N-cadherin expression is upregulated during further migration and differentiation into BBB phenotype yet rapidly diminishes after the functional barrier is established on E16. The localization of N-cadherin to the junctional cleft and additionally to the extrajunctional membrane may indicate that it not only participates in interendothelial adhesion but also might be considered a candidate for inductional signaling function. The developmental regulation of β-catenin and plakoglobin in endothelial cells of the pecten indicates a role in signal transduction of β-catenin during differentiation and for plakoglobin mainly in cell adhesion (data not shown). Pigmented cells show a different set of cadherins during development. Chicken R-cadherin and B-cadherin are differentially expressed in pigmented cells and may participate in mophogenetic processes leading to the establishment of the pecten structure (data not shown). A survey on the development of the pecten oculi is given in Fig. 14.

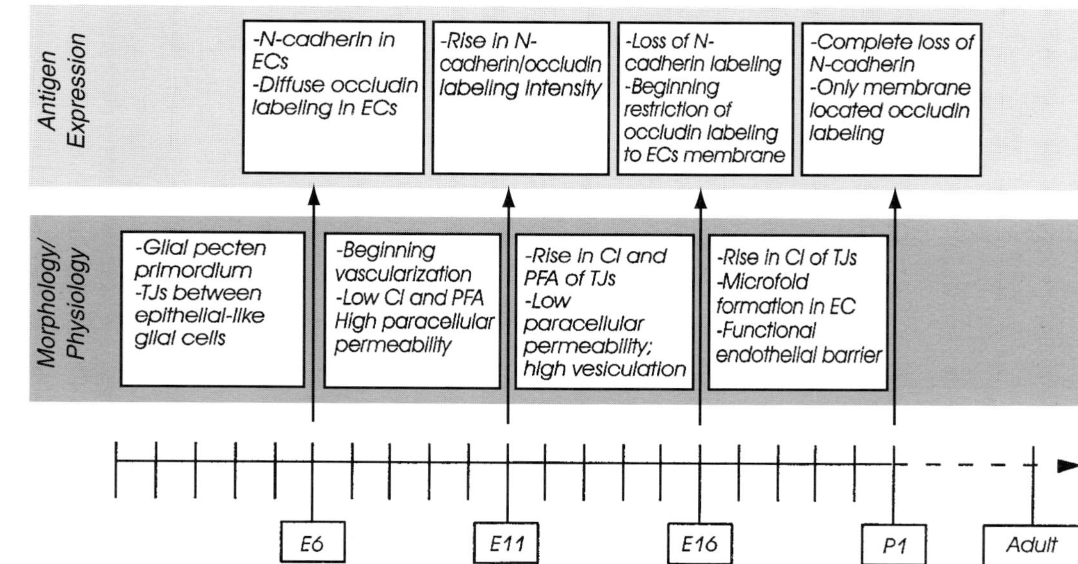

FIG. 14 Schedule of blood–retina barrier development in the pecten. ECs, endothelial cells; CI, complexity index of tight junctions (TJs); PFA, P-face association; E, embryonic day; P, posthatching day.

V. The Formation of the Vascular System in the Retina

After demonstrating that the pecten oculi is a vascular convolute in the avian eye that exhibits all features of an endothelial blood–retina barrier, we now will consider the possibility that the existence of the pecten plays a role in the avascularity of the avian retina. To this end, we will briefly describe the vasculature of the vertebrate retina, including new data on angiogenesis, and then consider more closely the metabolic interrelationships between the pecten and the avascular retina (Fig. 15).

A. The Vasculature of the Vertebrate Retina

Among the vertebrate groups, a great variability of retinal blood supply has developed. In principle, avascular and vascular retinae can be distinguished. The retinae of submammalian species (with the exception of the eel) are avascular, i.e., they do not contain intraretinal blood vessels. These retinae are nourished both from the choriocapillaris (which is lacking in the eel) at their outer surface and from the vitreous body at their inner surface. Both vascular systems have in common that they do not invade the neuroectodermal tissue. The vitreal vessels in teleosts are either bundled as a falciform process or form a widespread network of epiretinal capillaries (Collin, 1989) separated from the retina by the inner limiting membrane. In amphibians and reptiles, the inner retinal blood supply is similarly organized. The eyes of some reptiles, such as crocodiles and snakes, possess a vascular convolute called the "conus papillaris" which is similar to the pecten oculi of birds: A primordial tissue of glial origin is vascularized from the perineural plexus as well. By contrast, epiretinal blood vessels and the falciform process in fish, as well as vitreal vessels in the mammalian and human embryos, are formed as derivatives of the hyaloid arterial system (as the terminal portion of the ophthalmic artery). In more advanced developmental stages of the retina of most mammals, blood vessels penetrate the neuroectodermal tissue from the optic disc and form true intraretinal capillaries ["euangiotic" retina; for comprehensive reviews on retinal vasculature in vertebrates, see Chase (1982) and Stone and Maslim (1997)].

It has been hypothesized that there is a clear-cut correlation between the thickness of the retina and the state of vascularity. If the retinal thickness exceeds about 150 μm, it should be vascularized (Chase, 1982; Buttery *et al.,* 1991). It is believed that in thin retinae, the oxygen supply by the choriocapillaris is sufficient to fuel oxidative metabolism of the inner layers of the retina. Beyond a certain threshold of thickness (i.e., distance for oxygen diffusion), additional intraretinal blood vessels would be necessary to provide oxygen to the inner retina. Indeed, many data in the literature support such a relationship between retinal thickness and vascularization. However, there are also exceptions to this rule. For example, some vascularized mammalian retinae (e.g., that of the tree shrew) are thinner than, for instance, the avascular visual streak of the rabbit retina (Chao *et al.,* 1997). Furthermore, it has recently been shown that in normal guinea pig retina, which is avascular, the oxygen pressure at the inner surface of the retina is close to zero (Yu *et al.,* 1996). This is unexpected because such a low oxygen supply should evoke growth of blood vessels into the retina which, nevertheless, has never been observed in guinea pig retina.

Vascularization of mammalian retinae has been investigated in an abundance of studies which cannot be sufficiently summarized in this overview. Some of the most recent reviews are those of Chan-Ling (1994) and Stone and Maslim (1997). It is generally accepted that retinal vasculature codistributes with astroglial cells (Wolter, 1957; Schnitzer, 1987, 1988a,b; Stone and Dreher, 1987; Tout *et al.,* 1988; Trivino *et al.,* 1992; Gariano *et al.,* 1996; Chan-Ling, 1997). However, the codistribution commonly refers to the tangential dimension only; in the radial dimension, astrocytes codistribute with the innermost vessels in the nerve fiber and ganglion cell layers, whereas vessels in the inner nuclear layer are associated with Müller cell processes (Schnitzer, 1987). Therefore, the recent description of GFAP-positive astrocytes associated with blood vessels in the inner nuclear layer of the human retina is rather surprising (Trivino *et al.,* 1997).

The distribution of astrocytes in the retina may also shed some light on the mechanism of retinal vascularization. To address the question of which factors determine vascularization in the retina, it is important to distinguish between vasculogenesis and angiogenesis. The differentiation of angioblasts from mesoderm and the formation of a capillary plexus from growing and fusing angioblasts [called vasculogenesis by Risau *et al.* (1988) and Risau and Flamme (1995)] must strictly be separated from the angiogenic growth of new vascular sprouts from preexistent blood vessels (Risau, 1997). Several previous reports suggest that vascularization of the feline (Chan-Ling *et al.,* 1990), canine (McLeod *et al.,* 1987), rat (Shakib *et al.,* 1968), or primate retina (Ashton, 1970; Gariano *et al.,* 1994) involves both developmental mechanisms. In particular, blood vessel formation in the monkey and human retina occurs in two different successive steps, first by vasculogenesis in the nerve fiber layer and then by angiogenesis in the deeper capillary plexus (Gariano *et al.,* 1994). Remembering that in most (mammalian) species retinal astrocytes are confined to the nerve fiber layer, it might be concluded that astrocytes are involved in the process(es) of vasculogenesis rather than in angiogenic mechanisms. This implies that angiogenesis is independent

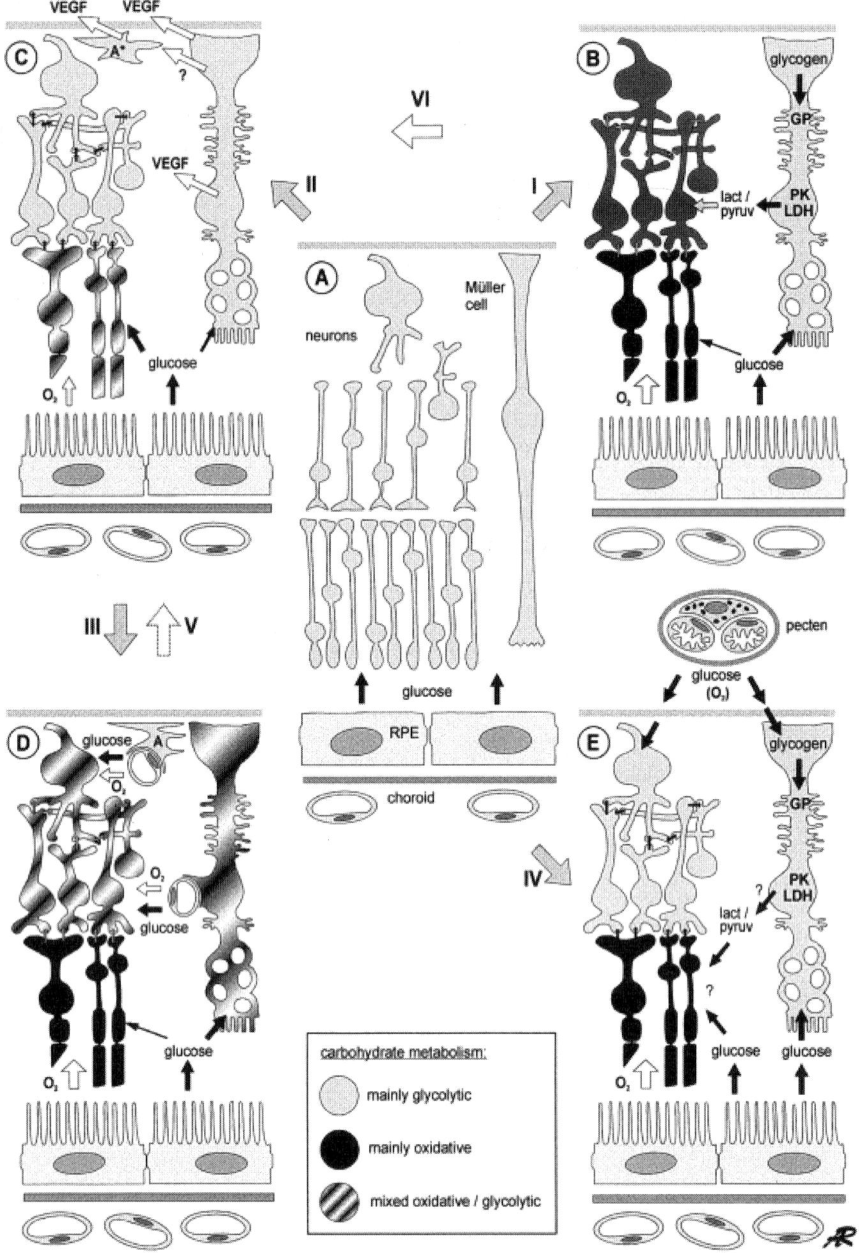

FIG. 15 Hypothetical scheme of how to build differently vascularized vertebrate retinae. In the middle (A), an embryonic retina is shown in which the mainly glycolytic metabolism of all the immature cells is sufficiently supported by glucose delivery from the choroid (via the

of astrocytes. This hypothesis is in line with the well-known observation that angiogenesis in the brain occurs from the perineural plexus before any astrocytes are differentiated (Risau and Lemmon, 1988; Engelhardt and Risau, 1995). According to this definition, the tangential spread of astroglial and endothelial precursors from the optic disc into the innermost layers of the retina corresponds to vasculogenesis, whereas ingrowth of capillaries into deeper retinal layers is due to angiogenesis.

Astrocytes or their precursors invade the retina from the optic nerve (Ling and Stone, 1988; Watanabe and Raff, 1988; Ling *et al.,* 1989). Ac-

---

retinal pigment epithelium; RPE). In the course of further differentiation, the energy demands of retinal neurons increase dramatically; one way to satisfy these demands is to "switch" to the more efficient oxidative metabolism (I). In many cases (B), the necessary oxygen (O_2) can be sufficiently provided by the choroid. This situation is observed in mature avascular retinae such as that of the guinea pig or the peripheral part of the rabbit retina. It is characterized by an obligatory oxidative metabolism of the neurons which is, for the most part, fueled by lactate/pyruvate (lact/pyruv) derived from the glycolytic metabolism of the Müller cells. A similar situation can be generated in species normally requiring an additional intraretinal vascularization if the choroidal oxygen supply is increased by postnatal breathing of pure oxygen (e.g., in premature infants). In the majority of mammalian species (with high spatial densities of neurons and/or complex neuronal circuits), the choroidal oxygen supply is insufficient to support an oxidative metabolism of the differentiating postnatal neurons (II, C). Thus, the cells are forced to maintain the inefficient glycolytic type of metabolism with a very high rate of glucose consumption. It is probably the resulting glucose deficiency which stimulates Müller cells and/or immigrating astrocytes (A*) to secrete vascular endothelial growth factor (VEGF) in order to evoke a vascularization of the retinal tissue. As a result, in most adult mammalian retinae (D), the combined choroidal and intraretinal glucose and oxygen supply allows for a sufficient oxidative (or, at least, mixed oxidative/glycolytic) energy metabolism of retinal neurons (and, perhaps dependent on the distance to blood vessels, also of many Müller cells). A transition from B to C may be enforced after cessation of the oxygen breathing of premature infants (VI). In such cases, the retina becomes hypoxic and responds by glial VEGF production and (pathological) neovascularization. A similar neovascularization occurs in mature retinae when ischemia (e.g., blood vessel occlusions) or insulin deficiency (diabetes mellitus) prevent sufficient glucose take into retinal glial cells (III: transition from D to C). In birds, the energetic demands of the retina are as high as those in mammals but another mechanism of metabolic support is realized (IV, E). The avian pecten oculi (and, possibly, also the reptilian conus papillaris) supplies large amounts of glucose (via the vitreous body) because of (i) a high glucose concentration in the avian blood plasma and (ii) a high density of glucose transporter molecules (isoform 1 = Glut1) in the endothelial cells of the pecten. This allows the maintenance of a mainly glycolytic metabolism in mature neurons and Müller cells, and thus avoids the necessity of an intraretinal vascularization. It should be pointed out that in mature retinae of all vertebrates, the (inner segments of) photoreceptor cells rely mainly on oxidative energy metabolism which is enabled by the particular choroidal oxygen supply. Their extremely active local oxygen consumption, however, prevents substantial oxygen diffusion from choroid into the inner retinal layers where, depending on species-specific energy demands, energetic failure may occur without additional intraretinal vascularization. A, mature astrocyte: LDH, lactate dehydrogenase; PK, pyruvate kinase.

cording to Huxlin *et al.* (1992), astrocyte immigration into the retina coincides with blood vessel formation, rendering it impossible to distinguish astroglial from vascular cell precursors (Gariano *et al.,* 1994). Whereas Chan-Ling and Stone (1991) present evidence for an early spread of mesenchymal vascular precursor cells which is followed by the migration of astrocytes, Ling *et al.* (1989), Jiang *et al.* (1994, 1995), Behzadian *et al.* (1995), and Stone *et al.* (1995) present data suggesting a primary spread of astrocytes into the retina, just ahead of newly formed vessels. Thus, most of the available data support the view that astroglial cell invasion precedes, and perhaps attracts, that of endothelial cell precursors. Avascular retinae, or avascular regions of an otherwise vascularized retina, are not entered by astrocytes during development. The failure of astrocyte immigration may be responsible for the avascularity of such retinae or retinal regions.

Another line of evidence also suggests that glial cells determine the extent of vascularization. It is well known that the VEGF is responsible for vascular growth, acting as a paracrine angiogenic factor during development and in tumors (Leung *et al.,* 1989; Breier *et al.,* 1992; Plate *et al.,* 1992; Millauer *et al.,* 1994; Folkman, 1995; Breier and Risau, 1996; Murata *et al.,* 1996; Beck and D'Amore, 1997; Provis *et al.,* 1997; Zhang and Stone, 1997). This growth factor is expressed and secreted by macroglial or glioma cells and binds to receptors on the endothelial cell surface. The binding of VEGF to its receptors is believed to be the main signal for angiogenesis (Millauer *et al.,* 1993; Breier and Risau, 1996). Interestingly, the level of VEGF expression is upregulated by hypoxia and downregulated by hyperoxia (Shweiki *et al.,* 1992; Ikeda *et al.,* 1995; Mukhopadhyay *et al.,* 1995; Donahue *et al.,* 1996). Thus, Stone *et al.* (1995) and Zhang and Stone (1997) postulate that hypoxia, due to the onset of neuronal activity, stimulates glial cells to secrete VEGF, thereby inducing angiogenesis. Inhibition of VEGF has been shown to suppress neovascularization of the retina (Aiello *et al.,* 1995; Aiello, 1997). Overexpression of VEGF results in pathological angiogenesis or hypervascularization, for example, in the diabetic retinopathy (Pe'er *et al.,* 1995; Amin *et al.,* 1997; Mathews *et al.,* 1997) or during embryonic development (Flamme *et al.,* 1995). Recently, the growth hormone was also shown to be important for hypoxia-induced neovascularization of the mouse retina (Smith *et al.,* 1997). Both inhibition of growth hormone secretion and expression of a growth hormone-antagonist transgene induced a reduction of neovascularization but not of VEGF or flk mRNA.

This "simple" VEGF-involving mechanism of hypoxia-induced vascularization cannot explain the observation that oxygen pressure in the inner retina of the guinea pig is near zero (Yu *et al.,* 1996) since the retina is not vascularized. This high degree of hypoxia should certainly be sufficient to induce angiogenesis, e.g., via VEGF expression by Müller cells. However, no blood vessels invade the retina of guinea pigs or of other species, such as the horse and the rabbit, in which a similar hypoxia of the inner retina seems to occur (Tillis *et al.,* 1988). This raises the question as to whether other mechanisms may be involved and/or a concerted action of several factors may be required for an induction of vasculogenesis or angiogenesis. Mandriota and Pepper (1997) described the induction of *in vitro* angiogenesis and upregulation of urokinase and tissue-type plasminogen activator, two of the hallmarks of VEGF activity, to be dependent on basic fibroblast growth factor (bFGF).

Thus, vascularization is certainly also controlled by bFGF and many other factors, such as cytokines, and probably by their mutual interaction(s) (Strömblad and Cheresh, 1996). In addition, the endowing of growing endothelial cells with adhesion molecules might be another important factor. Preliminary results showed that expression of different cadherins on the surface of endothelial and pigmented cells of the pecten is developmentally regulated. This might indicate that the expression pattern of cadherins correlates to the acquisition or loss of barrier properties and the capability to migrate, respectively.

Thus, the lack of blood vessels in the retina of the guinea pig, for example, might be due to an inherent lack or failure of any of the previously mentioned "links" (astrocyte immigration and synthesis of growth factors, and/or receptors) between hypoxia and induction of angiogenesis. This hypothesis appears to be rather unlikely, however, given the fact that most vertebrate retinae are avascular and that the vascularized mammalian retinae constitute an exception among vertebrates rather than the rule. This consideration supports the view that intraretinal vasculogenesis, which is preceded by the immigration of astrocytes, is a response to extraordinary metabolic needs and stimulated by metabolic deficiencies, and that in the retinae of most vertebrates either the metabolic demands are below the threshold of the angiogenetic mechanism(s) or other compensatory mechanisms are sufficient to fuel the metabolism of retinal cells. The latter case may apply to the retina of birds, which have a high body temperature and, thus, a very active cellular metabolism.

B. Is There a Role for the Pecten Oculi in the Avascularity of the Avian Retina?

Since the thickness of the avian retina clearly exceeds 150 μm, the hypothesis of Chase (1982) would predict that this tissue should be vascularized (as a mammalian retina of the same thickness would certainly be). However, there is a complete lack of blood vessels in the avian retina. This might be attributed to the presence of the pecten oculi: This vascular organ may provide enough substrates and oxygen to fuel an active oxidative metabolism of the inner retina (Wingstrand and Munk, 1965). This assumption would easily account for the avascularity of the avian retina: In these circumstances, there would be no retinal tissue hypoxia, no (enhanced) VEGF expression by Müller cells, and no retinal vasculogenesis/angiogenesis. However, there are several reasons to believe that the previously mentioned assumption is wrong. First, the operative deletion of the pecten does not cause a vascularization of the retina (Brach, 1975). Second, Müller cells of the normal chicken retina do express VEGF, and additional overexpression of VEGF does not induce any vascular sprouting (Schmidt and Flamme, 1998). Moreover, the avian VEGF receptor homolog Quek1/VEGFR2 has been shown to be expressed in the quail retina (Wilting *et al.*, 1997). Third, the cells of the inner retina lose (most of) their mitochondria during embryonic development (Hughes *et al.*, 1972; Buono and Sheffield, 1991a; Ruggiero and Sheffield, 1998), suggesting that there is no significant oxidative metabolism throughout most of the retinal tissue. It must be concluded that the avian retina fuels its metabolism largely by glycolysis. Indeed, biochemical experiments have shown that avian retinae have the highest glycolytic rate among all vertebrate tissues studied (Hughes *et al.*, 1972). Thus, the function of the pecten seems to be an efficient supply of glucose rather than of oxygen. This hypothesis fits well with the finding that pectinate endothelial cells express an extremely high density of glucose transporter molecules (Glut1) on their greatly enlarged membrane surface area. Moreover, the avian blood plasma contains significantly higher glucose concentrations than are found in mammals or other vertebrates studied (Krebs, 1972). This, together with the high glucose transport capacity of pectinate endothelial cells, appears to provide the basis for the very high glucose concentration found in the avian vitreous (Krebs, 1972; Fig. 15E). It should be pointed out that the latter is maintained despite a very high retinal glucose consumption rate. Finally, the significance of the pecten oculi for the retinal energy metabolism is corroborated by the observation that a necrotizing inflammation of the pecten obviously causes blindness (Raidal, 1997).

It may therefore be concluded that in birds the necessity of an intraretinal vascularization is prevented by the presence of the pecten oculi. By using the ample glucose supply of the avian bloodstream, this structure seems to pump vast amounts of glucose into the vitreous. This allows even the mature retinal cells to maintain the (predominantly) glycolytic mode of energy production which is characteristic of all embryonic tissues and prevents the development of a high oxygen demand (Figs. 15A and 15E). Buono and Sheffield (1991b) have shown that the aerobic isoform of the lactate dehydrogenase is distributed throughout the chicken retina, but that the anaerobic isoform of this enzyme is localized to the inner plexiform layer and the ganglion cell layer and increases during development. Whereas this metabolic situation seems to be inefficient from an economic point of view, it may have crucial functional advantages such as a high tolerance of hypoxia (which particularly may occur during the avian flight) and the avoidance of intraretinal blood vessels as optical obstacles.

C. Are There Metabolic Similarities between Avian and Avascular Mammalian Retinae?

Keeping in mind that intraretinal vascularization in most (but not all) mammals is an exception among vertebrates, the previously mentioned facts and hypotheses concerning avian retinal metabolism may also help to shed some light on the avascularity vs vascularity of mammalian retinae. In particular, we promote the idea that it is a (relative) lack of glucose rather than of oxygen which finally enforces an ingrowth of retinal blood vessels. There are two arguments in favor of this idea: (i) The avian retina is avascular, probably because it is well supplied by glucose, and (ii) mammalian retinae such as those of guinea pig (Yu *et al.*, 1996) and rabbit (Tillis *et al.*, 1988), which are almost anoxic throughout most of their thickness, never develop an intraretinal vascularization. Furthermore, at least during darkness (i.e., about half of each day), the inner plexiform (and nuclear) layer even of mature vascularized mammalian retinae is almost anoxic (Ahmed *et al.*, 1993; Yu *et al.*, 1994), but there is no further compensatory proliferation of blood vessels. On the other hand, there is a very high glycolytic rate even in vascularized retinae; most of their glucose utilization (but not their energy production) is due to anaerobic and aerobic glycolysis rather than respiration (Krebs, 1972; Winkler, 1995).

Whereas oxygen is supplied to the retina by rapid (passive) diffusion, glucose supply requires active (i.e., energy-consuming) transport across the membranes of endothelial and/or pigment epithelial cells. Thus, glucose transport is limited by both cellular energy stores and the number of transporter molecules available (e.g., dependent on cell surface area). Later, most cells confronted with a temporary lack of oxygen will first try to compensate by increasing glycolytic rate and, thus, glucose consumption. This means that hypoxia will be directly followed by a glucose deficiency before any ingrowth of blood vessels can occur. The problem of glucose deficiency is dramatically enhanced by the fact that a "switch" from respiration to glycolysis increases the load of the glucose-transporting mechanisms more than 10-fold.

Indeed, the neurons in vascularized retinae are able to perform such a switch: The rate of glycolysis is increased in darkness (due to the metabolic demands of the "dark current" of photoreceptor cells). If maintained *in vitro* with sufficient glucose supply, cat photoreceptors can produce almost normal light responses for as much as 1 h in complete anoxia (Noell, 1958). It might be speculated that mammalian retinae that become vascularized during ontogenesis have, from early developmental stages on, very high metabolic demands which cannot be fully satisfied even by the established vascularization (Figs. 15C and 15D). This in turn may require a maintenance of the (early in development dominating) glycolytic capacity of the neurons because the spatial density of neurons in such retinae, usually thicker than 150 μm in adulthood, is very high and the information processing very complex. These retinae will be characterized by a high rate of glucose utilization, and the retinal glucose supply will always be challenged. This situation is similar to that of the avian retina in which the problem is solved by the "glucose pump function" of the pecten oculi (Fig. 15E). Because this structure (and also the underlying high blood glucose level) is missing in mammalian eyes, intraretinal vascularization is required (Figs. 15C and 15D).

By contrast, the neurons of avascular mammalian retinae seem to develop an obligatory oxidative mode of energy production; if rabbit retinae are maintained *in vitro* under anoxic conditions, their photoreceptors completely lose physiological activity within a few minutes, irrespective of the glucose concentration (Noell, 1958). It might be speculated that in such retinae, the neuronal energy demands are low enough to be sufficiently fueled by the poor oxygen supply of the oxidative metabolism (Fig. 15B). Thus, the neuronal glucose consumption rate is (relatively) low, and there are no deficits in retinal glucose supply. Such a situation may well be maintained without any further retinal vascularization.

If the previously described assumptions are true, we must now search for a mechanism that links glucose deficiency with stimulation of blood vessel growth, probably via VEGF release (Fig. 15C). It has already been pointed out that VEGF is synthesized and released by the glial cells of the retina rather than by the neurons (Alou *et al.,* 1995; Stone *et al.,* 1995; Zhang and Stone, 1997). There are ample data supporting the view that these cells, even under aerobic conditions, predominantly perform glycolysis and supply neuronal cells with metabolites such as lactate (Dringen *et al.,* 1993; Poitry-Yamate *et al.,* 1995), which in the neuronal cell can be transformed by the lactate dehydrogenase into pyruvate fueling the neuronal Krebs cycle. This means that the metabolism of retinal glial cells crucially depends on a sufficient supply of glucose, whereas it may be rather independent of the available oxygen levels. Thus, their VEGF release may be directly triggered by glucose deficiency.

How does this hypothesis fit with the experimental observations? In cases in which the hypoxic neurons maintain a high level of glycolysis, they compete for the available glucose. This may be the reason for the normal postnatal vascularization of most mammalian retinae. If increased oxygen levels are supplied during the "critical" stages of development (e.g., oxygen breathing of premature newborn children), the retinal neurons can rely completely on the more economic oxidative metabolism, and there remains enough glucose to meet the needs of the glial cells. In these cases, the normal retinal vascularization fails to occur. A similar situation may be realized in certain mammalian retinae which have neurons with lower metabolic needs, such as in guinea pigs.

VI. Concluding Remarks

The literature on BBB provides an immense and inexhaustible body of data difficult to survey. It deals mainly with permeability properties of the vessel wall, problems of drug delivery into the brain, and aspects of metabolism and physiology at the interface between blood and brain under either normal or pathologic conditions. Fewer studies focus on the description of the morphology and biochemistry of the brain microvasculature. In this field, the literature is multiply overlapping with other cellular systems such as epithelia or peripheral endothelial cells. This may be due to the relative simplicity and the facilitated accessibility of epithelial tissue and/or peripheral blood vessels such as arteries and veins in comparison to brain capillaries. Another approach consists of the investigation of endothelial cell cultures, with the hope that basically these cells behave similarly to their analog *in vivo*. However, brain capillary endothelial cells are known to be extremely dependent on the brain microenvironment; if isolated from neural tissue, they immediately respond by alterations of the expression patterns of many antigens, such as glucose transporters, receptors, adhesion molecules, or of the barrier properties. Therefore, it is highly desirable to establish an *in vivo* model of the BBB that will allow study of its induction, maturation, and maintenance. The pecten oculi of the avian eye appears to fulfill all requirements for the investigation of BBB induction and maturation. Like a burning glass, this model focuses on several fascinating problems: (i) The migration of endothelial cells from a site where no barrier is established to a site where a barrier is established; (ii) the consecutive expression of barrier antigens and associated molecules (compounds of the cadherin/catenin system in the adherens junction as well as the tight junction complex), including their interplay; (iii) the question of why the blood–retina barrier switches in the pecten from the glial mode to the endothelial mode and whether this may reflect the phylogeny of the BBB which is also characterized by this switch; and (iv) the important problem of angiogenesis in the retina. One of the interesting questions is whether or not the pecten participates in the inhibition of vascular sprouting and/or astroglial migration in the avian retina. Furthermore, the presence and function of the pecten may have a great impact on metabolism and nutrition of the avian retina, including the glioneuronal interrelationships. The elucidation of these problems should shed some light on the regulation of angiogenesis in general as well as on the regulation of metabolic activities in the central nervous system.

Acknowledgments

Investigations considered in this overview were supported by the Graduiertenkolleg Neurobiologie of the University of Tübingen, by Grant Wo 215/11-3 (to H. G. and H. W.) from the Deutsche Forschungsgemeinschaft and by Grant 01KS9504, Project C5 (A. R.) from the Bundesministerium fuer Bildung und Wissenschaft (IZKF at the University of Leipzig). The authors gratefully acknowledge the generosity of Uwe Kniesel for providing Fig. 3. Figures 1 and 2 were produced with generous support of Christian F. Bardele and Horst Schoppmann. We acknowledge the gift of the anti-occludin- and the HT7-antibody by Drs. S. Tsukita (Kyoto, Japan) and W. Risau (Bad Nauheim, Germany), respectively.

References

Abbott, N. J., Hughes, C. C. W., Revest, P. A., and Greenwood, J. (1992). Development and characterization of a rat capillary endothelial culture. Towards an in vitro blood–brain barrier. *J. Cell Sci.* **103,** 23–38.

Ahmed, J., Braun, R. D., Dunn, R., and Linsenmeier, R. A. (1993). Oxygen distribution in the macaque retina. *Invest. Ophthalmol. Vis. Sci.* **34,** 516–521.

Aiello, L. P. (1997). Vascular endothelial growth factor. *Invest. Ophthalmol. Vis. Sci.* **38,** 1647–1652.

Aiello, L. P., Pierce, E. A., Foley, E. D., Takagi, H., Chen, H., Riddle, L., Ferrara, N., King, G. L., and Smith, L. E. H. (1995). Suppression of retinal neovascularization in vivo by inhibition of vascular endothelial growth factor (VEGF) using soluble VEGF-receptor chimeric proteins. *Proc. Natl. Acad. Sci. USA* **92,** 10457–10461.

Albrecht, U., Seulberger, H., Schwarz, H., and Risau, W. (1990). Correlation of blood–brain barrier function and HT 7 protein distribution in chick brain circumventricular organs. *Brain Res.* **535,** 49–61.

Alou, T., Hemo, J., Itin, A., Peer, A., Stone, J., and Keshet, E. (1995). Vascular endothelial growth factor acts as a survival factor for newly formed retinal vessels and has implications for retinopathy of prematurity. *Nature Med.* **1,** 1024–1029.

Amin, R. H., Frank, R. N., Kennedy, A., Eliott, D., Puklin, J. E., and Abrams, G. W. (1997). Vascular endothelial growth factor is present in glial cells of the retina and optic nerve of human subjects with nonproliferative diabetic retinopathy. *Invest. Ophthalmol. Vis. Sci.* **38,** 36–47.

Anderson, J. M., and Van Itallie, C. M. (1995). Tight junctions and the molecular basis for regulation of paracellular permeability. *Am. J. Physiol.* **269,** G467–G475.

Ando-Akatsuka, Y., Saitou, M., Hirase, T., Kishi, M., Sakaqkibara, A., Itoh, M., Yonemura, S., Furuse, M., and Tsukita, S. (1996). Interspecies diversity of the occludin sequence: cDNA cloning of human, mouse, dog, and rat-kangaroo homologues. *J. Cell Biol.* **133,** 43–48.

Arthur, F. E., Shivers, R. R., and Bowman, P. D. (1987). Astrocyte-mediated induction of tight junctions in brain capillary endothelium: An efficient in vitro model. *Dev. Brain Res.* **36,** 155–159.

Ashton, N. (1970). Retinal angiogenesis in the human embryo. *Br. Med. Bull.* **26,** 103–106.

Bachsich, P., and Gellért, A. (1935). Beiträge zur Kenntnis der Struktur und Funktion des Pecten oculi im Vogelauge. *Graefes Arch. Ophthalmol.* **133,** 448–460.

Bauer, H., Sonnleitner, U., Lametschwandtner, A., Steiner, M., Adam, H., and Bauer, H. C. (1995). Ontogenetic expression of the erythroid-type glucose transporter (Glut-1) in the telencephalon of the mouse: Correlation to the tightening of the blood–brain barrier. *Dev. Brain Res.* **86,** 317–325.

Bauer, H. C., Bauer, H., Lametschwandtner, A., Amberger, A., Ruiz, P., and Steiner, M. (1993). Neovascularization and the appearance of morphological characteristics of the blood–brain barrier in the embryonic mouse central nervous system. *Dev. Brain Res.* **75,** 269–278.

Bawa, S. R., and YashRoy, R. C. (1974). Structure and function of vulture pecten. *Acta Anat.* **89,** 473–480.

Beck, L., and D'Amore, P. A. (1997). Vascular development: Cellular and molecular regulation. *FASEB J.* **11,** 365–373.

Beck, W. D., Roberts, R. L., and Olson, J. J. (1986). Glial cells influence membrane-associated enzyme activity at the blood–brain barrier. *Brain Res.* **381,** 131–137.

Behrens, J., von Kries, J. P., Kühl, M., Bruhn, L., Wedlich, D., Grosschedl, R., and Birchmeier, W. (1996). Functional interaction of β-catenin with the transcription factor LEF-1. *Nature* **382,** 638–642.

Behzadian, M. A., Wang, X.-L., Jiang, B., and Caldwell, R. B. (1995). Angiostatic role of astrocytes: Suppression of vascular endothelial cell growth by TGF and other inhibitory factor(s). *Glia* **15,** 480–490.

Bertossi, M., Ribatti, D., Nico, B., Virgintino, D., Mancini, L., and Roncali, L. (1989). Computerized three-dimensional reconstruction of the developing blood–brain barrier. *Acta Neuropathol.* **79,** 48–51.

Bertossi, M., Roncali, L., Nico, B., Ribatti, D., Mancini, L., Virgintino, D., Fabiani, G., and Guidazzoli, A. (1993). Perivascular astrocytes and endothelium in the development of the blood–brain barrier in the optic tectum of the chick embryo. *Anat. Embryol.* **188,** 21–30.

Biegel, D., Spencer, D. D., and Pachter, J. S. (1995). Isolation and culture of human brain microvessel endothelial cells for the study of blood brain barrier properties in vitro. *Brain Res.* **692,** 183–189.

Blochmann, F., and von Husen, E. (1911). Ist der Pecten des Vogelauges ein Sinnesorgan? *Biol. Zentralbl.* **31,** 150–156.

Boado, R. J. (1996). Brain-derived peptides increase the expression of a blood–brain barrier GLUT-1 glucose transporter reporter gene. *Neurosci. Lett.* **220,** 53–56.

Boado, R. J., Wang, L., and Pardridge, W. M. (1994). Enhanced expression of the blood–brain barrier GLUT-1 glucose transporter gene by brain-derived factors. *Mol. Brain Res.* **22,** 259–267.

Bolz, S., Farrell, C. L., Dietz, K., and Wolburg, H. (1996). Subcellular distribution of glucose transporter (GLUT-1) during development of the blood–brain barrier in rats. *Cell Tissue Res.* **284,** 355–365.

Brach, V. (1975). The effect of intraocular ablation of the pecten oculi of the chicken. *Invest. Ophthalmol.* **14,** 166–168.

Braekevelt, C. R. (1984). Electron microscopic observations on the pecten of the nighthawk (*Chordeiles minor*). *Ophthalmologica* **189,** 211–220.

Breier, G., and Risau, W. (1996). The role of vascular endothelial growth factor in blood vessel formation. *Trends Cell Biol.* **6,** 454–456.

Breier, G., Albrecht, U., Sterrer, S., and Risau, W. (1992). Expression of vascular endothelial growth factor during embryonic angiogenesis and endothelial cell differentiation. *Development* **114,** 521–532.

Brightman, M. W., and Reese, T. S. (1969). Junctions between intimately apposed cell membranes in the vertebrate brain. *J. Cell Biol.* **40,** 648–677.

Buono, R. J., and Sheffield, J. B. (1991a). Changes in distribution of mitochondria in the developing chick retina. *Exp. Eye Res.* **53,** 187–198.

Buono, R. J., and Sheffield, J. B. (1991b). Changes in expression and distribution of lactate dehydrogenase isoenzymes in the developing chick retina. *Exp. Eye Res.* **53,** 199–204.

Butt, A. M., Jones, H. C., and Abbott, N. J. (1990). Electrical resistance across the blood–brain barrier in anaesthetized rats: A developmental study. *J. Physiol.* **429,** 47–62.

Buttery, R. G., Hinrichsen, C. F. L., Weller, W. L., and Haight, J. R. (1991). How thick should a retina be? A comparative study of mammalian species with and without intraretinal vasculature. *Vision Res.* **31,** 169–187.

Caley, D. W., and Maxwell, D. S. (1970). Development of the blood vessels and extracellular spaces during postnatal maturation of rat cerebral cortex. *J. Comp. Neurol.* **138,** 31–48.

Cancilla, P. A., and De Bault, L. E. (1983). Neutral amino acid transport properties of cerebral endothelial cells in vitro. *J. Neuropathol. Exp. Neurol.* **42,** 191–199.

Ceballos, G., and Rubio, R. (1995). Coculture of astroglial and vascular endothelial cells as apposing layers enhances the transcellular transport of hypoxanthine. *J. Neurochem.* **64,** 991–999.

Chan-Ling, T. (1994). Glial, neuronal, and vascular interactions in the mammalian retina. *Progr. Retina Eye Res.* **13,** 357–389.

Chan-Ling, T. (1997). Glial, vascular, and neuronal cytogenesis in whole-mounted cat retina. *Microsc. Res. Technol.* **36,** 1–16.

Chan-Ling, T., and Stone, J. (1991). Factors determining the migration of astrocytes into the developing retina. Migration does not depend on intact axons or patent vessels. *J. Comp. Neurol.* **303,** 375–386.

Chan-Ling, T., Halasz, P., and Stone, J. (1990). Development of retinal vasculature in the cat: Processes and mechanisms. *Curr. Eye Res.* **9,** 459–478.

Chao, T. I., Grosche, J., Friedrich, K. J., Biedermann, B., Francke, M., Pannicke, T., Reichelt, W., Wulst, M., Mühle, C., Pritz-Hohmeier, S., Kuhrt, H., Faude, F., Drommer, W., Kasper, M., Buse, E., and Reichenbach, A. (1997). Comparative studies on mammalian Müller (retinal glial) cells. *J. Neurocytol.* **26,** 439–454.

Chase, J. (1982). The evolution of retinal vascularization in mammals. *Ophthalmology* **89,** 1518–1525.

Collin, S. P. (1989). Topographic organization of the ganglion cell layer and intraocular vascularization in the retinae of two reef teleosts. *Vision Res.* **29,** 765–775.

Coomber, B. L., and Stewart, P. A. (1986). Three-dimensional reconstruction of vesicles in endothelium of blood–brain barrier versus highly permeable microvessels. *Anat. Rec.* **215,** 256–261.

Crozier, W. J., and Wolf, E. (1944a). Flicker response contours for the sparrow, and the theory of the avian pecten. *J. Gen. Physiol.* **27,** 315–324.

Crozier, W. J., and Wolf, E. (1944b). Modifications of the flicker response contour, and the signification of the avian pecten. *J. Gen. Physiol.* **27,** 287–313.

De Bault, L. E., and Cancilla, P. A. (1980). Gamma-glutamyl transpeptidase in isolated brain endothelial cells: induction by glial cells in vitro. *Science* **207,** 653–655.

Dehouck, M.-P., Méresse, S., Delorme, P., Fruchart, J.-C., and Cecchelli, R. (1990). An easier reproducible and mass-production method to study the blood–brain barrier in vitro. *J. Neurochem.* **54,** 1798–1801.

Dermietzel, R., and Krause, D. (1991). Molecular anatomy of the blood–brain barrier as defined by immunocytochemistry. *Int. Rev. Cytol.* **127,** 57–109.

Dermietzel, R., Krause, D., Kremer, M., Wang, C., and Stevenson, B. (1992). Pattern of glucose transporter (Glut 1) expression in embryonic brains is related to maturation of blood–brain barrier tightness. *Dev. Dyn.* **193,** 152–163.

Derouiche, A., and Frotscher, M. (1991). Astroglial processes around identified glutamatergic synapses contain glutamine synthetase. Evidence for transmitter degradation. *Brain Res.* **55,** 346–350.

Derouiche, A., and Rauen, T. (1995). Coincidence of L-glutamate/L-aspartate transporter (GLAST) and glutamine synthetase (GS) immunoreactions in retinal glia: Evidence for coupling of GLAST and GS in transmitter clearance. *J. Neurosci. Res.* **42,** 131–143.

Dieterich, C. E., Dieterich, H. J., Spycher, M. A., and Pfautsch, M. (1973). Fine structural observations of the pecten oculi capillaries of the chicken. *Z. Zellforsch. Mikr. Anat.* **146,** 473–489.

Donahue, M. L., Phelps, D. L., Watkins, R. H., LoMonaco, M. B., and Horowitz, S. (1996). Retinal vascular endothelial growth factor (VEGF) mRNA expression is altered in relation to neovascularization in oxygen induced retinopathy. *Curr. Eye Res.* **15,** 175–184.

Dringen, R., Gebhardt, R., and Hamprecht, B. (1993). Glycogen in astrocytes: Possible function as lactate supply for neighboring cells. *Brain Res.* **623,** 208–214.

Eichhorn, M., and Flügel, C. (1988). Histochemical demonstration of carbonic anhydrase and Na^+/K^+-ATPase in the pecten oculi of the fowl. *Exp. Eye Res.* **47,** 147–153.

El Hafny, B., Bourre, J. M., and Roux, F. (1996). Synergistic stimulation of gamma-glutamyl transpeptidase and alkaline phosphatase activities by retinoic acid and astroglial factors in immortalized rat brain microvessel endothelial cells. *J. Cell. Physiol.* **167,** 451–460.

Engelhardt, B., and Risau, W. (1995). Development of the blood–brain barrier. *In* "New Concepts of a Blood–Brain Barrier" (J. Greenwood *et al.,* eds.), pp. 11–31. Plenum, New York.

Fadool, J. M., and Linser, P. J. (1993). 5A11 antigen is a cell recognition molecule which is involved in neuronal–glial interactions in avian neural retina. *Dev. Dyn.* **196,** 252–262.

Farrell, C. L., and Pardridge, W. M. (1991). Blood brain barrier glucose transporter is asymmetrically distributed on brain capillary endothelial lumenal and ablumenal membranes. An electron microscopic immunogold study. *Proc. Natl. Acad. Sci. USA* **88,** 5779–5783.

Finnemann, S. C., Marmorstein, A. D., Neill, J. M., and Rodriguez Boulan, E. (1997). Identification of the retinal pigment epithelium protein RET-PE2 as CE-9/OX-47, a member of the immunoglobulin superfamily. *Invest. Ophthalmol. Vis. Sci.* **38,** 2366–2374.

Fischlschweiger, W., and O'Rahilly, R. (1966). The ultrastructure of the pecten oculi in the chick. *Acta Anat.* **65,** 561–578.

Flamme, I., von Reutern, M., Drexler, H. C. A., Syed-Ali, S., and Risau, W. (1995). Overexpression of vascular endothelial growth factor in the avian embryo induces hypervascularization and increased veascular permeability without alterations of embryonic pattern formation. *Dev. Biol.* **171,** 399–414.

Flamme, I., Frölich, T., and Risau, W. (1997). Molecular mechanisms of vasculogenesis and embryonic angiogenesis. *J. Cell. Physiol.* **173,** 206–210.

Folkman, J. (1995). Clinical applications of research on angiogenesis. *N. Engl. J. Med.* **333,** 1757–1763.

Forsyth, R. J. (1996). Astrocytes and the delivery of glucose from plasma to neurons. *Neurochem. Int.* **28,** 231–242.

Frank, R. N., Duta, S., and Mancini, M. A. (1987). Pericyte coverage is greater in the retinal than in the cerebral capillaries of the rat. *Invest. Ophthalmol. Vis. Sci.* **28,** 1086–1091.

Furuse, M., Hirase, T., Itoh, M., Nagafuchi, A., Yonemura, S., and Tsukita, S. (1993). Occludin: A novel integral membrane protein localizing at tight junctions. *J. Cell Biol.* **123,** 1777–1788.

Furuse, M., Itoh, M., Hirase, T., Nagafuchi, A., Yonemura, S., and Tsukita, S. (1994). Direct association of occludin with ZO-1 and its possible involvement in the localization of occludin at tight junctions. *J. Cell Biol.* **127,** 1617–1626.

Gariano, R. F., Iruela-Arispe, M. L., and Hendrickson, A. E. (1994). Vascular development in primate retina: Comparison of laminar plexus formation in monkey and human. *Invest. Ophthalmol. Vis. Sci.* **35,** 3442–3455.

Gariano, R. F., Sage, E. H., Kaplan, H. J., and Hendrickson, A. E. (1996). Development of astrocytes and their relation to blood vessels in fetal monkey retina. *Invest. Ophthalmol. Vis. Sci.* **37,** 2367–2375.

Gerhardt, H., Liebner, S., and Wolburg, H. (1996). The pecten oculi of the chicken as a new in vivo model of the blood–brain barrier. *Cell Tissue Res.* **285,** 91–100.

Gorovits, R., Avidan, N., Avisar, N., Shaked, I., and Vardimon, L. (1997). Glutamine synthetase protects against neuronal degeneration in injured retinal tissue. *Proc. Natl. Acad. Sci. USA* **94,** 7024–7029.

Grebenkämper, K., and Galla, H.-J. (1994). Translational diffusion measurements of a fluorescent phospholipid between MDCK-1 cells support the lipid model of the tight junctions. *Chem. Phys. Lipids* **71,** 133–143.

Guerin, J. L., and Bobilya, D. J. (1997). Hypothalamic extract influences a blood–brain barrier model of porcine brain capillary endothelial cells. *Neurochem. Res.* **22,** 321–326.

Gumbiner, B., Stevenson, B., and Grimaldi, A. (1988). The role of the cell adhesion molecule uvomorulin in the formation and maintenance of the epithelial junctional complex. *J. Cell Biol.* **107,** 1575–1587.

Hannah, R. S., and Nathaniel, E. J. H. (1974). The postnatal development of blood vessels in the substantia gelatinosa of rat cervical cord—An ultra-structural study. *Anat. Rec.* **178,** 691–710.

Hayashi, Y., Nomura, M., Yamagishi, S.-I., Harada, S.-I., Yamashita, J., and Yamamoto, H. (1997). Induction of various blood–brain barrier properties in non-neural endothelial cells by close apposition to co-cultured astrocytes. *Glia* **19,** 13–26.

Hirase, T., Staddon, J. M., Saitou, M., Ando-Akatsuka, Y., Itoh, M., Furuse, M., Fujimoto, K., Tsukita, S., and Rubin, L. L. (1997). Occludin as a possible determinant of tight junction permeability in endothelial cells. *J. Cell Sci.* **110,** 1603–1613.

Holash, J. A., Noden, D. M., and Stewart, P. A. (1993). Re-evaluating the role of astrocytes in blood–brain barrier induction. *Dev. Dyn.* **197,** 14–25.

Hossler, F. E., and Olson, K. R. (1984). Microvasculature of the avian eye: Studies on the eye of the duckling with microcorrosion casting, scanning electron microscopy, and stereology. *Am. J. Anat.* **170,** 205–221.

Huber, O., Bierkamp, C., and Kemler, R. (1996). Cadherins and catenins in development. *Curr. Opin. Cell Biol.* **8,** 685–691.

Hughes, J. T., Jerrome, D., and Krebs, H. A. (1972). Ultrastructure of the avian retina. An anatomical study of the retina of the domestic pigeon (*Columba liva*) with particular reference to the distribution of mitochondria. *Exp. Eye Res.* **14,** 189–195.

Hülsken, J., Behrens, J., and Birchmeier, W. (1994). Tumor-suppressor gene products in cell contacts: The cadherin-APC-*armadillo* connection. *Curr. Opin. Cell Biol.* **6,** 711–716.

Huxlin, K. R., Sefton, A. J., and Furby, J. H. (1992). The origin and development of retinal astrocytes in the mouse. *J. Neurocytol.* **21,** 530–544.

Ikeda, E., Achen, M. G., Breier, G., and Risau, W. (1995). Hypoxia-induced transcriptional activation and increased mRNA stability of vascular endothelial growth factor in C6 glioma cells. *J. Biol. Chem.* **270,** 19761–19766.

Janzer, R. C., and Raff, M. C. (1987). Astrocytes induce blood–brain barrier properties in endothelial cells. *Nature* **325,** 253–256.

Jiang, B., Liou, G. I., Behzadian, M. A., and Caldwell, R. B. (1994). Astrocytes modulate retinal vasculogenesis: Effects on fibronectin expression. *J. Cell Sci.* **107,** 2499–2508.

Jiang, B., Behzadian, M. A., and Caldwell, R. B. (1995). Astrocytes modulate retinal vasculogenesis: Effects on endothelial cell differentiation. *Glia* **15,** 1–10.

Joo, F. (1993). The blood–brain barrier in vitro: The second decade. *Neurochem. Int.* **23,** 499–521.

Kachar, B., and Reese, T. S. (1982). Evidence for the lipidic nature of tight junction strands. *Nature* **296,** 464–466.

Kauth, H., and Sommer, H. (1953). Das Ferment Kohlensäureanhydratase im Tierkörper. IV. Über die Funktion des Pecten im Vogelauge. *Biol. Zentralbl.* **72,** 196–209.

Kiama, S. G., Bhattacharjee, J., Maina, J. N., and Weyrauch, K. D. (1994). A scanning electron microscope study of the pecten oculi of the black kite (Milvus migrans): Possible involvement of melanosomes in protecting the pecten against damage by ultraviolet light. *J. Anat.* **185,** 637–642.

Kniesel, U., and Wolburg, H. (1999). The tight junctions of the blood–brain barrier. *Cell. Mol. Neurobiol.,* in press.

Kniesel, U., Risau, W., and Wolburg, H. (1996). Development of blood–brain barrier tight junctions in the rat cortex. *Dev. Brain Res.* **96,** 229–240.

Krebs, H. A. (1972). The Pasteur effect and the relations between respiration and fermentation. *In* "Essays in Biochemistry" (F. Dickens and P. N. Campbell, eds.), pp. 1–34. Academic Press, New York.

Kumagai, A. K., Glasgow, B. J., and Pardridge, W. M. (1994). Glut1 glucose transporter expression in the diabetic and non-diabetic human eye. *Invest. Ophthalmol. Vis. Sci.* **35,** 2887–2894.

Lampugnani, M. G., and Dejana, E. (1997). Interendothelial junctions: Structure, signalling and functional roles. *Curr. Opin. Cell Biol.* **9,** 674–682.

Lampugnani, M. G., Corada, M., Andriopoulou, P., Esser, S., Risau, W., and Dejana, E. (1997). Cell confluence regulates tyrosine phosphorylation of adherens junction components in endothelial cells. *J. Cell Sci.* **110,** 2065–2077.

Lane, N. J., Reese, T. J., and Kachar, B. (1992). Structural domains of the tight junctional intramembrane fibrils. *Tissue Cell* **24,** 291–300.

Lane, N. J., Revest, P. A., Whytock, S., and Abbott, N. J. (1995). Fine-structural investigation of rat brain microvascular endothelial cells: Tight junctions and vesicular structures in freshly isolated and cultured preparations. *J. Neurocytol.* **24,** 347–360.

Laterra, J., and Goldstein, G. W. (1991). Astroglial-induced in vitro angiogenesis. Requirements for RNA and protein synthesis. *J. Neurochem.* **57,** 1231–1239.

Laterra, J., Guérin, C. J., and Goldstein, G. W. (1990). Astrocytes induce neural microvascular endothelial cells to form capillary-like structures in vitro. *J. Cell Physiol.* **144,** 204–215.

Latker, C. H., and Beebe, D. C. (1984). Developmental changes in the blood–ocular barriers in chicken embryos. *Exp. Eye Res.* **39,** 401–414.

Leiner, M. (1951). Über die Bedeutung des Pecten im Vogelauge. *Zool. Anz. Suppl.* **15,** 117–123.

Leung, D. W., Cachianes, G., Kuang, W.-J., Goeddel, D. V., and Ferrara, N. (1989). Vascular endothelial growth factor is a secreted angiogenic mitogen. *Science* **246,** 1306–1309.

Liebner, S., Gerhardt, H., and Wolburg, H. (1997). Maturation of the blood–retina barrier in the developing pecten oculi of the chicken. *Dev. Brain Res.* **100,** 205–219.

Ling, T., and Stone, J. (1988). The development of astrocytes in the cat retina: Evidence of migration from the optic nerve. *Dev. Brain Res.* **44,** 73–86.

Ling, T., Mitrofanis, J., and Stone, J. (1989). Origin of retinal astrocytes in the rat: Evidence of migration from the optic nerve. *J. Comp. Neurol.* **286,** 345–352.

Linser, P. J., Trapido-Rosenthal, H. G., and Orona, E. (1997). Glutamine synthetase is a glial-specific marker in the olfactory regions of the lobster (Panulirus argus) nervous system. *Glia* **20,** 275–283.

Linser, P., and Moscona, A. A. (1984). Variable CAII compartmentalization in vertebrate retina. *Ann. N. Y. Acad. Sci.* **429,** 430–446.

Lobrinus, J. A., Juillerat-Jeanneret, L., Darekar, P., Schlosshauer, B., and Janzer, R. C. (1992). Induction of the blood–brain barrier specific HT7 and neurothelin epitopes in endothelium cells of the chick chorioallantoic vessels by a soluble factor derived from astrocytes. *Dev. Brain Res.* **70,** 207–212.

Maher, F., Vannucci, S. J., and Simpson, J. A. (1994). Glucose transporter proteins in brain. *FASEB J.* **8,** 1003–1011.

Mandriota, S. J., and Pepper, M. S. (1997). Vascular endothelial growth factor-induced in vitro angiogenesis and plasminogen activator expression are dependent on endogenous basic fibroblast growth factor. *J. Cell Sci.* **110,** 2293–2302.

Mann, I. C. (1924). The function of the pecten. *Br. J. Ophthalmol.* **8,** 209–226.

Mathews, M. K., Merges, C., McLeod, D. S., and Lutty, G. A. (1997). Vascular endothelial growth factor and vascular permeability changes in human diabetic retinopathy. *Invest. Ophthalmol. Vis. Sci.* **38,** 2729–2741.

Maxwell, K., Berliner, J. A., and Cancilla, P. A. (1987). Induction of gamma-glutamyl transpeptidase in cultured cerebral endothelial cells by a product released by astrocytes. *Brain Res.* **410,** 309–314.

Maxwell, K., Berliner, J. A., and Cancilla, P. A. (1989). Stimulation of glucose analogue uptake by cerebral microvessel endothelial cells by a product released by astrocytes. *J. Neuropathol. Exp. Neurol.* **48,** 69–80.

McCarthy, K. M., Skare, I. B., Stankewich, M. C., Furuse, M., Tsukita, S., Rogers, R. A., Lynch, R. D., and Schneeberger, E. E. (1996). Occludin is a functional component of the tight junction. *J. Cell Sci.* **109,** 2287–2298.

McLeod, D. S., Lutty, G. A., Wajer, S. D., and Flower, R. W. (1987). Visualization of a developing vasculature. *Microvasc. Res.* **33,** 257–269.

Méresse, S., Dehonk, M.-P., Delorme, P., Bensaid, M., Tauber, J.-P., Delbart, C., Fruchart, J.-C., and Cecchelli, R. (1989). Bovine brain endothelial cells express tight junctions and monoamine oxidase activity in long-term culture. *J. Neurochem.* **53,** 1363–1371.

Michaelson, I. C. (1954). "Retinal Circulation in Man and Animals." Charles Thomas, Springfield, IL.

Millauer, B., Wizigmann-Voos, S., Schnürch, H., Martinez, R., Moller, N. P. H., Risau, W., and Ullrich, A. (1993). High affinity VEGF binding and developmental expression suggest flk-1 as a major regulator of vasculogenesis and angiogenesis. *Cell* **72,** 835–846.

Millauer, B., Shawver, L. K., Plate, K. H., Risau, W., and Ullrich, A. (1994). Glioblastoma growth inhibited *in vivo* by a dominant-negative flk-1 mutant. *Nature* **367,** 576–579.

Molenaar, M., van de Wetering, M., Oosterwegel, M., Peterson-Maduro, J., Godsave, S., Korinek, V., Roose, J., Destrée, O., and Clevers, H. (1996). XTcf-3 transcription factor mediates β-catenin-induced axis formation in Xenopus embryos. *Cell* **86,** 391–399.

Mollgard, K., and Saunders, N. R. (1986). The development of the human blood–brain and blood–CSF barriers. *Neuropathol. Appl. Neurobiol.* **12,** 337–358.

Moses, M. A., Klagsbrun, M., and Shing, Y. (1995). The role of growth factors in vascular cell development and differentiation. *Int. Rev. Cytol.* **161,** 1–48.

Mukhopadhyay, D., Tsiokas, L., Zhou, X. M., Foster, D., Brugge, J. S., and Sukhatme, V. P. (1995). Hypoxic induction of human vascular endothelial growth factor expression through c-src activation. *Nature* **375,** 577–581.

Murata, T., Nakagawa, K., Kalil, A., Ishibashi, T., Inomata, H., and Sueishi, K. (1996). The temporal and spatial vascular endothelial growth factor expression in retinal vasculogenesis of rat neonates. *Lab. Invest.* **74,** 68–77.

Muruganandam, A., Herx, L. M., Monette, R., Durkin, J. P., and Stanimirovic, D. B. (1997). Development of immortalized human cerebromicrovascular endothelial cell line as an in vitro model of the human blood–brain barrier. *FASEB J.* **11,** 1187–1197.

Mustonen, T., and Alitalo, K. (1995). Endothelial receptor tyrosine kinases involved in angiogenesis. *J. Cell Biol.* **129,** 895–898.

Nabeshima, S., Reese, T. S., Landis, D. M. D., and Brightman, M. W. (1975). Junctions in the meninges and marginal glia. *J. Comp. Neurol.* **164,** 127–170.

Nagy, Z., Peters, H., and Hüttner, I. (1984). Fracture faces of cell junctions in cerebral endothelium during normal and hyperosmotic conditions. *Lab. Invest.* **50,** 313–322.

Nico, B., Cantino, D., Bertossi, M., Ribatti, D., Sassoe, M., and Roncali, L. (1992). Tight endothelial junctions in the developing microvasculature. A thin section and freeze-fracture study in the chick embryo optic tectum. *J. Submicrosc. Cytol. Pathol.* **24,** 85–96.

Nielsen, S., Nagelhus, E. A., Amiry-Moghaddam, M., Bourque, C., Agre, P., and Ottersen, O. P. (1997). Specialized membrane domains for water transport in glial cells: High resolution immunogold cytochemistry of aquaporin-4 in rat brain. *J. Neurosci.* **17,** 171–180.

Noell, W. K. (1958). Studies on visual cell viability and differentiation. *Ann. N. Y. Acad. Sci.* **72,** 337–361.

Noske, W., and Hirsch, M. (1986). Morphology of tight junctions in the ciliary epithelium of rabbits during arachidonic acid-induced breakdown of the blood–aqueous barrier. *Cell Tissue Res.* **245,** 405–412.

Nusse, R. (1997). A versatile transcriptional effector of wingless signaling. *Cell* **89,** 321–323.

O'Rahilly, R., and Meyer, D. B. (1961). The development and histochemistry of the pecten oculi. *In* "The Structure of the Eye" (G. K. Smelser, ed.), pp. 207–219. Academic Press, New York.

Orsulic, S., and Peifer, M. (1996). An in vitro structure–function study of armadillo, the β-catenin hologue, reveals both separate and overlapping regions of the protein required for cell adhesion and for wingless signaling. *J. Cell Biol.* **134,** 1283–1300.

Pardridge, W. M., Boado, R. J., and Farrell, C. R. (1990). Brain-type glucose transporter (Glut-1) is selectively localized to the blood–brain barrier. *J. Biol. Chem.* **265,** 18035–18040.

Pe'er, J., Shweiki, D., Itin, A., Hemo, I., Gnessin, H., and Keshet, E. (1995). Hypoxia-induced expression of vascular endothelial growth factor by retinal cells is a common factor in neovascularizing ocular diseases. *Lab. Invest.* **72,** 638–645.

Peters, A., Palay, S. L., and Webster, H. D. F. (1991). "The Fine Structure of the Nervous System. Neurons and Their Supporting Cells." Oxford Univ. Press, New York.

Pettigrew, J. D., Wallman, J., and Wildsoet, C. F. (1990). Saccadic oscillations facilitate ocular perfusion from the avian pecten. *Nature* **343,** 362–363.

Pinto da Silva, P., and Kachar, B. (1982). On tight junction structure. *Cell* **28,** 441–450.

Plate, K. H., Breier, G., Weich, H. A., and Risau, W. (1992). Vascular endothelial growth factor is a potential tumour angiogenesis factor in human gliomas *in vivo*. *Nature* **359,** 845–848.

Poitry-Yamate, C. L., Poitry, S., and Tsacopoulos, M. (1995). Lactate released by Müller glial cells is metabolized by photoreceptors from mammalian retina. *J. Neurosci.* **15,** 5179–5191.

Provis, J. M., Leech, J., Diaz, C. M., Penfold, P. L., Stone, J., and Keshet, E. (1997). Development of the human retinal vasculature: Cellular relations and VEGF expression. *Exp. Eye Res.* **65,** 555–568.

Raidal, S. R. (1997). Bilateral necrotizing pectenitis causing blindness in a rainbow lorikeet (*Trichoglossus haematodus*). *Avian Pathol.* **26,** 871–876.

Rajasekaran, A. K., Hojo, M., Huima, T., and Rodriguez-Boulan, E. (1996). Catenins and zonula occludens-1 form a complex during early stages in the assembly of tight junctions. *J. Cell Biol.* **132,** 451–464.

Rascher, G., and Wolburg, H. (1997). The tight junctions of the leptomeningeal blood–cerebrospinal fluid barrier during development. *J. Brain Res.* **38,** 525–540.

Raub, T. J., Kuentzel, S. L., and Sawada, G. A. (1992). Permeability of bovine brain microvessel endothelial cells in vitro. Barrier tightening by a factor released from astroglioma cells. *Exp. Eye Res.* **199,** 330–340.

Raviola, E., and Raviola, G. (1967). A light and electron microscopic study of the pecten of the pigeon eye. *Am. J. Anat.* **130,** 427–462.

Reese, T. S., and Karnovsky, M. J. (1967). Fine structural localization of a blood–brain barrier to exogenous peroxidase. *J. Cell Biol.* **34,** 207–217.

Risau, W. (1991). Induction of the blood–brain barrier endothelial cell differentiation. *Ann. N. Y. Acad. Sci.* **633,** 405–419.

Risau, W. (1997). Mechanisms of angiogenesis. *Nature* **386,** 671–674.

Risau, W., and Flamme, I. (1995). Vasculogenesis. *Annu. Rev. Cell Dev. Biol.* **11,** 73–91.

Risau, W., and Lemmon, V. (1988). Changes in the vascular extracellular matrix during embryonic vasculogenesis and angiogenesis. *Dev. Biol.* **125,** 441–450.

Risau, W., Hallmann, R., Albrecht, U., and Henke-Fahle, S. (1986). Brain induces the expression of an early cell surface marker for blood–brain barrier-specific endothelium. *EMBO J.* **5,** 3179–3184.

Risau, W., Sariola, H., Zerwes, H.-G., Sasse, J., Ekblom, P., Kemler, R., and Doetschman, T. (1988). Vasculogenesis and angiogenesis in embryonic–stem-cell-derived embryoid bodies. *Development* **102,** 471–478.

Rodriguez-Peralta, L. A. (1975). Hematic and fluid barriers of the retina and vitreous body. *J. Comp. Neurol.* **132,** 109–124.

Rohen, J. (1955). Arterio-venöse Anastomosen in der Orbita bei Vögeln. *Morphol. Jahrbuch* **95,** 364–383.

Romanoff, A. L. (1960). The organ of special sense. *In* "The Avian Embryo" (A. L. Romanoff, ed.), pp. 396–404. Macmillan, New York.

Rubin, L. L. (1991). The blood–brain barrier in and out of cell culture. *Curr. Opin. Neurobiol.* **1,** 360–363.

Rubin, L. L. (1992). Endothelial cells: Adhesion and tight junctions. *Curr. Opin. Cell Biol.* **4,** 830–833.

Rubin, L. L., Hall, D. E., Porter, S., Barbu, K., Cannon, C., Horner, H. C., Janatpour, M., Liaw, C. W., Manning, K., Morales, J., Tanner, L. J., Tomaselli, K. J., and Bard, F. (1991). A cell culture model of the blood–brain barrier. *J. Cell Biol.* **115,** 1725–1736.

Ruggiero, F. P., and Sheffield, J. B. (1998). The use of avidin as a probe for the distribution of mitochondrial carboxylases in developing chick retina. *J. Histochem. Cytochem.* **46,** 177–183.

Saitou, M., Ando-Akatsuka, Y., Itoh, M., Furuse, M., Inazawa, J., Fujimoto, K., and Tsukita, S. (1997). Mammalian occludin in epithelial cells: Its expression and subcellular distribution. *Eur. J. Cell Biol.* **73,** 222–231.

Sato, T. N., Tozawa, Y., Deutsch, U., Wolburg-Buchholz, K., Fujiwara, Y., Gendron-Maguire, M., Gridley, T., Wolburg, H., Risau, W., and Qin, Y. (1995). Distinct roles of the receptor tyrosine kinases tie-1 and tie-2 in blood vessel formation. *Nature* **376,** 70–74.

Schlosshauer, B., and Herzog, K.-H. (1990). Neurothelin: An inducible cell surface glycoprotein of blood–brain barrier-specific endothelial cells and distinct neurons. *J. Cell Biol.* **110,** 1261–1274.

Schmidt, M., and Flamme, I. (1998). The in vivo activity of vascular endothelial growth factor isoforms in the avian embryo. *Growth Factors* **15,** 183–197.

Schnitzer, J. (1987). Retinal astrocytes: Their restriction to vascularized parts of the mammalian retina. *Neurosci Lett.* **78,** 29–34.

Schnitzer, J. (1988a). Astrocytes in mammalian retina. *Prog. Retina Res.* **7,** 209–231.

Schnitzer, J. (1988b). The development of astrocytes and blood vessels in the postnatal rabbit retina. *J. Neurocytol.* **17,** 433–449.

Schulze, C., and Firth, J. A. (1992). Interendothelial junctions during blood–brain barrier development in the rat. Morphological changes at the level of individual tight junctional contacts. *Dev. Brain Res.* **69,** 85–96.

Schulze, C., and Firth, J. A. (1993). Immunohistochemical localization of adherens junction components in blood–brain barrier microvessels of the rat. *J. Cell Sci.* **104,** 773–782.

Schulze, C., Smales, C., Rubin, L. L., and Staddon, J. M. (1997). Lysophophatidic acid increases tight junctional permeability in cultured brain endothelial cells. *J. Neurochem.* **68,** 991–1000.

Seaman, A. R., and Storm, H. (1963). A correlated light and electron microscopic study of the pecten oculi of the domestic fowl (*Gallus domesticus*). *Exp. Eye Res.* **2,** 163–172.

Seulberger, H., Lottspeich, F., and Risau, W. (1990). The inducible blood–brain barrier specific molecule HT7 is a novel immunoglobuline-like cell surface glycoprotein. *EMBO J.* **9,** 2151–2158.

Seulberger, H., Unger, C. M., and Risau, W. (1992). Ht 7, neurothelin, basigin, gp 42 and OX-47;md;Many names for one developmentally regulated immunoglobulin-like surface glycoprotein on blood–brain barrier endothelium, epithelial tissue barriers and neurons. *Neurosci. Lett.* **140,** 93–97.

Shakib, M., DeOliveira, L. F., and Henkind, P. (1968). Development of retinal vessels: II: Earliest stages of vessel formation. *Invest. Ophthalmol. Vis. Sci.* **7,** 689–700.

Shivers, R. R. (1979). The blood–brain barrier of a reptile, Anolis carolinensis. A freeze-fracture study. *Brain Res.* **169,** 221–230.

Shivers, R. R., Arthur, F. E., and Bowman, P. D. (1988). Induction of gap junctions and brain endothelium-like tight junctions in cultured bovine endothelial cells. Local control of cell specialization. *J. Submicrosc. Cytol. Pathol.* **20,** 1–14.

Shweiki, D., Itin, A., Soffer, D., and Keshet, E. (1992). Vascular endothelial growth factor induced by hypoxia may mediate hypoxia-initiated angiogenesis. *Nature* **359,** 843–845.

Smith, B. J., Smith, S. A., and Braekevelt, C. R. (1996). Fine structure of the pecten oculi of the barred owl (Strix varia). *Histol. Histopathol.* **11,** 89–96.

Smith, L. E. H., Kopchick, J. J., Chen, W., Knapp, J., Kinose, F., Daley, D., Foley, E., Smith, R. G., and Schaeffer, J. M. (1997). Essential role of growth hormone in ischemia-induced retinal neovascularization. *Science* **276,** 1706–1709.

Spatz, M., Kawai, N., Merkel, N., Bembry, J., and McCarron, R. M. (1997). Functional properties of cultured endothelial cells derived from large microvessels of human brain. *Am. J. Physiol.* **272,** C231–C239.

Stanness, K. A., Westrum, L. E., Fornaciari, E., Mascagni, P., Nelson, J. A., Stenglein, S. G., Myers, T., and Janigro, D. (1997). Morphological and functional characterization of an in vitro blood–brain barrier model. *Brain Res.* **771,** 329–342.

Stewart, P. A., and Coomber, B. L. (1986). Astrocytes and the blood–brain barrier. *Astrocytes* **1,** 311–328.

Stewart, P. A., and Hayakawa, E. M. (1987). Interendothelial junctional changes underlie the developmental "tightening" of the blood–brain barrier. *Dev. Brain Res.* **32,** 271–281.

Stewart, P. A., and Hayakawa, K. (1994). Early ultrastructural changes in blood–brain barrier vessels of the rat embryo. *Dev. Brain Res.* **78,** 25–34.

Stewart, P. A., and Tuor, U. J. (1994). Blood–eye barriers in the rat: Correlation of ultrastructure with function. *J. Comp. Neurol.* **340,** 566–576.

Stone, J., and Dreher, Z. (1987). Relationship between astrocytes, ganglion cells and vasculature of the retina. *J. Comp. Neurol.* **255,** 35–49.

Stone, J., and Maslim, J. (1997). Mechanisms of retinal angiogenesis. *Prog. Retina Eye Res.* **16,** 157–181.

Stone, J., Itin, A., Alon, T., Peer, J., Gnessin, H., Chan-Ling, T., and Keshet, E. (1995). Development of retinal vasculature is mediated by hypoxia-induced vascular endothelial growth factor (VEGF) expression by neuroglia. *J. Neurosci.* **15,** 4738–4747.

Strömblad, S., and Cheresh, D. A. (1996). Cell adhesion and angiogenesis. *Trends Cell Biol.* **6,** 462–468.

Suzuki, F., and Nagano, T. (1991). Three-dimensional model of tight junction fibrils based on freeze-fracture images. *Cell Tissue Res.* **264,** 381–384.

Tao-Cheng, J.-H., and Brightman, M. W. (1987). Cell membrane interactions between astroglia and brain endothelium in vitro. *In* "The Biochemical Pathology of Astrocytes" (M. D. Norenberg, L. Hertz, and A. Schousboe, eds.), p. 2. A. R. Liss, New York.

Tao-Cheng, J.-H., Nagy, Z., and Brightman, M. W. (1987). Tight junctions of brain endothelium in vitro are enhanced by astroglia. *J. Neurosci.* **7,** 3293–3299.

Tillis, T. N., Muray, D. L., Schmidt, G. J., and Weiter, J. J. (1988). Preretinal oxygen changes in the rabbit under conditions of light and dark. *Invest. Ophthalmol. Vis. Sci.* **29,** 988–991.

Tontsch, U., and Bauer, H. C. (1991). Glial cells and neurons induce blood brain barrier related enzymes in cultured cerebral endothelial cells. *Brain Res.* **539,** 247–253.

Tout, S., Ashwell, K., and Stone, J. (1988). The development of astrocytes in the albino rabbit retina and their relationship to retinal vasculature. *Neurosci. Lett.* **90,** 241–247.

Trivino, A., Ramirez, J. M., Ramirez, A. I., Salazar, J. J., and Garcia-Sanchez, J. (1992). Retinal perivascular astroglia. An immunoperoxidase study. *Vision Res.* **32,** 1601–1608.

Trivino, A., Ramirez, J. M., Ramirez, A. I., Salazar, J. J., and Garcia-Sanchez, J. (1997). Comparative study of astrocytes in human and rabbit retina. *Vision Res.* **37,** 1707–1711.

Tsacopoulos, M., and Magistretti, P. J. (1996). Metabolic coupling between glia and neurons. *J. Neurosci.* **16,** 877–885.

Tsukita, S., Furuse, M., and Itoh, M. (1996). Molecular dissection of tight junctions. *Cell Struct. Funct.* **21,** 381–385.

Tucker, R. (1975). The surface of the pecten oculi in the pigeon. *Cell Tissue Res.* **157,** 457–465.

Uehara, M., Oomori, S., Kitagawa, H., and Ueshima, T. (1990). The development of the pecten oculi in the chick. *Jpn. J. Vet. Sci.* **52,** 503–512.

Van Deurs, B., and Koehler, J. K. (1979). Tight junctions in the choroid plexus epithelium. A freeze-fracture study including complementary replicas. *J. Cell Biol.* **80,** 662–673.

Verbavatz, J.-M., Ma, T., Gobin, R., and Verkman, A. S. (1997). Absence of orthogonal arays in kidney, brain and muscle from transgenic knockout mice lacking water channel aquaporin-4. *J. Cell Sci.* **110,** 2855–2860.

Vinores, S. A. (1995). Assessment of blood–retinal integrity. *Histol Histopathol.* **10,** 141–154.

Wakai, S., and Hirokawa, N. (1978). Development of the blood–brain barrier to horseradish peroxidase in the chick embryo. *Cell Tissue Res.* **195,** 195–203.

Watanabe, T., and Raff, M. C. (1988). Retinal astrocytes are immigrants from the optic nerve. *Nature* **332,** 834–837.

Welsch, U. (1972). Enzymhistochemische und feinstrukturelle Beobachtungen am Pecten oculi von Taube (*Columba livia*) und Lachmöve (*Larus ridibundus*). *Z. Zellforsch. Mikr. Anat.* **132,** 231–244.

Wiesinger, H. (1995). Glia-specific enzyme systems. In "Neuroglia" (H. Kettenmann and B. R. Ransom, eds.), pp. 488–499. Oxford Univ. Press, New York.

Wilting, J., Eichmann, A., and Christ, B. (1997). Expression of the avian VEGF receptor homolgues Quek1 and Quek2 in blood–vascular and lymphatic endothelial and non-endothelial cells during quail embryonic development. *Cell Tissue Res.* **288,** 207–223.

Wingstrand, K. G., and Munk, O. (1965). The pecten oculi of the pigeon with particular regard to its function. *Biol. Skr. Dan. Vid. Selsk.* **14,** 1–64.

Winkler, B. S. (1995). A quantitative assessment of glucose metabolism in the isolated rat retina. *In* "Vision et Adaptation" (Y. Christen, M. Doly, and M.-T. Droy-Lefaix, eds.), pp. 78–96. Elsevier, Amsterdam.

Wolburg, H. (1995). Orthogonal arrays of intramembranous particles: A review with special reference to astrocytes. *J. Brain Res.* **36,** 239–258.

Wolburg, H., and Risau, W. (1995). Formation of the blood–brain barrier. *In* "Neuroglia" (H. Kettenmann and B. R. Ransom, eds.), pp. 763–776. Oxford Univ. Press, New York.

Wolburg, H., Neuhaus, J., Kniesel, U., Krauss, B., Schmid, E.-M., Öcalan, M., Farrell, C., and Risau, W. (1994). Modulation of tight junction structure in blood–brain barrier endothelial cells. Effects of tissue culture, second messengers and cocultured astrocytes. *J. Cell Sci.* **107,** 1347–1357.

Wolter, J. R. (1957). Perivascular glia of the blood vessels of the human retina. *Am. J. Ophthalmol.* **44,** 766–773.

Woods, D. F., and Bryant, P. J. (1993). Apical junctions and cell signalling in epithelia. *J. Cell Sci.* **17,** 171–181.

Yew, D. T. (1978). The origin and initial development of the pecten oculi. *Anat. Anz.* **143,** 383–387.

Yoshida, Y., Yamada, M., Wakabayashi, K., and Ikuta, F. (1988). Endothelial fenestrae in the rate fetal cerebrum. *Dev. Brain Res.* **44,** 211–219.

Yu, D.-Y., Cringle, S. J., Alder, V. A., Su, E.-N., and Yu, P. K. (1994). Intraretinal oxygen distribution in rats as a function of systemic blood pressure. *Am. J. Physiol.* **267,** H2498–H2507.

Yu, D.-Y., Cringle, S. J., Alder, V. A., Su, E.-N., and Yu, P. K. (1996). Intraretinal oxygen distribution and choroidal regulation in the avascular retina of the guinea pig. *Am. J. Physiol.* **270,** H965–H973.

Zhang, Y., and Stone, J. (1997). Role of astrocytes in the control of developing retinal vessels. *Invest. Ophthalmol. Vis. Sci.* **38,** 1653–1666.

Bacteriorhodopsin

Janos K. Lanyi
Department of Physiology and Biophysics, University of California,
Irvine, California 92697-4560

Bacteriorhodopsin is a seven-transmembrane helical protein that contains all-*trans* retinal. In this light-driven pump, a reaction cycle initiated by photoisomerization to 13-*cis* causes translocation of a proton across the membrane. Local changes in the geometry of the protonated Schiff base and the proton acceptor Asp85, and the proton conductivities of the half channels that lead from this active site to the two membrane surfaces, interact so as to allow timely proton transfers that result in proton release on the extracellular side and proton uptake on the cytoplasmic one. The details of the steps in this photocycle, and the underlying principles that ensure unidirectionality of the movement of a proton across the protein, provide strong clues to how ion pumps function.
KEY WORDS: Bacteriorhodopsin, Retinal protein, Retinal isomerization, Proton pump, Proton translocation, Halobacteria.

I. Introduction

Bacteriorhodopsin is the simplest of the pumps that generate transmembrane electrochemical potential for protons. Unlike changes in substrate-binding energy during chemical reactions as in ion-motive ATPases and the NADH/NADP transhydrogenase, electron transfer after photoexcitation and during redox reactions as in the photosynthetic reaction center, cytochrome oxidase, and the cytochrome bc complex, in bacteriorhodopsin the proton transport is driven directly by pK_a changes initiated by light-dependent isomerization of the retinal chromophore. The transport is thus based entirely on acid-base reactions, in what is probably the smallest imaginable (26 kDa) functional membrane-spanning protein. Bacteriorhodopsin is the best understood ionic pump and has become the paradigm for proton transport systems as well as for the structure of membrane proteins in general and G protein-linked receptors in particular.

International Review of Cytology, Vol. 187
0074-7696/99 $30.00

Bacteriorhodopsin is an integral membrane protein whose 248 residues are arranged in seven transmembrane helical segments (A–G) that enclose a binding cavity for the all-*trans* retinal chromophore (Henderson *et al.*, 1990; Grigorieff *et al.*, 1996; Kimura *et al.*, 1997a,b; Pebay-Peyroula *et al.*, 1997) (Fig. 1, see color plate). The retinal is bound via a protonated Schiff base to Lys216 near the center of helix G and lies nearly parallel to the plane of the membrane. A naturally occurring extended two-dimensional hexagonal lattice ("purple membrane") of trimers of this protein has made it possible to determine its three-dimensional electron-density map at 3.5–7 Å (Henderson *et al.*, 1986, 1990; Baldwin *et al.*, 1988) and more recently at 3 Å (Kimura *et al.*, 1997a,b) resolution. Three-dimensional crystals produced by condensation of a cubic lipid-detergent phase (Landau and Rosenbusch, 1996) have yielded a 2.5 Å structure (Pebay-Peyroula *et al.*, 1997). These structures are the points of departure in all attempts to describe the transport mechanism.

Photoisomerization of the retinal from 13-*trans*,15-*anti* (the "light-adapted" chromophore) to 13-*cis*, 15-*anti* (Aton *et al.*, 1977; Braiman and Mathies, 1982) initiates a reaction cycle that, through a series of thermal steps, translocates a proton across the membrane. Spectroscopic changes in the visible have identified the intermediates that accumulate in photostationary states during this "photocycle" at cryogenic temperatures (Lozier *et al.*, 1975; Becher *et al.*, 1978; Shichida *et al.*, 1983). Photoexcitation with short light pulses made it also possible to follow the interconversions of these states at ambient temperature and in real time during a single turnover (Xie *et al.*, 1987; Gerwert *et al.*, 1990; Zimányi *et al.*, 1989; Milder *et al.*, 1991; Braiman *et al.*, 1991; Ames and Mathies, 1990). The retinal and its configurational transformations have been described in studies by a large variety of spectroscopic methods, including visible, UV, resonance Raman, FTIR, and NMR spectroscopy, whereas changes in protein residues have been revealed by FTIR and UV spectroscopy and by studies utilizing site-specific mutations. The proton exchange between the protein and the aqueous medium has been followed using pH-indicator dyes, either free in the bulk or covalently bound to the protein surface. During the past few years a considerable body of work utilizing these methods firmly established the outlines of a coherent mechanistic and thermodynamic model. For additional information on bacteriorhodopsin and other bacterial rhodopsins the reader is directed to earlier reviews (Spudich and Bogomolni, 1988; Henderson *et al.*, 1990; Mathies *et al.*, 1991; Lanyi, 1992, 1994, 1993, 1997; Oesterhelt *et al.*, 1992; Rothschild, 1992; Ebrey, 1993; Khorana, 1993; Lanyi and Váró, 1995; Maeda, 1995; Haupts *et al.*, 1997).

The fact that bacteriorhodopsin is a light-driven pump is not essential to understand how the proton is transported across the membrane. The only light-dependent event in the photocycle of bacteriorhodopsin is the initial isomerization of the retinal. All subsequent steps are thermal relaxations, much as in the translocation cycles of ion pumps driven by chemical transformations. This allows conclusions to be drawn from its mechanism that could have general relevance for ion pumps.

II. The Photocycle

The photocycle contains numerous intermediate states characterized by their absorption maxima in the visible and UV, as well in the infrared, designated as J, K, L, M, N, and O.[1] In most cases substates have also been described (K and KL, M_1 and M_2, etc.). To some extent the nomenclature is arbitrary and based on convention. It generally follows the shifts in the absorption maxima in the visible (The photointermediates are often designed with the measured or estimated wavelength maxima as subscripts, e.g., K_{600}, L_{550}, M_{410}, N_{560}, and O_{640}.) These may reflect well understood changes in molecular properties, as in the case of the far blue-shifted M state in which the reason for the shift is that the Schiff base is unprotonated. In other cases, ascertaining the identity of the intermediate in question is based on direct information on such properties as the protonation state of residues or the isomeric state of the retinal. The named intermediates seem to arise and decay in a linear sequence after photoisomerization of the retinal in the K state, although some proposed models contain branching. The lifetimes of these states range from femtoseconds to milliseconds. Under physiological conditions the turnover of the cycle is under a few tens of milliseconds.

A. Spectroscopy of Intermediate States

As the initial BR state, each of the intermediates has a single broad, asymmetric absorption band in the visible. The maxima are between about 550 and 620 nm, except M, which has a maximum at about 410 nm. The J, K, and O states are red shifted and L and N are blue shifted relative to BR. In the UV, all intermediates between the initial few and the last

[1] In this chapter, J, K, L, M, N, and O are the photointermediates of the bacteriorhodopsin photocycle; BR is the initial state. Superscripts, where given, refer to the net protonation of the protein relative to the initial state. Subscripts for M substrates refer to pre- and postswitch states (i.e., M_1 and M_2) or an M with changed N-like protein conformation (i.e., M_N). D, E, N, R, Q, T, S, V, Y, and F designate aspartate, glutamate, asparagine, arginine, glutamine, threonine, serine, valine, tyrosine, and phenylalanine residues, respectively. Proteins with residue replacements are designated with the wild-type residue, its number, and the replacement, e.g., D96N.

intermediate in the cycle exhibit a band near 335 nm, the "*cis*-peak," that reflects the 13-*cis* isomeric state of the retinal (Kuschmitz and Hess, 1982; Drachev *et al.,* 1987).

FTIR difference spectra contain many bands that reveal the state of ionizable protein residues. However, the spectra are dominated by changes in the retinal, most prominently by a pair of negative and positive bands from the ethylenic stretch vibration. The frequency of this mode is strongly correlated with the absorption maximum in the visible (Aton *et al.,* 1977). Another feature common to many intermediates is in the single carbon–carbon bond stretch (the "fingerprint") region between 1100 and 1300 cm^{-1}. In all but the O state it consists of several negative and positive bands from the depletion of the all-*trans* isomer and the presence of the 13-*cis* isomeric state. Some of the features that more specifically identify the intermediates are the following. In the K state large amplitude hydrogen-out-of-plane (HOOP) modes between 800 and 1000 cm^{-1} indicate that the retinal skeleton is twisted. The L state is characterized by a complex spectral feature near 1740 cm^{-1} due to perturbation of Asp96 (including possibly its deprotonation) and Asp115 (Braiman *et al.,* 1988a; Gerwert *et al.,* 1989; Maeda *et al.,* 1992b; Braiman *et al.,* 1991). The M minus BR difference spectrum contains a positive C—O stretch band at 1762 cm^{-1} due to the protonation of Asp85 and exhibits greatly decreased positive chromophore band intensities, e.g., for C–C stretch at 1186 cm^{-1} (Braiman *et al.,* 1988a, 1991; Souvignier and Gerwert, 1992). In the N state the 1762 cm^{-1} band shifts to 1755 cm^{-1} and a negative band at 1742 cm^{-1} appears from deprotonation of Asp96 (Braiman *et al.,* 1988a, 1991; Pfefferlé *et al.,* 1991; Maeda *et al.,* 1992b), whereas strong bands, positive at 1655 cm^{-1} and negative at 1670 cm^{-1}, due to amide I vibrations reflect protein backbone changes (Braiman *et al.,* 1987, 1991; Ormos, 1991; Ormos *et al.,* 1992; Perkins *et al.,* 1992). The recovery of the amplitude in the fingerprint region in N reveals that the Schiff base is now protonated. However, under some conditions these characteristics of the N state appear before the Schiff base is reprotonated. This state is termed the M_N intermediate (Sasaki *et al.,* 1992). In the O state the 1755 cm^{-1} positive band persists but the 1742 cm^{-1} negative band is absent because Asp96 is reprotonated. The downshifted ethylenic stretch band at 1505 cm^{-1} is useful in following the O intermediate (Souvignier and Gerwert, 1992). The fingerprint region reveals that the retinal is no longer 13-*cis.* As in the K state, large-amplitude HOOP bands indicate twist of the retinal chain.

B. Kinetics

The kinetics of a reaction sequence with as many intermediate states as in the bacteriorhodopsin photocycle is difficult to solve (Nagle, 1991). An exact solution is virtually precluded by the considerable overlap of the spectra of most of the intermediates with one another and the spectrum of the initial state, and the fact that several interconversions have similar rate constants. In general, the solution of the rate equations of a complex reaction sequence is derived from the matrix of the measurable relaxation constants and amplitudes that are model-dependent functions of the desired elementary rate constants. Global analyses of the time course of absorption changes at selected wavelengths identified these phenomenological parameters (Nagle *et al.*, 1982, 1995; Maurer *et al.*, 1987; Nagle, 1991; Lozier *et al.*, 1992; Müller *et al.*, 1991; Souvignier and Gerwert, 1992) but did not uniquely define a photocycle model. However, it is clear that a single sequence containing only unidirectional reactions is not consistent with the data (Nagle *et al.*, 1982). The models proposed to account for the observed multiple rise and decay time constants include parallel photocycles (Hanamoto *et al.*, 1984; Dancsházy *et al.*, 1988; Birge, 1990; Balashov *et al.*, 1991; Drachev *et al.*, 1992; Eisfeld and Stockburger, 1992; Pusch *et al.*, 1992), branched photocycles (Sherman *et al.*, 1979; Beach and Fager, 1985; Drachev *et al.*, 1992), single unbranched photocycles but with one or more reversible reactions (Váró *et al.*, 1990; Gerwert *et al.*, 1990; Ames and Mathies, 1990; Váró and Lanyi, 1991b,c; Zimányi and Lanyi, 1993), and a two-photon cycle in which the slower decaying M is produced by photoreaction of the N intermediate (Kouyama *et al.*, 1988; Fukuda and Kouyama, 1992b). Spectroscopic data alone do not provide a sufficient basis to fully decide among these alternatives, and it is now clear that solution of the kinetics requires additional and independent information on the structure and ionization state of residues and identification of the sequence of proton release and uptake during the photocycle.

From such an indirect approach a concensus based on time-resolved FTIR (Gerwert *et al.*, 1990; Souvignier and Gerwert, 1992; Hessling *et al.*, 1993, 1997; Riesle *et al.*, 1996; Dioumaev and Braiman, 1997; Zscherp and Heberle, 1997) and resonance Raman (Terner *et al.*, 1979a,b; Fodor *et al.*, 1988a,b; Ames *et al.*, 1989; Ames and Mathies, 1990; Alshuth and Stockburger, 1986; Rath *et al.*, 1993; Althaus *et al.*, 1995; Smith *et al.*, 1987) spectra, as well as kinetics derived from visible spectroscopy (Váró and Lanyi, 1991b,c; Zimányi *et al.*, 1992b; Milder *et al.*, 1991; Zimányi and Lanyi, 1993), has developed in favor of the simplest viable photocycle model: a single reaction sequence with reversible reactions, BR-hv $\rightarrow$ K $\leftrightarrow$ L $\leftrightarrow$ M $\leftrightarrow$ N $\leftrightarrow$ O $\rightarrow$ BR. However, this scheme is clearly an oversimplification. A model with one M intermediate does not fit the data in the 10 to 100-μs time range in which the L to M interconversion takes place. Introducing another intermediate between L and M yielded calculated spectra for some of the putative intermediates that were obviously mixtures and thus did not solve the problem (Lozier *et al.*, 1992). On the other hand, the introduction of two sequential M substates (M_1 and M_2), on purely kinetic grounds, gave a simple and reasonable explanation for the fact that the concentration of L tends to zero at the time when M reaches its maximal concentration despite what appears to be an earlier established L to M equilibrium (Váró and Lanyi, 1991b; Zimányi and Lanyi, 1993). As discussed later, the existence of more than one M state is now supported by more direct observations and constitutes the cornerstone of the transport mechanism.

The proposal of back reactions in the linear scheme has elicited a great deal of controversy. Their presence was tested by pertubation of the equilibrium of intermediate states they produce. If the kinetics of the recovery includes regeneration of the intermediate depleted by the perturbation, at the expense of the other intermediate in question and at the rate at which the intermediate was produced in the first place, the back reaction may be considered as proven. Such an approach was first used for the putative N $\leftrightarrow$ O equilibrium. Since the transient concentration of the O state is lower at higher temperatures, an infrared pulse was used to heat the sample for a few milliseconds after O was formed by a first flash (Chernavskii *et al.*, 1989). The concentration of O first decreased then increased, indicating that it was in equilibrium with another (the N) state. Similarly, the proposed M $\leftrightarrow$ N equilibrium was probed by depleting M with a second, blue flash after it was formed (Druckmann *et al.*, 1993; Zimányi *et al.*, 1993). It recovered with a time constant similar to its formation, again indicating that M was in equilibrium with another (also the N) state.

These two-flash experiments were made technically easier by the fact that the M $\leftrightarrow$ N and the N $\leftrightarrow$ O equilibria are established on a millisecond time scale. The proposed L $\leftrightarrow$ M equilibrium is produced in the order of 0.1 ms, and the evidence for its existence is of another kind. If such an equilibrium occurs during the photocycle the M state will coexist with L until there is a unidirectional reaction that converts the L + M mixture to the next state. In the D96N mutant, in which for kinetic reasons the N and O intermediates do not accumulate, at pH < 6 a significant amount of L was found to be present for hundreds of milliseconds as the initial state recovered from M (Zimányi *et al.*, 1992b). In some mutants, such as V49A, the amount of L was dominant over M, and the kinetics suggested that the reason for this was that the L $\leftrightarrow$ M equilibrium was shifted in favor of L (Brown *et al.*, 1994a).

III. Structure of the Protein

The seven-transmembrane helices enclose a cavity that spans the width of the membrane. It is occupied by the retinal and nearly all of the buried protonable residues. The Schiff base is about midway across the protein and divides the interhelical cavity into an extracellular (EC) region containing numerous charged residues and a cytoplasmic (CP) region containing mostly hydrophobic residues (Henderson *et al.*, 1990; Grigorieff *et al.*, 1996; Kimura *et al.*, 1997a,b; Pebay-Peyroula *et al.*, 1997). The putative pathways of the protons through the two peripheral regions are sometimes referred to as the extracellular and cytoplasmic "half-channels," respectively. From a functional point of view these regions can be termed proton release and proton uptake domains.

A. The Extracellular Proton Release Domain

The extracellular domain has a complex structure. The small magnitude of the ^{15}N isotropic shift of the Schiff base, compared to those measured in model systems with various anions in solution, can be explained only with the existence of a diffuse counter-ion in which the negative charge is not localized on a single group but rather distributed over a hydrogen-bonded network (De Groot *et al.*, 1989, 1990). This network contains the charged residues Asp85, Asp212, and Arg82 and probably bound water molecules (Hildebrandt and Stockburger, 1984), which together provide a delocalized negative charge to balance the positive charge of the Schiff base (Dér *et al.*, 1991). The net charge of this complex is zero, consistent with its buried location in the protein. That the interaction between the protein counter-ion and the Schiff base is unusually weak is indicated by two additional facts. First, the absorption maximum in the visible is considerably red shifted from the 440-nm maximum of model compounds where the counterion, e.g., chloride, is at van der Waals distance from a protonated Schiff base. This "opsin shift" arises to a large extent (although not entirely) from the weakness of the Schiff base–counter-ion interaction (Kakitani *et al.*, 1985; Nakanishi *et al.*, 1980; Warshel, 1978). Second, the C—N stretch frequency of the Schiff base is unusually low (Alshuth and Stockburger, 1986; Smith *et al.*, 1984). The presence of bound water molecules, hydrogen-bonded to both Schiff base and Asp85, is suggested by the L minus BR difference FTIR spectra for the wild-type and D85N proteins (Maeda *et al.*, 1992a, 1994; Kandori *et al.*, 1995; Yamazaki *et al.*, 1995a,b, 1996). Difference bands in the 3450–3750 cm^{-1} region that originate from changed O–H stretch frequencies of water in the L intermediate are strongly affected by mutation of residues 85 and 212.

Strong coulombic influence of Asp85, Arg82, and Asp212 on the Schiff base is indicated by shifts of the absorption maximum of the chromophore, as well as changes of the Schiff base pK_a and the deprotonation kinetics, upon site-directed mutations of these residues. Replacement of either

Asp212 (Needleman *et al.*, 1991) or Asp85 (Mogi *et al.*, 1988; Subramaniam *et al.*, 1990; Otto *et al.*, 1990; Miercke *et al.*, 1991; Zimányi *et al.*, 1992a; Brown *et al.*, 1993a; Turner *et al.*, 1993) causes a red shift in the absorption maximum of the chromophore, and replacement of Arg82 causes a blue shift (Subramaniam *et al.*, 1990; Brown *et al.*, 1993a), as expected from electrostatic through-space interaction of these negatively and positively charged residues with the Schiff base (Honig *et al.*, 1979; Baasov *et al.*, 1987; Baasov and Sheves, 1986). However, Asp85 and Asp212 are clearly not equivalent in this regard. Replacement of Asp85 with asparagine shifts the absorption maximum of the chromophore by 47 nm (Zimányi *et al.*, 1992a; Brown *et al.*, 1993a) to the red and lowers the pK_a of the Schiff base from well above 10 to about 7 (Otto *et al.*, 1990) or 8.4 (Brown *et al.*, 1993a), but similar replacement of Asp212 results in a red shift of only about 15 nm and the pK_a remains unchanged (Subramaniam *et al.*, 1990; Needleman *et al.*, 1991). The importance of the negative charge of Asp85 as the principal component of the Schiff base counter-ion is demonstrated by a newly acquired pH dependence of the spectral properties of the recombinant protein in which this residue is replaced by histidine (Subramaniam *et al.*, 1992).

Much FTIR evidence suggests that by becoming protonated in the K state and deprotonated again in the M state, Tyr185 plays a key role in the early events of the photocycle (Rothschild *et al.*, 1986; Braiman *et al.*, 1988b). According to NMR and UV–Raman experiments, however, the protein does not contain a tyrosinate up to pH 10 (Herzfeld *et al.*, 1990; Ames *et al.*, 1990), and none form during the photocycle (Ames *et al.*, 1992). A recent suggestion is that Tyr185 forms a polarizable hydrogen bond with Asp212, and the bond becomes stronger after photoexcitation (Rothschild *et al.*, 1990). This is based partly on the appearance of a positive band at 1738 cm^{-1} in M of the wild-type protein (Braiman *et al.*, 1988a; Rothschild *et al.*, 1990), interpreted as partial protonation of Asp212, and partly on the disappearance of FTIR bands assigned to Tyr185 in D212N (Rothschild *et al.*, 1990). On the other hand, replacing Asp212 with asparagine (Needleman *et al.*, 1991), or Tyr185 with phenylalanine (Mogi *et al.*, 1987), was found to decrease transport but not by more than about two-thirds, indicating that although located at the active site these residues are not indispensable for proton transfer. From site-specific mutants the 1738-cm^{-1} band has been assigned recently to perturbation of Asp115 (Sasaki *et al.*, 1994), a residue not in the vicinity of the Schiff base.

The blue shift of the absorption maximum of the deprotonated Schiff base provides a method for spectroscopic titration of the unphotolysed chromophore. The pK_a of a Schiff base in model compounds is near 7 when in solution (Baasov and Sheves, 1986), but in bacteriorhodopsin it is about 13 (Druckmann *et al.*, 1982). The strongly elevated pK_a reflects the high free energy of the uncompensated charge of the counter-ion when the Schiff base is deprotonated (Sheves *et al.*, 1986) and might also be influenced by water molecules that form hydrogen bonds (Gat and Sheves, 1993). Its marked lowering upon replacement of Asp85 with neutral residues is consistent with this observation. As expected from the critical role of Asp85, the Schiff base of halorhodopsin, a related retinal protein with a threonine residue at the position equivalent to Asp85, has a nearly unperturbed pK_a of 7.5 (Lanyi, 1986).

Titration of Asp85 toward acid pH results in a red shift of the maximum of the chromophore that transforms the purple protein into a blue one, similarly to replacement of this aspartate with uncharged residues (Fischer and Oesterhelt, 1979; Mowery *et al.*, 1979; Edgerton *et al.*, 1980; Váró and Lanyi, 1989). When the ionic strength is sufficient to eliminate effects of surface charge on the local pH, the purple-to-blue shift occurs at a pH of about 2.5 (Moore *et al.*, 1978; Fischer and Oesterhelt, 1979; Mowery *et al.*, 1979; Kimura *et al.*, 1984; Ariki *et al.*, 1987; Dun;tı;ach *et al.*, 1988; Jonas and Ebrey, 1991). This anomalously low pK_a is the consequence of stabilization of the aspartate anion by Arg82 because replacement of Arg82 with glutamine or alanine raises the apparent pK_a of Asp85 to about 7 (Subramaniam *et al.*, 1990; Thorgeirsson *et al.*, 1991; Brown *et al.*, 1993a). As expected from the different shapes of aspartate and glutamate that will place them in different environments in the protein, the pK_a for the color transition is also different in the D85E mutant and higher than that in wild-type (Lanyi *et al.*, 1992; Greenhalgh *et al.*, 1992). The pK_a of Asp212 is less certain but appears to be well below that of Asp85 because Asp212 remains anionic at a pH low enough to form the blue chromophore (Metz *et al.*, 1992).

An additional method of titrating Asp85 uses the rate of retinal isomeric equilibration ("dark adaptation"). The "light-adapted" chromophore that contains 100% all-*trans* retinal relaxes in the dark over tens of minutes, or hours depending on conditions, to a mixture of all-*trans* and 13-*cis,* 15-*syn* (Kalisky *et al.*, 1977; Maeda *et al.*, 1977; Sperling *et al.*, 1977, 1979; Harbison *et al.*, 1984; Scherrer *et al.*, 1989). The rate of this process is proportional to the fractional protonation of Asp85 (Balashov *et al.*, 1993). This method, as well as the more direct spectroscopic titration, revealed that Asp85 protonates with two pK_as and suggested a model in which an unknown protonatable group interacts with Asp85 so as to influence its pK_a (Balashov *et al.*, 1995, 1996; Richter *et al.*, 1996a,b). The nature of this interaction is that near neutral pH either one or the other residue can be protonated but not both. In the D85E and Y185F mutants the anomalous behavior of Asp85 could also be demonstrated by conventional spectroscopic titration (Richter *et al.*, 1996b). The significance of this phenomenon is that when Asp85 becomes protonated by the Schiff base in the photocycle, the unknown residue will dissociate and will be the candidate for the group that releases a proton to the extracellular surface. First Glu204 (Brown *et al.*, 1995a; Richter *et al.*, 1996a) and later Glu194 (Balashov *et al.*, 1997; Dioumaev *et al.*, 1998) were shown to be involved in this effect: In the E204Q and E194Q mutants the anomaly in the titration of Asp85 disappeared, and in the E204D mutant it was modified. From the structure it is apparent that these residues could constitute a proton transfer chain from the vicinity of Asp85 and the extracellular surface. Indeed, replacement of either Glu204 (Brown *et al.*, 1995a) or Glu194 (Balashov *et al.*, 1997; Dioumaev *et al.*, 1998) interfered with proton release.

Glu9 and particularly Glu74 are closer to the extracellular surface than the residues described previously. Their involvement in proton transfer to the surface is questionable since the E9A and the E74C mutants are not perturbed in proton release. Glu194, on the other hand, not only is essential for proton release but also, when changed to an aspartate, was found to become transiently protonated during proton release (Dioumaev *et al.*, 1998). The proton release pathway appears to pass through Glu204 and Glu194 and then to the surface.

B. The Cytoplasmic Proton Uptake Domain

While the cytoplasmic surface carries a high negative charge from Asp36, Asp38, Asp102, Asp104, and numerous acidic residues on the C-terminal tail, the interior of this half of the protein contains mostly nonprotonable or uncharged residues. The important exceptions are Asp96 and Arg227. Since Asp96 is the internal proton donor to the Schiff base during the photocycle, the nearly unchanged rate of the Schiff base reprotonation to alkaline pH indicates that the pK_a of this residue is at least 10 (Zimányi *et al.*, 1993). An unusually high pK_a for Asp96 is also suggested by the fact that the typical 1742-cm^{-1} negative FTIR band due to deprotonation of its COOH group persists at least up to pH 9 or 10 (Engelhard *et al.*, 1985; Gerwert *et al.*, 1989; Braiman *et al.*, 1991; Pfefferlé *et al.*, 1991; Maeda *et al.*, 1992b). Direct titration of Asp96 in the infrared had shown the pK_a to be as high as 11 (Száraz *et al.*, 1994). This high proton affinity must be due to the hydrophobic environment of the cytoplasmic domain. It ensures that Asp96 will stay protonated and function as a proton donor throughout the physiological pH range (Miller and Oesterhelt, 1990).

However, the pK_a of Asp96 must be greatly lowered during the photocycle if it is to act as a proton donor. Diffraction of the two-dimensional lattice of the purple membrane indicates that the cytoplasmic end of helix F tilts away from the center of the protein in the N state (Kamikubo *et al.*, 1996; Vonck, 1996), but this occurs already in the M state when it is stabilized by lower water activity or other treatments or in the D96N mutant (Subramaniam *et al.,* 1993; Nakasako *et al.,* 1991; Kamikubo *et al.,* 1997; Dencher *et al.,* 1989; Han *et al.,* 1994). The effects of hydrostatic pressure, osmotic agents on the proton exchange between Schiff base and Asp96 (Váró and Lanyi, 1991a, 1995; Cao *et al.,* 1991), and kinetic analysis of a cooperativity in the rigid membrane lattice of the purple membrane (Váró *et al.,* 1996) suggested that the opening of a cleft at the cytoplasmic surface affects Asp96 through increased hydration of its environment. Transfer of a proton from Asp96 to the Schiff base must take place through a rather hydrophobic region. A narrow channel that may contain a few water molecules could provide a hydrogen-bonded chain over a 12 Å distance, but there is little or no experimental evidence for this (Pebay-Peyroula *et al.,* 1997).

The decrease in the pK_a of Asp96, and the ensuing protonation of the Schiff base, is followed by recovery of the protein conformation (Brown *et al.,* 1995b) and thus reprotonation of Asp96, but from the cytoplasmic surface (Zimányi *et al.,* 1993).

IV. Proton Release Mechanism

A. Protonation of Asp85 by the Retinal Schiff Base

Destabilization of the Schiff base proton in the L intermediate leads to its transfer to the anionic Asp85. Evidence indicating that Asp85 is the proton acceptor is the following: (i) Deprotonation of the Schiff base is accompanied by the simultaneous appearance of a positive FTIR band at 1762 cm^{-1} assigned to the COOH vibration of Asp85 (Braiman *et al.,* 1988a, 1991); (ii) the Schiff base remains protonated during the photocycle when Asp85 is replaced with a nonprotonable group (Stern *et al.,* 1989; Otto *et al.,* 1990) or at such a low pH that Asp85 is protonated from the start (Mowery *et al.,* 1979; Váró and Lanyi, 1989); (iii) the Schiff base deprotonates much more rapidly when Asp85 is replaced with glutamate (Butt *et al.,* 1989; Lanyi *et al.,* 1992; Greenhalgh *et al.,* 1992); and (iv) proton transport is reactivated in the otherwise inactive D85C mutant when a carboxylate residue is introduced by reaction of Cys85 with iodoacetic acid (Greenhalgh *et al.,* 1992).

The free energy difference between the L state (protonated Schiff base and deprotonated Asp85) and the M state (deprotonated Schiff base and protonated Asp85) can be estimated by summing the ΔGs of the protonation of Asp85 in the wild-type protein and the deprotonation of the Schiff base in the D85N or D85T proteins (Brown *et al.,* 1993a). This calculation yields a ΔG of 30–33 kJ/mol, which is lower than the free energy retained after absorption of the photon (Birge *et al.*, 1991), allowing for overcoming the free energy gap and for losses. It corresponds to an expected effective ΔpK_a of 5.3–5.7 pH units between donor and acceptor.

The equilibrium constant K = [L]/[M] calculated from the kinetics at room temperature is in fact about 4 (Váró and Lanyi, 1991b), indicating that the pK_a difference between proton donor and acceptor in L is narrowed to 0.6. Thus, the photoisomerization either lowered the pK_a of the Schiff base by about 5 pH units or the pK_a of Asp85 is raised to create the conditions in which the proton transfer which produces the M state can take place. It is very likely that both occur. It has been argued from the C—O stretch frequency of Asp85 and the increased hydrophobicity in its environment that the pK_a of this residue rises considerably during the photocycle (Braiman *et al.*, 1996).

It is reasonable that the pK_a of the protonated Schiff base would be lowered in L. First, quantum chemical calculations suggest that the *trans*-to-*cis* rotation of the C_{13}–C_{14} double bond evident in the L state disrupts the π system of the retinal chain. This will decrease electron density on the Schiff base nitrogen and destabilize the proton (Orlandi and Schulten, 1979; Tavan *et al.*, 1985). Second, because the distal part of the chain and the β-ionone ring of the retinal are fixed by three flanking tryptophan residues (Henderson *et al.*, 1990), isomerization of the C_{13}–C_{14} bond displaces the Schiff base (Dencher *et al.*, 1992a) to a different, possibly more electronegative environment. Indeed, an increased deuterium shift of the Schiff base C—N frequency (22–24 cm^{-1} in L vs 16 cm^{-1} in BR) indicates that the hydrogen bond between the Schiff base and its counter-ion becomes stronger (Alshuth and Stockburger, 1986; Smith *et al.*, 1984). Third, *ab initio* calculations of the magnitude and sign of the ΔG between a model protonated Schiff base–aspartate ion pair (i.e., before proton transfer) and the corresponding neutral pair (i.e., after proton transfer) (Scheiner and Hillenbrand, 1985; Scheiner and Duan, 1991) indicate that it is critically dependent on the geometry of the hydrogen bond and the polarizability of the environment. These will be changed after photoexcitation. Consistently, the pK_a of the Schiff base in model compounds is strongly dependent on the orientation of a sterically fixed counter-ion (Gat and Sheves, 1993). Finally, a decrease in the amplitude of the C_{15}–H in-plane vibration at 1303 cm^{-1} suggests that in the L state the retinal skeleton is twisted so as to remove a steric conflict between the hydrogens on C_{12} and C_{15} (Maeda *et al.*, 1991; Pfefferlé *et al.*, 1991). Such a strain would further disrupt the extended π system along the retinal skeleton and contribute to lowering the pK_a of the Schiff base (Schulten and Tavan, 1978; Orlandi and Schulten, 1979; Fahmy *et al.*, 1989).

The nature of the change in the geometry of the Schiff base relative to Asp85 and the other nearby residues is not clear. Comparison of the linear dichroism of bacteriorhodopsins containing either retinal or 3,4-dehydroretinal defined the initial direction of the N → H bond as pointing toward the exterior surface (Lin and Mathies, 1989). If proton transfer is to take place from the Schiff base to Asp85 the interaction between the Schiff base and its complex counterion must change so as to reorient the C—N–H bond more directly toward Asp85. Molecular dynamics calculations suggested that the interaction of the Schiff base with its counter-ion is stabilized by water molecules in the retinal binding pocket (Zhou *et al.,* 1993). FTIR spectra indicated that the hydrogen-bonding properties of one or a few bound water molecules in L depend on Asp85. The L state is normally characterized by a negative band at 3642 cm^{-1} that is sensitive to both D_2O and $H_2{}^{18}O$ and indicates disappearance of weakly bound water (Maeda *et al.,* 1992a). It is replaced by a small positive band at about 3652 cm^{-1} (free water) and a large broad absorption increase in the 3450–3560 cm^{-1} region (in part strongly bound water). These changes disappear in the M state, implicating the protonated Schiff base as a participant in hydrogen bonding with the water. They are also absent in L state of D85N, and the remaining features in this spectral region are no longer changed in $H_2{}^{18}O$, thus also implicating Asp85 in the hydrogen bonding with water (Maeda *et al.,* 1994). These findings, together with perturbation of Asp96 in L (Braiman *et al.,* 1988a, 1991; Maeda *et al.,* 1992b) and the increase of the deuterium effect on the C—N frequency (Smith *et al.,* 1984; Alshuth and Stockburger, 1986), suggest large-scale rearrangement of the hydrogen bonds of Asp85, the Schiff base, bound water, and Asp96. The result appears to be stronger hydrogen bonding within both proton channels, but in particular on the extracellular side so as to create the conditions for proton transfer from the Schiff base to Asp85.

B. Proton Release to the Extracellular Surface

The transfer of the Schiff base proton to Asp85 during the L to M reaction takes place inside the protein, but at approximately the same time a proton appears on the extracellular surface of the membrane. The release of protons after flash excitation, as well as the subsequent uptake on the cytoplasmic side, has been followed by measuring transient absorption changes of pH-indicator dyes, either in the bulk or covalently bound at the protein surface. Dyes in the bulk are nitrophenol (Lozier *et al.,* 1976; Drachev *et al.,* 1984), pyranine (Grzesiek and Dencher, 1986; Heberle and Dencher, 1990, 1992a; Otto *et al.,* 1989, 1990; Cao *et al.,* 1993b), bromocresol green (Dencher and Wilms, 1975; Mitchell and Rayfield, 1986), and phenol red and chlorophenol red (Váró and Lanyi, 1990b; Zimányi *et al.,* 1992b). Dyes covalently bound are fluorescein conjugated to residue Lys129 (on the extracellular end of helix D) as succinimidyl ester or isothiocyanate (Heberle and Dencher, 1992a) and iodoacetamidofluorescein conjugated to engineered cysteine residues at locations of interest (Alexiev *et al.,* 1994a,b; 1995; Scherrer *et al.,* 1994). Dyes located on the surface show the proton release to be at 20–100 μs (faster in the detergent-solubilized protein), i.e., roughly concurrent with the L to M reaction. In the wild type the temperature dependence reveals that deprotonation of the Schiff base is kinetically linked to the release of the proton (Alexiev *et al.,* 1994a), but under some conditions, and especially in some mutants, the release may occur considerably after the Schiff base deprotonation (Heberle *et al.,* 1993; Cao *et al.,* 1995). The latter observations suggested that the two processes are not directly, or even necessarily, coupled.

Dyes located in the bulk detect the released proton on a time scale of several hundred microseconds. The transfer of the proton from the surface to the bulk is therefore considerably delayed (Heberle and Dencher, 1992a,b; Scherrer *et al.,* 1992). As expected from model studies (Gutman *et al.,* 1985), proton exchange between the surface layer and the bulk is measurably accelerated, for kinetic reasons, when a mobile buffer is added (Drachev *et al.,* 1984; Grzesiek and Dencher, 1986). Fixed buffering groups at the surface effectively conduct the released protons laterally along the membrane surface, and they reach fluoresceine covalently linked to a residue on the cytoplasmic surface long before they are detected by pyranine in the bulk (Heberle and Dencher, 1992a; Alexiev *et al.,* 1994a). This is true even when the dye is on the opposite side of the membrane, necessitating that the released proton travel along the surface, pass around the edge of the membrane sheet, and reach the dye (Alexiev *et al.,* 1995; Heberle *et al.,* 1994).

At pH $\geq$ 6 the direction of the absorption changes of the dyes shows that proton release precedes proton uptake. Since the release on the extracellular side occurs approximately during the rise of M and the uptake on the cytoplasmic side occurs approximately during the decay of N, the finding that there is approximately one proton lost transiently from the protein as it passes through the M and N states (Grzesiek and Dencher, 1986; Heberle and Dencher, 1992a; Váró and Lanyi, 1990b) is readily interpreted as vectorial release and uptake. This proton loss can also be observed during steady illumination of purple membrane sheets. A pH decrease measured with a glass electrode detects protons released in the medium during the photostationary state that develops and in amounts consistent with the accumulation of M and N (Garty *et al.,* 1977; Fischer and Oesterhelt, 1979; Takeuchi *et al.,* 1981; Váró and Lanyi, 1990b).

At pH below 6, however, the proton kinetics in the medium are dramatically changed. Bromocresol green (Dencher and Wilms, 1975), chlorophenol red (Váró and Lanyi, 1990b; Zimányi *et al.,* 1992b), as well as pyranine

(Cao *et al.*, 1993a) detect not proton release but rather transient proton uptake and at a later time in the photocycle, approximately coincident with the accumulation of the O state. Similarly, conductivity change after flash excitation revealed delayed proton disappearance at lower pH rather than the rapid release observed at higher pH (Marinetti and Mauzerall, 1983). Consistent with this, the direction of measured pH change during photostationary states is reversed at lower pH (Garty *et al.*, 1977; Fischer and Oesterhelt, 1979; Takeuchi *et al.*, 1981; Váró and Lanyi, 1990b). Under these conditions, therefore, the protein gains rather than loses a proton transiently in the photocycle. The dye kinetics indicate that the proton uptake step occurs at about the same time as at higher pH but the release step is delayed until the O to BR reaction. The proton release occurs under these conditions after the proton uptake. This effect of pH on the proton release must reflect the protonation state of the release site and can be brought about independent of pH, or with different pH dependence, when the release pathway is perturbed by mutation.

Because proton release at pH > 6 appears approximately concurrent with the rise of M, it has long been associated with the L to M reaction in a mechanistic sense. However, lowering the pH to 6 and below causes the appearance of an $M_2 \rightarrow M_1$ back reaction rather than an increase in the rate of the $M_1 \rightarrow L$ back reaction (Zimányi *et al.*, 1992b). This suggests that the proton release is connected to interconversion of M substates rather than directly to deprotonation of the Schiff base. According to this mechanism the proton release should be concurrent with the second phase of the M rise (with the time constant of the $M_1^{(0)} \leftrightarrow M_1^{(-1)}$ reaction) but not with the first (with the time constant of the $L^{(0)} \leftrightarrow M_1^{(0)}$ reaction). This is confirmed by the time course of the response of a surface-bound dye (Heberle and Dencher, 1992b). Proton release exhibits zero amplitude during the first phase of M rise and occurs concurrently with the second phase.

C. The Proton Transfer Chain between the Schiff Base and the Surface

The late decay of the C—O stretch band of Asp85 indicates that this residue remains protonated until the recovery of BR at the end of the photocycle (Braiman *et al.*, 1988a, 1991; Gerwert *et al.*, 1990; Pfefferlé *et al.*, 1991; Müller *et al.*, 1991; Souvignier and Gerwert, 1992; Bousché *et al.*, 1992). The proton released when Asp85 is protonated must originate from another groups with access to the extracellular surface. This group, originally termed XH (Zimányi *et al.*, 1992b), must have a high enough pK_a in BR to keep it protonated up to at least pH 9 in order to maintain it as a source of protons but will be lowered in L so as to cause its timely deprotonation.

At first, the most obvious candidate for XH seemed to be Arg82. On the basis of the observed coulombic interaction between the anionic Asp85 and the protonated Arg82, it might be expected that protonation of Asp85 would lower the pK_a of Arg82, as required if Arg82 were XH. Indeed, replacement of Arg82 with glutamine or alanine changes the early proton release near pH 7 to delayed proton uptake (Otto *et al.,* 1990; Balashov *et al.,* 1992; Cao *et al.,* 1995); this is similar to the behavior of the wild-type protein at lower pH. Its replacement with a lysine changed the pK_a for release (Govindjee *et al.,* 1996). Presumably, when XH is rendered nonfunctional proton exchange with the bulk proceeds at any pH as it would normally only in the low pH pathway. On the other hand, the pK_a of arginine in proteins is usually >10. Titration of recombinant bacteriorhodopsins with various site-specific replacements of Asp85, Arg82, or both allowed calculation of the pK_a of Arg82 during protonation changes of the Schiff base and Asp85 (Brown *et al.,* 1993a). The result of this calculation based on a coulombic model is that the pK_a of Arg82 is 13.8 in the initial state and decreases to only 11.5 in the M state. This precludes Arg82 being XH.

As discussed previously, the interaction between the protonation states of Asp85 and Glu204 (Brown *et al.,* 1995a; Richter *et al.,* 1996a; Sampogna and Honig, 1996; Balashov *et al.,* 1995, 1996) and Glu194 (Balashov *et al.,* 1997; Dioumaev *et al.,* 1998) observed in the dark would account for proton release during the photocycle if these residues were to transiently dissociate and protonate, respectively, after protonation of Asp85. A negative C—O stretch band from Glu204 (or Asp204 in the E204D mutant), which is expected if the proton originates from dissociation of residue 204, is observable, but its amplitude is rather small (Brown *et al.,* 1995a). On the other hand, when Glu194 is replaced with an aspartate, the transient protonation of this residue during the photocycle is clearly detectable. The C—O stretch band of a protonated aspartate appears at 1720 cm^{-1} closely correlated with the protonation of Asp85, and at pH > 7 it decays in a few milliseconds, at which time covalently bound fluoresceine detects a proton at the surface (Dioumaev *et al.,* 1998). At pH < 7 it decays with a longer time constant that corresponds to the end of the photocycle (i.e., the pK_a of proton release is about 7), and under these conditions fluoresceine also detects the proton at this later time. The pK_a of Asp194 in the unphotolyzed protein, as determined with solid-state ^{13}C NMR, is well below the physiological level (about 3) and consistent with the prediction that it is anionic as a proton acceptor in the proton release chain. The amplitude of the 1720-cm^{-1} band exhibits an apparent pK_a of about 9, which corresponds to the calculated pK_a of Glu204. Above pH 9 protonation of Asp194 does not occur. Presumably, these observations were possible because the deprotonation of Asp194 is slower than the deprotonation of Glu204, and its protonated form can accumulate. Thus, the results strongly suggest that Glu204, or a site that is dependent on the protonation state of Glu204, is the source of the released proton, and Glu194 gains this proton and then dissociates, releasing it to the extracellular surface.

V. Reprotonation of the Schiff Base

Time-resolved FTIR spectra (Gerwert *et al.,* 1990; Bousché *et al.,* 1991; Braiman *et al.,* 1991; Souvignier and Gerwert, 1992) indicate that the appearance of the negative 1742-cm^{-1} band, which indicates deprotonation of Asp96, is coincident with the reprotonation of the Schiff base. Thus, proton loss from Asp96 and proton gain by the Schiff base are described by a single kinetic process. That Asp96 is the internal proton donor to the Schiff base is also strongly supported by results with mutated proteins: Replacement of Asp96 with nonprotonable residues greatly slows reprotonation of the Schiff base and makes it dependent on pH, although proton transport still occurs (Holz *et al.,* 1989; Gerwert *et al.,* 1989; Tittor *et al.,* 1989; Otto *et al.,* 1989; Miller and Oesterhelt, 1990; Miercke *et al.,* 1991; Cao *et al.,* 1991). Because a buried proton donor with suitable pK_a and access is lacking under these conditions, the Schiff base is reprotonated directly from the cytoplasmic surface.

Since the distance between Asp96 and the Schiff base is about 12 Å (Henderson *et al.,* 1990), proton transfer between them would seem not to be feasible unless facilitated by an intervening hydrogen-bonded chain. A hydrogen-bonded chain of a few water molecules arranged in single file has been suggested for proton conduction in the interior of proteins in general (Nagle and Tristram-Nagle, 1983; Merz and Zundel, 1981) and for bacteriorhodopsin (Nagle and Tristram-Nagle, 1983; Schulten and Schulten, 1986; Zhou *et al.,* 1993), and it has been calculated to be feasible in energy-minimized structures (Nonella *et al.,* 1991; Zhou *et al.,* 1993). Strongly bound water near the Schiff base was detected by neutron diffraction (Papadopoulos *et al.,* 1990), but an extended hydrogen-bonded chain of water molecules does not seem to be present in the cytoplasmic region of the unphotolyzed protein. Possibly, it could be formed after an appropriate protein conformational change in the photocycle (Dencher *et al.,* 1992a,b; Fukuda and Kouyama, 1992b).

A protonation equilibrium between Asp96 and the Schiff base that is not far from unity was suggested by the appearance of a second M decay time constant higher than pH 8.5 with linear pH dependency such as expected for the subsequent reprotonation of Asp96 (Otto *et al.,* 1989) and found for the decay of N (Kouyama *et al.,* 1988). Under some conditions the Schiff base deprotonation is described adequately by the scheme M $\leftrightarrow$ N $\rightarrow$ BR (Otto *et al.,* 1989; Váró and Lanyi, 1990a; Cao *et al.,* 1991; Ames and Mathies, 1990; Souvignier and Gerwert, 1992). The existence of a significant thermal N $\rightarrow$ M back reaction is supported by the results of double-perturbation experiments. Depletion of the M state with a second (blue) flash is followed by partial recovery of M with the time constant of the M $\leftrightarrow$ N equilibration reaction. Two parallel M states with different decay time constants are thus ruled out (Druckmann *et al.,* 1993; Zimányi *et al.,* 1993; Brown *et al.,* 1993b). However, there are conditions in which the decay of N is measurably slower than the second decay component of M (Fukuda and Kouyama, 1992a; Zimányi *et al.,* 1993). This would be consistent with an M $\leftrightarrow$ N equilibrium only in a scheme which contained two sequential N states with similar spectra and connected by a unidirectional reaction. The existence of two N substates had been postulated (Mathies *et al.,* 1991; Milder, 1991) on the grounds that N differs from the O intermediate in both the isomeric configuration of the retinal and the protonation state of Asp96, and thus the N to O reaction might in principle be resolved into the reprotonation and reisomerization steps. This is supported by the pH dependence of the N $\rightarrow$ O reaction (Ames and Mathies, 1990). This scheme is confirmed by direct evidence (Zimányi *et al.,* 1993), which indicates that, as expected, the proton uptake is during the interconversion of two consecutive N intermediates.

The pK_a of the Schiff base is about 8 at this time in the photocycle (Brown and Lanyi, 1996). The pK_a of Asp96, on the other hand, is above 10 (Ormos, 1991). An equilibrium between M_2 and N that does not lie far toward M_2 can be established only if the pK_a's of the Schiff base and Asp96 approach one another. That the pK_a of Asp96 is lowered at this time in the photocycle, and independently of the protonation state of the Schiff base, is suggested by the photoreaction of the D212N mutant (Cao *et al.,* 1993b). In the blue form of this recombinant protein (at pH > 7) the Schiff base remains protonated after photoexcitation, and the photocycle in the neutral pH range is described by the scheme BR-hv $\rightarrow$ K $\leftrightarrow$ L $\leftrightarrow$ N $\rightarrow$ BR. The state N is different from L in that it has a small shift in the absorption maximum in the visible, a negative 1742-cm^{-1} band that is the C—O stretch of the deprotonated Asp96, and the appearance of amide I and II bands indicative of a protein backbone change, as in N of the wild type. The pH indicator dye pyranine detects the release of a proton at a time well after the L $\rightarrow$ N reaction but before the decay of N. Neither the FTIR changes nor the proton release are observed in the double-mutant D212N/D96N. The results thus indicate that Asp96 deprotonates even in the absence of its normal proton acceptor, the unprotonated Schiff base.

Net proton translocation is not detected, however, and the kinetics of the photovoltage produced (Moltke *et al.,* 1992) indicates that the subsequent proton uptake is on the same side as the release, i.e., most likely the cytoplasmic side where Asp96 is located. It is reasonable to suppose that such lowering of the pK_a of Asp96 also takes place in the wild-type photocycle, but that here the proton is captured directly by the Schiff base.

Destabilization of the protonated Asp96 must be caused by a change in its dielectric environment. The protein conformation change that includes tilt of the cytoplasmic end (Thorgeirsson *et al.,* 1997) of helix F away from the center of the molecule, detected by various crystallographic methods (Nakasako *et al.,* 1991; Kataoka *et al.,* 1994; Kamikubo *et al.,* 1996, 1997; Vonck, 1996; Brown *et al.,* 1997; Oka *et al.,* 1997; Subramaniam *et al.,* 1993, 1997; Han *et al.,* 1994; Dencher *et al.,* 1989; Sass *et al.,* 1997) and spin–spin exchange measurement (Thorgeirsson *et al.,* 1997), would cause increased binding of water near Asp96. A role for water is suggested by the findings that dehydration affects specifically the Asp96 to Schiff base proton transfer (Váró and Lanyi, 1991a; Cao *et al.,* 1991) and the conformational change at the cytoplasmic surface. Increased hydration will lower the pK_a of Asp96. Replacement of the neighboring residue T46 with valine causes marked acceleration of the reprotonation of the Schiff base and the slowing of the subsequent proton uptake (Marti *et al.,* 1991), suggesting that interaction with this residue also modulates the pK_a of Asp96. Importantly, the proton conformation with the tilted helix F is assumed by the unphotolyzed protein, provided that the Schiff base–counter-ion interaction is weakened. This occurs in the D85N mutant at a pH level higher than 9 in which the Schiff base is unprotonated and in D85N/D96N at neutral pH (Kataoka *et al.,* 1994; Brown *et al.,* 1997). Since the retinal isomeric state is unaffected by the conformation change and vice versa, the driving force for the conformation shift is likely to be loss of coulombic interaction at the active site (Brown *et al.,* 1997).

The rate of proton transfer from D96 to the Schiff base is determined largely by the enthalpy cost of separating the proton from the aspartate anion (Cao *et al.,* 1991). The rate is about seven orders of magnitude slower than predicted for proton conduction via a string of water molecules, e.g., in gramicidin. This would correspond to an additional barrier of 40 kJ/mol. The activation enthalpy for M decay after replacing Asp96 with asparagine, which eliminates the transition state ion pair, is indeed lowered by about 40 kJ/mol (Tittor *et al.,* 1989; Miller and Oesterhelt, 1990; Cao *et al.,* 1991). As expected from such a model, the hydration state of the protein has a strong influence on the proton transfer in the wild-type protein but not in D96N. Withdrawal of bound water from the protein, either by lowering the vapor pressure over deposited films (Váró and Lanyi, 1991a) or by adding osmotically active solutes to purple membrane suspensions (Cao *et al.,* 1991), specifically inhibits the $M_2 \leftrightarrow N$ equilibration reaction in the photocycle.

VI. Proton Uptake at the Cytoplasmic Surface and Recovery of the Initial State

From FTIR and resonance Raman spectra it is evident that reprotonation of Asp96 and reisomerization of the retinal from 13-*cis* to all-*trans* are both associated with the $N \rightarrow O$ chromophore reaction. Resolution of this transition into the $N^{(-1)} \rightarrow N^{(0)} \rightarrow O \rightarrow BR$ sequence at high pH (Zimányi *et al.,* 1993) and into the $N^{(0)} \rightarrow O^{(0)} \rightarrow O^{(+1)} \rightarrow BR$ sequence at low pH (Cao *et al.,* 1993a), however, indicates that the two processes are not necessarily coincident.

Proton uptake at pH 8.5, measured with pyranine, occurs after the rise of N but well before its decay, i.e., during the process described as the $N^{(-1)}$ to $N^{(0)}$ reaction (Zimányi *et al.,* 1993). The time-dependent absorption changes at 410 and 570 nm between pH 9 and 11 allowed calculation of the pH dependencies of the apparent rate constants. As expected, if it refers to proton uptake, the $N^{(-1)} \rightarrow N^{(0)}$ reaction was found to decrease with increasing pH, whereas its back reaction was not pH dependent (Zimányi *et al.,* 1993). The calculated pK_a for the proton uptake is about 11. Thus, in contrast with the proton release to the extracellular side that has a pK_a within the physiological range and allows for two alternative pathways, the pK_a for proton uptake at the cytoplasmic side is well above the pH range for the proton transport. Since it is not certain that the proton taken up directly reprotonates Asp96, this pK_a might refer to the high proton affinity of a cytoplasmic proton transfer complex comprising perhaps Thr46 and bound water rather than the regained proton affinity of Asp96 in $N^{(-1)}$. However, such an intermediate site might not exist since the pK_a of Asp96 in the unphotolyzed state is about 11 (Száraz *et al.,* 1994).

At lower pH (e.g., pH 4) the proton uptake measured with pyranine lags behind the rise of the O state (Cao *et al.,* 1993a). Therefore, the same kind of argument that suggested two N states at high pH also requires two O states at low pH. The N decay pathway under these conditions is thus described by the scheme $N^{(0)} \leftrightarrow O^{(0)} \leftrightarrow O^{(+1)} \rightarrow BR$.

Thus, it appears that the relationship of the proton uptake and/or the reprotonation of Asp96 to the reisomerization of the retinal to all-*trans* is not the same under all conditions. The reisomerization, as reflected in the photocycle by the $N \rightarrow O$ chromophore transition, appears at different times in the low and high pH pathways. In the high pH pathway it follows proton uptake, whereas in the low pH pathway it precedes it. This might be a consequence of the different protonation state of the extracellular proton release complex, i.e., of the net charge in the region of the Schiff base, in the two pathways. Indeed, theory suggests (Warshel and Ottolenghi, 1979; Tavan *et al.,* 1985) that a more negative environment for the Schiff base should raise the barriers to bond rotations in the retinal. There is, in fact, such a connection between the barrier for isomerization and the charge environment of the Schiff base. When the pH was low enough to protonate Asp85 (Ohno *et al.,* 1977; Balashov *et al.,* 1992), or the anionic Asp85 was replaced with a neutral residue (Turner *et al.,* 1993), equilibration of the 13-*cis* and all-*trans* chromophores became unusually rapid. At acid pH the thermal equilibration of these isomeric states was found to be proportional to the protonated state of Asp85 (Balashov *et al.,* 1993).

The kinetics of proton uptake from the cytoplasmic surface is simplified when Asp96 is replaced with a nonprotonable group. The M decay is described by a single exponential with a pH-dependent rate because proton transfer to the Schiff base is now directly from the bulk. When azide, a weak acid, is added the rate of the protonation becomes rapid (Tittor *et al.,* 1989, 1994b; Cao *et al.,* 1991; Le Coutre *et al.,* 1995), and the biphasic kinetics that develops allows determination of the pK_a of the Schiff base at this stage (8.3; Brown and Lanyi, 1996). Without azide, however, the M decay is so slow that N scarcely accumulates or does not accumulate at all (Cao *et al.,* 1993b), i.e., the rate of reisomerization is not slowed correspondingly by replacement of Asp96. Proton transfer to the Schiff base is hindered by what appears to be an increased entropic barrier (Tittor *et al.,* 1989; Miller and Oesterhelt, 1990; Cao *et al.,* 1991). Resolution of the proton trajectory into two segments, between the Schiff base and residue 96 (about 12 Å) and between residue 96 and the cytoplasmic surface (about 6 Å), and comparison of the rates and the activation parameters indicated that the entropically unfavorable consequence of replacing Asp96 is on the capture of the proton at the opening of the cytoplasmic channel (Cao *et al.,* 1991). This suggests another role for Asp96: Its negative charge in the N state sustains a protein conformation appropriate for effective entry of a proton into the cytoplasmic channel. This function of Asp96 is strongly influenced by the nearby residues Thr46 and Ser226 since their replacement slows proton uptake by about two orders of magnitude (Marti *et al.,* 1991; Brown *et al.,* 1994b).

The initial state is recovered as Asp85 is deprotonated in the final O → BR reaction (Bousché *et al.,* 1992). Asp85 transfers its proton either to the extracellular proton release complex (at high pH) or to the bulk (at low pH). Resonance Raman (Smith *et al.,* 1983) indicates that minor relaxations in the retinal, and therefore most likely in the protein, accompany this internal proton transfer. Various amino acid replacements in the extracellular region of the protein affected deprotonation of Asp85 in the same way, whether the deprotonation was during the O → BR reaction or during a pH jump in the dark. The good correspondence of the rate constants for these reactions over three orders of magnitude strongly suggested that the last step in the photocycle is limited by deprotonation of Asp85.

VII. The Reprotonation Switch

According to the alternating access hypothesis, ion pumps cycle between two conformations that determine access to one or the other side of the membrane but not to both at the same time (Tanford, 1982; Kalisky *et al.*, 1981; Lanyi, 1993, 1995; Oesterhelt *et al.*, 1992). In bacteriorhodopsin the "access" refers to proton exchange between the Schiff base and either Asp85 on the extracellular side or Asp96 on the cytoplasmic side. The change of access from one of these residues to the other is called the reprotonation (or protonation) switch (Nagle and Mille, 1981; Schulten *et al.*, 1984; Henderson *et al.*, 1990; Váró and Lanyi, 1991c; Fodor *et al.*, 1988a; Mathies *et al.*, 1991; Oesterhelt *et al.*, 1992; Zimányi *et al.*, 1992a). In principle the switch might be accomplished in two alternative ways: by changes of proton affinity in which proton transfer is controlled (i) by changes of the pK_a's of the respective donor and acceptor groups or (ii) by changes of local geometry in which proton transfer is controlled through altered bond angles and distances. In the first case, the geometry between proton donor and acceptor must allow proton transfer throughout, but in the second case the geometry itself constitutes the switch.

The logical place for the switch in the photocycle to occur is after deprotonation of the Schiff base but before its reprotonation, i.e., in the M state. Therefore, it is important to establish the identities of the immediate preswitch and postswitch states. Since the switch appears to be a multistep process, more than two states involved with the switch are expected. Indeed, there is a considerable amount of evidence for the existence of a multitude of M substates, as reviewed later, and some evidence argues for their functional role in the reprotonation switch.

First, the kinetic scheme for time-resolved difference spectra in the visible required postulating two sequential M substates, termed M_1 and M_2 connected by a unidirectional reaction (Váró and Lanyi, 1990a, 1991b). The introduction of the $M_1 \rightarrow M_2$ reaction (ratio of forward/reverse rates > 200 at pH 7, although it is much less at lower pH) explained why the concentration of L tended to zero as the concentration of M reached a maximum, even though the calculated $L \leftrightarrow M$ equilibrium with a single M would have predicted the persistence of considerable amounts of L in a mixture with M. Determining the concentration of L at this time in the photocycle was complicated by the accumulation of N because L and N have similar absorption (Váró and Lanyi, 1991b) and resonance Raman (Ames and Mathies, 1990) spectra. Confirming the results with the D96N protein, under conditions in which for kinetic reasons the accumulation of N was eliminated, provided important support to this interpretation of wild-type kinetics (Váró and Lanyi, 1991b; Zimányi and Lanyi, 1993).

Second, in purple membrane containing wild-type protein the maximum of M is 411–412 nm regardless of whether the M is the putative M_1 or M_2, but under some conditions the kinetically defined M substates exhibit different absorption maxima. M_2 is always blue shifted relative to M_1. The maximum of M_2 is a few nanometers lower than that of M_1 in the detergent-solubilized (monomeric) wild-type protein (Váró and Lanyi, 1991b; Milder *et al.,* 1991; Subramaniam *et al.,* 1991). Upon the D115N residue replacement the difference in λ_{max} under these conditions becomes as much as 18 nm (Váró *et al.,* 1992). Distinguishing the maxima of M substates in the purple membrane lattice is easier at pH 10, at which the rise of M is much faster and M_1 accumulates to concentrations comparable to that of M_2. Determined under these conditions, the maximum of M_1 in D96N, and in double mutants containing the D96N mutation, is like that in wild-type M_1 but that of M_2 is blue shifted by 7 nm (Zimányi *et al.,* 1992a; Cao *et al.,* 1995). Such a shift would be expected if hydrogen bonding of the Schiff base were diminished or broken. Thus, it appears that replacing Asp96 with an asparagine affects the maxima of M_1 and M_2 specifically, which is expected if M_1 and M_2 were respectively preswitch and postswitch states, with the Schiff base hydrogen bonded to acceptor/donors first in M_1 and then in M_2.

Third, the quantum yield and rate of a blue flash-induced photo-back reaction of M to BR changes during the lifetime of M in such a way as to suggest the existence of two distinct M substates from measurement of either absorption change (Druckmann *et al.,* 1992) or photoelectric effect (Dickopf and Heyn, 1997). The rate of conversion of the first to the second M and its pH dependence agreed well with that predicted from the kinetics of L, suggesting that these experiments detect the proposed M_1 and M_2.

Fourth, photoacoustic measurements indicated that a large decrease of enthalpy occurs between proton release and uptake (Ort and Parson, 1979; Garty *et al.,* 1982). This suggested a strong decrease of entropy and thus a large protein conformational change during the lifetime of M (Váró and Lanyi, 1991c). Recent studies with better time resolution localized a large part of this enthalpy decrease at 80–90 μs (Rohr *et al.,* 1992), i.e., at about the time constant of the proposed $M_1 \rightarrow M_2$ transition.

Finally, photoelectric measurements of bacteriorhodopsin expressed in *Xenopus laevis* were performed to test for M_1 and M_2 with blue light-dependent quenching of the photocurrent (Nagel *et al.,* 1998). An externally applied electrical potential had the effect of driving M_2 back to M_1, consistent with the idea that these M substates are in equilibrium and that they are connected by an electrogenic reaction, i.e., proton release.

FTIR spectra of bacteriorhodopsin films in which the decay of M was greatly slowed, i.e., in D96N at pH 10 and 276 K (Sasaki *et al.,* 1992) or in glucose-dehydrated wild-type protein (Perkins *et al.,* 1992), showed that the amide bands at 1670 (negative) and 1650 (positive) cm^{-1} (which originate from peptide bond vibrations) as well as the shift of the COOH frequency of Asp85 from 1761 to 1755 cm^{-1}, otherwise observed only in N, appear in the virtual absence of chromophore bands due to protonated Schiff base. The FTIR changes usually associated with protein changes in N can thus occur before the Schiff base is reprotonated and indicate the existence of a late M state different in its protein conformation from the earlier one. This late M, termed M_N (Sasaki *et al.,* 1992), is unlikely to be M_2. In the wild-type photocycle the amide bands arise virtually concurrently with the $M_2 \rightarrow N$ reaction (Braiman *et al.,* 1991; Gerwert *et al.,* 1990; Souvignier and Gerwert, 1992) which takes place much later than the $M_1 \rightarrow M_2$ reaction, and in D96N the M_N state coexists with its precursor M state in a constant ratio throughout the recovery of BR (Sasaki *et al.,* 1992). In view of these results, M_N is more likely to be a transient state between M_2 and N. The rationale for M_N is that in this state the pK_a of Asp96 is lowered. This idea is supported by the findings that (i) in D96N the FTIR band due to Asn96 shows a shift at about this time in the photocycle consistent with changed hydrogen bonding (Sasaki *et al.,* 1992) and (ii) in the photocycle of the blue form of the D212N protein the amide bands appear and the pK_a of Asp96 is lowered at the same time even though under these conditions a deprotonated Schiff base is not formed (Cao *et al.,* 1993b).

The molecular events that underlie these M substates, and therefore the protonation switch, certainly include the pK_a changes that make proton transfers in the EC or CP directions available. The coupling of the protonation of Asp85 and the proton release site in the extracellular proton release chain (described previously) ensures that the pK_a of Asp85 rises so as to block its deprotonation (Balashov *et al.,* 1995 ,1996; Richter *et al.,* 1996a,b). Once a proton is released to the EC, surface proton exchange with the EC half-channel is no longer possible. On the other hand, as described previously, deprotonation of the Schiff base causes large-scale protein conformational changes at the CP surface that appear to result in increased hydration of this region, and thus the lowering of the pK_a of Asp96 and probably increased proton conductivity to the CP surface. Reprotonation of the Schiff base is therefore from the CP side.

Because transport is possible in mutants in which Asp85 or Asp96 or both are replaced (Tittor *et al.*, 1994a,1997; Haupts *et al.*, 1997; Ganea *et al.*, 1998), the switch can obviously function without the pK_a changes of these residues. According to the isomerization/switch/transfer (IST) model (Haupts *et al.*, 1997; Tittor *et al.*, 1997), the switch step is independent of isomerization and proton transfer. Invoked in order to explain specifically the observations of blue light-induced CP to EC proton transport and blue plus green light-induced EC to CP proton transport in Asp85 mutants (Tittor *et al.*, 1994a), and to reconcile them with the transport in the wild type, this model postulates that S and T are in kinetic competition after photoisomerization. In some cases S occurs before T and in others the opposite occurs. An attractive candidate for the S step would be rotation of the C_{14}–C_{15} single bond of the retinal (Gerwert and Siebert, 1986), but this is contradicted by vibrational spectra of L, M, and N (Fodor *et al.*, 1988a,b).

Spectroscopic study of the photocycle of the D85N/D96N mutant suggested, however, that the switch event is complex and depends on both the local geometry near the Schiff base and the proton conductivities of the half-channels that lead to the two membrane surfaces. Protonation and deprotonation of the Schiff base were studied after pH jump without illumination and in the photocycle of the unprotonated Schiff base, in the visible and the infrared. The results suggested a hypothesis for the proton transfer switch different from the IST model. Here, the essential point is that in the metastable 13-*cis,* 15-*anti* photoproduct, but not in the stable all-*trans* isomeric state, access flickers between the EC and CP directions. The direction of proton transfer is decided both by this local access and by the presence of a suitable donor or acceptor group (in the wild type) or the proton conductivity in the half-channels (in D85N/D96N). In the wild-type transport cycle the concurrent local access in the extracellular and cytoplasmic directions during the lifetime of the metastable 13-*cis,* 15-*anti* state enables the changing pK_a's of the proton acceptor and donor to determine the direction of the proton transfers. Proton transfer from the Schiff base to Asp85 in the EC direction is followed by reprotonation by Asp96 from the CP direction because proton release to the EC surface raises the pK_a of Asp85 and a large-scale protein conformation change lowers the pK_a of Asp96. Since in D85N/D96N when the retinal is in the stable all-*trans,* 15-*anti* form access of the Schiff base in *locked* (in the EC directions), in this protein reisomerization, rather than changes in the proton conductivities of the EC and CP half-channels, provides the switch function. With this mechanism, the various modes of transport reported for Asp85 mutants (CP to EC direction with blue light and EC to CP direction with blue plus green light) are understood in the same terms as in the wild type.

VIII. Dissipation of Free Energy in the Photocycle

All the thermal reactions of the photocycle are driven by the excess free energy retained in the K state after absorption of a photon. Its amount is probably close to the approximately 50 kJ/mol excess enthalpy measured calorimetrically (Birge *et al.,* 1991). Understanding how this ΔG is transformed and dissipated in the photocycle will reveal how free energy in the retinal is transferred first to protein residues and then to protons so as to produce an electrochemical gradient across the membrane. Rough outlines of the thermodynamics of the photocycle between K and the last intermediate O have been reconstructed (Váró and Lanyi, 1991c) from photoacoustic measurements of the enthalpy changes (i.e., calorimetric enthalpies) and the temperature dependencies of the rate constants (i.e., van't Hoff activation enthalpies and entropies).

Some of the photocycle reactions appear to proceed near equilibrium, i.e., at close to 0 ΔG. This is kinetically optimal for internal reactions of enzymes in general (Albery and Knowles, 1976). Under physiological conditions the two reactions associated with proton exchange between the protein and the two aqueous phases and the internal proton transfer at the last photocycle step dissipate nearly all the excess free energy. At a physiological pH (about 7) the overall $M_1^{(0)} \rightarrow M_2^{(-1)}$ reaction exhibits an apparent equilibrium constant $K = [M_2^{(-1)}]/[M_1^{(0)}]$ not less than 200, which corresponds to a ΔG of at least -15 kJ/mol (Váró and Lanyi, 1991b). Of this amount about 7.5 kJ/mol is calculated to be lost dissipatively at the reprotonation switch, and the rest, which is pH dependent, is accounted for by proton release to the bulk on the extracellular side, i.e., 5.7 kJ/mol for every pH unit above the pK_a of the release complex (Zimányi *et al.,* 1992b). The other pH-dependent step at which free energy is lost in the $N^{(-1)} \rightarrow N^{(0)}$ (or at low pH the $O^{(0)} \rightarrow O^{(+1)}$) reaction (Zimányi *et al.,* 1993). Here, free energy is lost by proton uptake on the cytoplasmic side because the pH under most conditions is well below the pK_a of the proton uptake group. This pK_a is about 11 (Zimányi *et al.,* 1993), and ΔG will amount to -5.7 kJ/mol for every pH unit below 11. From the initial 50 kJ/mol available in K and the three calculated ΔGs, the free energy that remains for driving the O $\rightarrow$ BR reaction is estimated to be no more than 15 kJ/mol. This corresponds to an [BR]/[O] equilibrium constant of several hundred, consistent with the fact that no O state can be detected in coexistence with unphotolyzed bacteriorhodopsin.

The activation enthalpies associated with the observable rate constants, and the photoacoustic determination of an approximately 80 kJ/mol enthalpy decrease between proton release and uptake (Ort and Parson, 1979; Garty *et al.,* 1982), allowed reconstruction of the enthalpy cycle (Váró and

Lanyi, 1991c). The assumption that the entropy of the K state does not differ greatly from that of BR, and a reasonable although somewhat arbitrary equal apportion of the free energy changes between the $M_a \rightarrow M_2$ and O → BR reactions, allowed, in turn, reconstruction of the entropy cycle (Váró and Lanyi, 1991c). The two cycles indicate that the retained free energy in the system changes from ΔH to - $T{\cdot}\Delta S$ (i.e., from excess enthalpy to negative entropy) at the reprotonation switch. Enthalpy is converted to entropy at the switch because until M_1 the excess free energy resides in the chromophore, mainly as retinal bond torsions and the lowered proton affinity of the Schiff base relative to Asp85, but in M_2 and the subsequent states the high Schiff base pK_a recovers and the remaining free energy is transformed into the excess negative entropy of a restricted protein conformation. Relaxation of this conformation drives the completion of the photocycle. Consistent with this, removing the protein from the motionally restricted environment of the purple membrane lattice caused little or not change in the enthalpies and entropies of the photocycle reactions before the switch but resulted in large changes after the switch (Váró and Lanyi, 1991d). It is significant that these changes in the detergent-solubilized protein, as well as in purple membranes which contain residue replacements that perturb noncovalent bonds assumed to play roles in proton transfer between the Schiff base and D96 (e.g., T46V) (L. Brown, Y. Cao, R. Needleman and J. Lanyi, unpublished data), always include decreased activation enthalpies and more negative activation entropies. It appears therefore that the protein under these conditions becomes more flexible but less ordered (and/or binds less ordered water). Although the transition states of the reactions are reached at less enthalpic cost, the protein must pass through a greater number of conformational states (and/or organize more water).

At the switch the enthalpy of the system decreases below the initial level so that the M_2 to BR portion of the photocycle contains mainly endothermic reactions. This explains the well-known fact that the transient accumulation of the O intermediate, unlike the others, is greater at higher temperatures (Li *et al.,* 1984; Váró and Lanyi, 1991c; Chizhov *et al.,* 1992).

IX. Summary

Despite many unsolved problems, the mechanism and energetics of the light-driven proton transport are now basically understood. Energy captured during photoexcitation, and retained in the form of bond rotations and strains of the retinal, is transformed into directed changes in the pK_a's of vectorially arranged proton transfer groups. The framework for the spatial and temporal organization of these changes is provided by the protein near the retinal Schiff base. The transport is completed by proton transfer among three essential groups in three domains stacked approximately parallel to the membrane plane: (i) the anionic Asp85 that is included in a complex of residues on the extracellular side also containing Arg82, Asp212, and water, (ii) the protonated Schiff base; and (iii) the protonated Asp96 that is included in a complex of residues on the cytoplasmic side also containing Thr46. Other neighboring polar groups and water bound elsewhere which play a role in the transport do so either by further influencing the pK_a's of the three protonable groups or by providing passive pathways for proton transfer.

The Schiff base proton, destabilized after photoexcitation mainly by distortion of the π-electron system along the retinal skeleton, is transferred to the initially low pK_a group Asp85 located on the extracellular side. Proton release to the extracellular surface causes increase of the pK_a of Asp85 and prevents return of the transferred proton. The proton of the high pK_a group Asp96, with access to the cytoplasmic side, is destabilized by a protein conformational change through rearrangement of bound water and becomes transferred to the Schiff base. These internal events are coupled to proton release and uptake at the two aqueous surfaces. The charge of the extracellular hydrogen-bonded complex is redistributed upon protonation of Asp85, and if the pH is above the pK_a of the complex a proton is released to the bulk via Glu204 and Glu194. After reprotonation of the Schiff base the pK_a of the cytoplasmic hydrogen-bonded complex is raised well above the pH and Asp96 regains a proton from the bulk. If the pH is lower than the pK_a for proton release, the release is delayed until the end of the photocycle. In either sequence there is net transfer of a proton from the cytoplasmic to the extracellular phase.

The transfer of excess free energy from the chromophore to the protein, and finally to the transported proton, is described by a characteristic thermodynamic cycle. At physiological pH the excess enthalpy retained in the form of local perturbation near the active site (the retinal Schiff base) drives proton transfer to Asp85 and release on the extracellular side. The resulting protein conformation changes and its eventual recovery causes reprotonation of the Schiff base by Asp96 and its reprotonation from the cytoplasmic surface, respectively. ΔG is transformed into proton electrochemical potential as a proton is released at a pH higher than the pK_a of the extracellular proton release complex and a proton is taken up at a pH lower than the pK_a of the cytoplasmic proton uptake complex.

References

Albery, W. J., and Knowles, J. R. (1976). Evolution of enzyme function and development of catalytic efficiency. *Biochemistry* **15,** 5631–5640.

Alexiev, U., Marti, T., Heyn, M. P., Khorana, H. G., and Scherrer, P. (1994a). Covalently bound pH-indicator dyes at selected extracellular or cytoplasmic sites in bacteriorhodopsin. 2. Rotational orientation of helices D and E and kinetic correlation between M formation and proton release in bacteriorhodopsin micelles. *Biochemistry* **33,** 13693–13699.

Alexiev, U., Marti, T., Heyn, M. P., Khorana, H. G., and Scherrer, P. (1994b). Surface charge of bacteriorhodopsin detected with covalently bound pH indicators at selected extracellular and cytoplasmic sites. *Biochemistry* **33,** 298–306.

Alexiev, U., Mollaaghababa, R., Scherrer, P., Khorana, H. G., and Heyn, M. P. (1995). Rapid long-range proton diffusion along the surface of the purple membrane and delayed proton transfer into the bulk. *Proc. Natl. Acad. Sci. USA* **92,** 372–376.

Alshuth, T., and Stockburger, M. (1986). Time-resolved resonance Raman studies on the photochemical cycle of bacteriorhodopsin. *Photochem. Photobiol.* **43,** 55–66.

Althaus, T., Eisfeld, W., Lohrmann, R., and Stockburger, M. (1995). Application of Raman spectroscopy to retinal proteins. *Israel J. Chem.* **35,** 227–252.

Ames, J. B., and Mathies, R. A. (1990). The role of back-reactions and proton uptake during the N $\rightarrow$ O transition in bacteriorhodopsin's photocycle: A kinetic resonance Raman study. *Biochemistry* **29,** 7181–7190.

Ames, J. B., Fodor, S. P. A., Gebhard, R., Raap, J., van den Berg, M. M., Lugtenburg, J., and Mathies, R. A. (1989). Bacteriorhodopsin's M_{412} intermediate contains a 13-*cis*, 14-s-*trans*, 15-*anti*-retinal Schiff base chromophore. *Biochemistry* **28,** 3681–3687.

Ames, J. B., Bolton, S. R., Netto, M. M., and Mathies, R. A. (1990). Ultraviolet resonance Raman spectroscopy of bacteriorhodopsin: Evidence against tyrosinate in the photocycle. *J. Am. Chem. Soc.* **112,** 9007–9009.

Ames, J. B., Ros, M., Raap, J., Lugtenburg, J., and Mathies, R. A. (1992). Time-resolved ultraviolet resonance Raman studies of protein structure: Application to bacteriorhodopsin. *Biochemistry* **31,** 5328–5334.

Ariki, M., Magde, D., and Lanyi, J. K. (1987). Metal ion binding sites of bacteriorhodopsin. *J. Biol. Chem.* **262,** 4947–4951.

Aton, B., Doukas, A. G., Callender, R. H., Becher, B., and Ebrey, T. G. (1977). Resonance Raman studies of the purple membrane. *Biochemistry* **16,** 2995–2999.

Baasov, T., and Sheves, M. (1986). Alteration of pK_a of the bacteriorhodopsin protonated Schiff base. A study with model compounds. *Biochemistry* **25,** 5249–5258.

Baasov, T., Friedman, N., and Sheves, M. (1987). Factors affecting the C—N stretching in protonated retinal Schiff-base: A model study for bacteriorhodopsin and visual pigments. *Biochemistry* **26,** 3210–3217.

Balashov, S., Govindjee, R., Kono, M., Lukashov, E., Ebrey, T. G., Feng, Y., Crouch, R. K., and Menick, D. R. (1992). Arg82ala mutant of bacteriorhodopsin expressed in H. halobium: Drastic decrease in the rate of proton release and effect on dark adaptation. *In* "Structures and Functions of Retinal Proteins" (J.L. Rigaud, ed.), pp. 111–114. Libbey Eurotext, Montrouge.

Balashov, S. P., Govindjee, R., and Ebrey, T. G. (1991). Red shift of the purple membrane absorption band and the deprotonation of tyrosine residues at high pH. Origin of the parallel photocycles of trans-bacteriorhodopsin. *Biophys. J.* **60,** 475–490.

Balashov, S. P., Govindjee, R., Kono, M., Imasheva, E., Lukashev, E., Ebrey, T. G., Crouch, R. K., Menick, D. R., and Feng, Y. (1993). Effect of the arginine-82 to alanine mutation in bacteriorhodopsin on dark adaptation, proton release, and the photochemical cycle. *Biochemistry* **32,** 10331–10343.

Balashov, S. P., Govindjee, R., Imasheva, E. S., Misra, S., Ebrey, T. G., Feng, Y., Crouch, R. K., and Menick, D. R. (1995). The two pK_a's of aspartate-85 and control of thermal isomerization and proton release in the arginine-82 to lysine mutant of bacteriorhodopsin. *Biochemistry* **34,** 8820–8834.

Balashov, S. P., Imasheva, E. S., Govindjee, R., and Ebrey, T. G. (1996). Titration of aspartate-85 in bacteriorhodopsin: What it says about chromophore isomerization and proton release. *Biophys. J.* **70,** 473–481.

Balashov, S. P., Imasheva, E. S., Ebrey, T. G., Chen, N., Menick, D. R., and Crouch, R. K. (1997). Glutamate-194 to cysteine mutation inhibits fast light-induced proton release in bacteriorhodopsin. *Biochemistry* **36,** 8671–8676.

Baldwin, J. M., Henderson, R., Beckmann, E., and Zemlin, F. (1988). Images of purple membrane at 2.8 A resolution obtained by cryo-electron microscopy. *J. Mol. Biol.* **202,** 585–591.

Beach, J. M., and Fager, R. S. (1985). Evidence for branching in the photocycle of bacteriorhodopsin and concentration changes of late intermediate forms. *Photochem. Photobiol.* **41,** 557–562.

Becher, B., Tokunaga, F., and Ebrey, T. G. (1978). Ultraviolet and visible absorption spectra of the purple membrane protein and the photocycle intermediates. *Biochemistry* **17,** 2293–2300.

Birge, R. R. (1990). Nature of the primary photochemical events in rhodopsin and bacteriorhodopsin. *Biochim. Biophys. Acta Bio-Energetics* **1016,** 293–327.

Birge, R. R., Cooper, T. M., Lawrence, A. F., Masthay, M. B., Zhang, C.-F., and Zidovetzki, R. (1991). Revised assignment of energy storage in the primary photochemical event in bacteriorhodopsin. *J. Am. Chem. Soc.* **113,** 4327–4328.

Bousché, O., Braiman, M. S., He, Y.-W., Marti, T., Khorana, H. G., and Rothschild, K. J. (1991). Vibrational spectroscopy of bacteriorhodopsin mutants. Evidence that Asp-96 deprotonates during the M → N transition. *J. Biol. Chem.* **266,** 11063–11067.

Bousché, O., Sonar, S., Krebs, M. P., Khorana, H. G., and Rothschild, K. J. (1992). Time-resolved Fourier transform infrared spectroscopy of the bacteriorhodopsin mutant Tyr-185-Phe: Asp-96 reprotonates during O formation, As-85 and Asp-212 deprotonate during O decay. *Photochem. Photobiol.* **56,** 1085–1095.

Braiman, M. S., and Mathies, R. A. (1982). Resonance Raman spectra of bacteriorhodopsin's primary photoproduct: Evidence for a distorted 13-cis retinal chromophore. *Proc. Natl. Acad. Sci. USA* **79,** 403–407.

Braiman, M. S., Ahl, P. L., and Rothschild, K. J. (1987). Millisecond Fourier-transform infrared difference spectra of bacteriorhodopsin's M_{412} photoproduct. *Proc. Natl. Acad. Sci. USA* **84,** 5221–5225.

Braiman, M. S., Mogi, T., Marti, T., Stern, L. J., Khorana, H. G., and Rothschild, K. J. (1988a). Vibrational spectroscopy of bacteriorhodopsin mutants: Light-driven proton transport involves protonation changes of aspartate residues 85,96, and 212. *Biochemistry* **27,** 8516–8520.

Braiman, M. S., Mogi, T., Stern, L. J., Hackett, R. D., Chao, B. H., and Khorana, H. G. (1988b). Vibrational spectroscopy of bacteriorhodopsin mutants: I. Tyrosine-185 protonates and deprotonates during the photocycle. *Proteins* **3,** 219–229.

Braiman, M. S., Bousché, O., and Rothschild, K. J. (1991). Protein dynamics in the bacteriorhodopsin photocycle: Submillisecond Fourier transform infrared spectra of the L, M, and N photointermediates. *Proc. Natl. Acad. Sci. USA* **88,** 2388–2392.

Braiman, M. S., Dioumaev, A. K., and Lewis, J. R. (1996). A large photolysis-induced pK_a increase of the chromophore counterion in bacteriorhodopsin: Implications for ion transport mechanisms of retinal proteins. *Biophys. J.* **70,** 939–947.

Brown, L. S., and Lanyi, J. K. (1996). Determination of the transiently lowered pK_a of the retinal Schiff base during the photocycle of bacteriorhodopsin. *Proc. Natl. Acad. Sci. USA* **93,** 1731–1734.

Brown, L. S., Bonet, L., Needleman, R., and Lanyi, J. K. (1993a). Estimated acid dissociation constants of the Schiff base, asp-85 and arg-82 during the bacteriorhodopsin photocycle. *Biophys. J.* **65,** 124–130.

Brown, L. S., Zimányi, L., Ottolenghi, M., Needleman, R., and Lanyi, J. K. (1993b). Photoreaction of the N intermediate of bacteriorhodopsin and its relationship to the decay kinetics of the M intermediate. *Biochemistry* **32,** 7679–7685.

Brown, L. S., Gat, Y., Sheves, M., Yamazaki, Y., Maeda, A., Needleman, R., and Lanyi, J. K. (1994a). The retinal Schiff base–counterion complex of bacteriorhodopsin: Changed geometry during the photocycle is a cause of proton transfer to aspartate 85. *Biochemistry* **33,** 12001–12011.

Brown, L. S., Yamazaki, Y., Maeda, M., Sun, L., Needleman, R., and Lanyi, J. K. (1994b). The proton transfers in the cytoplasmic domain of bacteriorhodopsin are facilitated by a cluster of interacting residues. *J. Mol. Biol.* **239,** 401–414.

Brown, L. S., Sasaki, J., Kandori, H., Maeda, A., Needleman, R., and Lanyi, J. K. (1995a). Glutamic acid 204 is the terminal proton release group at the extracellular surface of bacteriorhodopsin. *J. Biol. Chem.* **270,** 27122–27126.

Brown, L. S., Váró, G., Needleman, R., and Lanyi, J. K. (1995b). Functional significance of a protein conformation change at the cytoplasmic end of helix F during the bacteriorhodopsin photocycle. *Biophys. J.* **69,** 2103–2111.

Brown, L. S., Kamikubo, H., Zimányi, L., Kataoka, M., Tokunaga, F., Verdegem, P., Lugtenburg, J., and Lanyi, J. K. (1997). A local electrostatic change is the cause of the large-scale protein conformation shift in bacteriorhodopsin. *Proc. Natl. Acad. Sci. USA* **94,** 5040–5044.

Butt, H.-J., Fendler, K., Bamberg, E., Tittor, J., and Oesterhelt, D. (1989). Aspartic acids 96 and 85 play a central role in the function of bacteriorhodopsin as a proton pump. *EMBO J.* **8,** 1657–1663.

Cao, Y., Váró, G., Chang, M., Ni, B., Needleman, R., and Lanyi, J. K. (1991). Water is required for proton transfer from aspartate 96 to the bacteriorhodopsin Schiff base. *Biochemistry* **30,** 10972–10979.

Cao, Y., Brown, L. S., Needleman, R., and Lanyi, J. K. (1993a). Relationship of proton uptake on the cytoplasmic surface and the reisomerization of the retinal in the bacteriorhodopsin photocycle: An attempt to understand the complex kinetics of the protons and the N and O intermediates. *Biochemistry* **32,** 10239–10248.

Cao, Y., Váró, G., Klinger, A. L., Czajkowsky, D. M., Braiman, M. S., Needleman, R., and Lanyi, J. K. (1993b). Proton transfer from asp-96 to the bacteriorhodopsin Schiff base is caused by decrease of the pKa of asp-96 which follows a protein backbone conformation change. *Biochemistry* **32,** 1981–1990.

Cao, Y., Brown, L. S., Sasaki, J., Maeda, A., Needleman, R., and Lanyi, J. K. (1995). Relationship of proton release at the extracellular surface to deprotonation of the Schiff base in the bacteriorhodopsin photocycle. *Biophys. J.* **68,** 1518–1530.

Chernavskii, D. S., Chizhov, I. V., Lozier, R. H., Murina, T. M., Prokhorov, A. M., and Zubov, B. V. (1989). Kinetic model of bacteriorhodopsin photocycle: Pathway from M state to bR. *Photochem. Photobiol.* **49,** 649–653.

Chizhov, I., Engelhard, M., Chernavskii, D. S., Zubov, B., and Hess, B. (1992). Temperature and pH sensitivity of the O_{640} intermediate of the bacteriorhodopsin photocycle. *Biophys. J.* **61,** 1001–1006.

Dancsházy, Z., Govindjee, R., and Ebrey, T. G. (1988). Independent photocycles of the spectrally distinct forms of bacteriorhodopsin. *Proc. Natl. Acad. Sci. USA* **85,** 6358–6361.

De Groot, H. J. M., Harbison, G. S., Herzfeld, J., and Griffin, R. G. (1989). Nuclear magnetic resonance study of the Schiff base in bacteriorhodopsin: Counterion effects on the ^{15}N shift anisotropy. *Biochemistry* **28,** 3346–3353.

De Groot, H. J. M., Smith, S. O., Courtin, J., Van den Berg, E., Winkel, C., Lugtenburg, J., Griffin, R. G., and Herzfeld, J. (1990). Solid-state ^{13}C and ^{15}N NMR study of the low pH forms of bacteriorhodopsin. *Biochemistry* **29,** 6873–6883.

Dencher, N. A., and Wilms, M. (1975). Flash photometric experiments on the photochemical cycle of bacteriorhodopsin. *Biophys. Struct. Mech.* **1,** 259–271.

Dencher, N. A., Dresselhaus, D., Zaccai, G., and Büldt, G. (1989). Structural changes in bacteriorhodopsin during proton translocation revealed by neutron diffraction. *Proc. Natl. Acad. Sci. USA* **86,** 7876–7879.

Dencher, N. A., Heberle, J., Büldt, G., Höltje, H.-D., and Höltje, M. (1992a). What do neutrons, x-ray synchrotron radiation, optical pH-indicators, and mutagenesis tell us about the light-driven proton pump bacteriorhodopsin? *In* "Membrane Proteins: Structures, Interactions and Models." (A. Pullman, J. Jortner, and B. Pullman, eds.), pp. 69–84. Kluwer, Dordrecht.

Dencher, N. A., Heberle, J., Büldt, G., Höltje, H.-D., and Höltje, M. (1992b). Active and passive proton transfer steps through bacteriorhodopsin are controlled by a light-triggered hydrophobic gate. *In* "Structures and Functions of Retinal Proteins" (J. L. Rigaud, ed.), pp. 213–216. Libbey Eurotext, Montrouge.

Dér, A., Száraz, S., Tóth-Boconádi, R., Tokaji, Z., Keszthelyi, L., and Stoeckenius, W. (1991). Alternative translocation of protons and halide ions by bacteriorhodopsin. *Proc. Natl. Acad. Sci. USA* **88,** 4751–4755.

Dickopf, S., and Heyn, M. P. (1997). Evidence for the first phase of the reprotonation switch of bacteriorhodopsin from time-resolved photovoltage and flash photolysis experiments on the photoreversal of the M-intermediate. *Biophys. J.* **73,** 3171–3181.

Dioumaev, A. K., and Braiman, M. S. (1997). Nano- and microsecond time-resolved FTIR spectroscopy of the halorhodopsin photocycle. *Photochem. Photobiol.* **66,** 755–763.

Dioumaev, A. K., Richter, H. T., Brown, L. S., Tanio, M., Tuzi, S., Saitô, H., Kimura, Y., Needleman, R., and Lanyi, J. K. (1998). Existence of a proton transfer chain in bacteriorhodopsin: Participation of Glu-194 in the release of protons to the extracellular surface. *Biochemistry* **37,** 2496–1506.

Drachev, L. A., Kaulen, A. D., and Skulachev, V. P. (1984). Correlation of photochemical cycle, H^+ release and uptake, and electrical events in bacteriorhodopsin. *FEBS Lett.* **178,** 331–335.

Drachev, L. A., Skulachev, V. P., Kaulen, A. D., and Zorina, V. V. (1987). The mechanism of H^+ transfer by bacteriorhodopsin—The properties and the function of intermediate R. *FEBS Lett.* **226,** 139–144.

Drachev, L. A., Kaulen, A. D., and Komrakov, A. Y. (1992). Interrelations of M intermediates in bacteriorhodopsin photocycle. *FEBS Lett.* **313,** 248–250.

Druckmann, S., Ottolenghi, M., Pande, A., Pande, J., and Callender, R. H. (1982). Acid-base equilibrium of the Schiff base in bacteriorhodopsin. *Biochemistry.* **21,** 4953–4959.

Druckmann, S., Friedman, N., Lanyi, J. K., Needleman, R., Ottolenghi, M., and Sheves, M. (1992). The back photoreaction of the M intermediate in the photocycle of bacteriorhodopsin: Mechanism and evidence for two M species. *Photochem. Photobiol.* **56,** 1041–1047.

Druckmann, S., Heyn, M. P., Lanyi, J. K., Ottolenghi, M., and Zimányi, L. (1993). Thermal equilibration between the M and N intermediates in the photocycle of bacteriorhodopsin. *Biophys. J.* **65,** 1231–1234.

Dunäch, M., Seigneuret, M., Rigaud, J.-L., and Padros, E. (1988). Influence of cations on the blue to purple transition of bacteriorhodopsin. Comparison of Ca^{2+} and Hg^{2+} binding and their effect on the surface potential. *J. Biol. Chem.* **263,** 17378–17384.

Ebrey, T. G. (1993). Light energy transduction in bacteriorhodopsin. *In* "Thermodynamics of Membranes, Receptors and Channels" (M. Jackson, ed.), pp. 353–387. CRC Press, New York.

Edgerton, M. E., Moore, T. A., and Greenwood, C. (1980). Investigations into the effect of acid on the spectral and kinetic properties of purple membrane from *Halobacterium halobium.* *Biochem. J.* **189,** 413–420.

Eisfeld, W., and Stockburger, M. (1992). Optical transient studies on the photochemical cycle of bacteriorhodopsin. In "Structures and Functions of Retinal Proteins" (J. L. Rigaud, ed.), pp. 139–142. Libbey Eurotex, Montrouge.

Engelhard, M., Gerwert, K., Hess, B., Kreutz, W., and Siebert, F. (1985). Light-driven protonation changes of internal aspartic acids of bacteriorhodopsin: An investigation by static and time-resolved infrared difference spectroscopy using [4-^{13}C] aspartic acid labeled purple membrane. *Biochemistry.* **24,** 400–407.

Fahmy, K., Siebert, F., Grossjean, M. F., and Tavan, P. (1989). Photoisomerization in bacteriorhodopsin studied by FTIR, linear dichroism and photoselection experiments combined with quantum chemical theoretical analysis. *J. Mol. Struct.* **214,** 257–288.

Fischer, U., and Oesterhelt, D. (1979). Chromophore equilibria in bacteriorhodopsin. *Biophys. J.* **28,** 211–230.

Fodor, S. P., Ames, J. B., Gebhard, R., van der Berg, E. M., Stoeckenius, W., Lugtenburg, J., and Mathies, R. A. (1988a). Chromophore structure in bacteriorhodopsin's N intermediate: Implications for the proton pumping mechanism. *Biochemistry* **27,** 7097–7101.

Fodor, S. P., Pollard, W. T., Gebhard, R., van den Berg, E. M., Lugtenburg, J., and Mathies, R. A. (1988b). Bacteriorhodopsin's L_{550} intermediate contains a C_{14}–C_{15} s-*trans*-retinal chromophore. *Proc. Natl. Acad. Sci. USA* **85,** 2156–2160.

Fukuda, K., and Kouyama, T. (1992a). Formation and decay of bacteriorhodopsin's N intermediate: Softening the protein conformation with alcohols affects intra-protein proton transfer and retinal isomerization. *Photochem. Photobiol.* **56,** 1057–1062.

Fukuda, K., and Kouyama, T. (1992b). Photoreaction of bacteriorhodopsin at high pH: Origins of the slow decay component of M. *Biochemistry* **31,** 11740–11747.

Ganea, C., Tittor, J., Bamberg, E., and Oesterhelt, D. (1998). Chloride- and pH-dependent proton transport by BR mutant D85N. *Biochim. Biophys. Acta Bio-Membr.* **1368,** 84–96.

Garty, H., Klemperer, G., Eisenbach, M., and Caplan, S. R. (1977). The direction of light-induced pH changes in purple membrane suspensions. Influence of pH and temperature. *FEBS Lett.* **81,** 238–242.

Garty, H., Caplan, S. R., and Cahen, D. (1982). Photoacoustic photocalorimetry and spectroscopy of *Halobacterium halobium* purple membranes. *Biophys. J.* **37,** 405–415.

Gat, Y., and Sheves, M. (1993). A mechanism for controlling the pK_a of the retinal protonated Schiff base in retinal proteins. A study with model compounds. *J. Am. Chem. Soc.* **115,** 3772–3773.

Gerwert, K., and Siebert, F. (1986). Evidence for light-induced 13-cis, 14-s-cis isomerization in bacteriorhodopsin obtained by FTIR difference spectroscopy using isotopically labeled retinals. *EMBO J.* **5,** 805–811.

Gerwert, K., Hess, B., Soppa, J., and Oesterhelt, D. (1989). Role of aspartate-96 in proton translocation by bacteriorhodopsin. *Proc. Natl. Acad. Sci. USA* **86,** 4943–4947.

Gerwert, K., Souvignier, G., and Hess, B. (1990). Simultaneous monitoring of light-induced changes in protein side-group protonation, chromophore isomerization, and backbone motion of bacteriorhodopsin by time-resolved Fourier-transform infrared spectroscopy. *Proc. Natl. Acad. Sci. USA* **87,** 9774–9778.

Govindjee, R., Misra, S., Balashov, S. P., Ebrey, T. G., Crouch, R. K., and Menick, D. R. (1996). Arginine-82 regulates the pK_a of the group responsible for the light-driven proton release in bacteriorhodopsin. *Biophys. J.* **71,** 1011–1023.

Greenhalgh, D. A., Subramaniam, S., Alexiev, U., Otto, H., Heyn, M. P., and Khorana, H. G. (1992). Effect of introducing different carboxylate-containing side chains at position 85 on chromophore formation and proton transport in bacteriorhodopsin. *J. Biol. Chem.* **267,** 25734–25738.

Grigorieff, N., Ceska, T. A., Downing, K. H., Baldwin, J. M., and Henderson, R. (1996). Electron-crystallographic refinement of the structure of bacteriorhodopsin. *J. Mol. Biol.* **259,** 393–421.

Grzesiek, S., and Dencher, N. A. (1986). Time-course and stoichiometry of light-induced proton release and uptake during the photocycle of bacteriorhodopsin. *FEBS Lett.* **208,** 337–342.

Gutman, M., Nachliel, E., and Gershon, E. (1985). Effect of buffer on kinetics of proton equilibration with a protonable group. *Biochemistry* **24,** 2937–2941.

Han, B.-G., Vonck, J., and Glaeser, R. M. (1994). The bacteriorhodopsin photocycle: Direct structural study of two substates of the M-intermediate. *Biophys. J.* **67,** 1179–1186.

Hanamoto, J. H., Dupuis, P., and El-Sayed, M. A. (1984). On the protein (tyrosine)–chromophore (protonated Schiff base) coupling in bacteriorhodopsin. *Proc. Natl. Acad. Sci. USA* **81,** 7083–7087.

Harbison, G. S., Smith, S. O., Pardoen, J. A., Winkel, C., Lugtenburg, J., Herzfeld, J., Mathies, R. A., and Griffin, R. G. (1984). Dark-adapted bacteriorhodopsin contains 13-*cis,* 15-*syn* and all-*trans,* 15-*anti* retinal Schiff bases. *Proc. Natl. Acad. Sci. USA* **81,** 1706–1709.

Haupts, U., Tittor, J., Bamberg, E., and Oesterhelt, D. (1997). General concept for ion translocation by halobacterial retinal proteins: The isomerization/switch/transfer (IST) model. *Biochemistry* **36,** 2–7.

Heberle, J., and Dencher, N. A. (1990). Bacteriorhodopsin in ice: Accelerated proton transfer from the purple membrane surface. *FEBS Lett.* **277,** 277–280.

Heberle, J., and Dencher, N. A. (1992a). Surface-bound optical probes monitor proton translocation and surface potential changes during the bacteriorhodopsin photocycle. *Proc. Natl. Acad. Sci. USA* **89,** 5996–6000.

Heberle, J., and Dencher, N. A. (1992b). The surface of the purple membrane: A transient pool for protons? *In* "Structures and Functions of Retinal Proteins" (J. L. Rigaud, ed.), pp. 221–224. Libbey Eurotext, Montrouge.

Heberle, J., Oesterhelt, D., and Dencher, N. A. (1993). Decoupling of photo- and proton cycle in the Asp85 → Glu mutant of bacteriorhodopsin. *EMBO J.* **12,** 3721–3727.

Heberle, J., Riesle, J., Thiedemann, G., Oesterhelt, D., and Dencher, N. A. (1994). Proton migration along the membrane surface and retarded surface to bulk transfer. *Nature* **370,** 379–382.

Henderson, R., Baldwin, J. M., Downing, K. H., Lepault, J., and Zemlin, F. (1986). Structure of the purple membrane from *Halobacterium halobium:* Recording, measurement and evaluation of electron micrographs at 3.5 A resolution. *Ultramicroscopy* **19,** 147–178.

Henderson, R., Baldwin, J. M., Ceska, T. A., Zemlin, F., Beckmann, E., and Downing, K. H. (1990). Model for the structure of bacteriorhodopsin based on high-resolution electron cryo-microscopy. *J. Mol. Biol.* **213,** 899–929.

Herzfeld, J., Das Gupta, S. K., Farrar, M. R., Harbison, G. S., McDermott, A. E., Pelletier, S. L., Raleigh, D. P., Smith, S. O., Winkel, C., Lugtenburg, J., and Griffin, R. G. (1990). Solid-state ^{13}C NMR study of tyrosine protonation in dark-adapted bacteriorhodopsin. *Biochemistry* **29,** 5567–5574.

Hessling, B., Souvignier, G., and Gerwert, K. (1993). A model-independent approach to assigning bacteriorhodopsin's intramolecular reactions to photocycle intermediates. *Biophys. J.* **65,** 1929–1941.

Hessling, B., Herbst, J., Rammelsberg, R., and Gerwert, K. (1997). Fourier transform infrared double-flash experiments resolve bacteriorhodopsin's M_1 to M_2 transition. *Biophys. J.* **73,** 2071–2080.

Hildebrandt, P., and Stockburger, M. (1984). Role of water in bacteriorhodopsin's chromophore: Resonance Raman study. *Biochemistry* **23,** 5539–5548.

Holz, M., Drachev, L. A., Mogi, T., Otto, H., Kaulen, A. D., Heyn, M. P., Skulachev, V. P., and Khorana, H. G. (1989). Replacement of aspartic acid-96 by asparagine in bacteriorhodopsin slows both the decay of the M intermediate and the associated proton movement. *Proc. Natl. Acad. Sci. USA* **86,** 2167–2171.

Honig, B., Ebrey, T. G., Callender, R. H., Dinur, U., and Ottolenghi, M. (1979). Photoisomerization, energy storage, and charge separation: A model for light energy transduction in visual pigments and bacteriorhodopsin. *Proc. Natl. Acad. Sci. USA* **76,** 2503–2507.

Jonas, R., and Ebrey, T. G. (1991). Binding of a single divalent cation directly correlates with the blue-to-purple transition in bacteriorhodopsin. *Proc. Natl. Acad. Sci. USA* **88,** 149–153.

Kakitani, H., Kakitani, T., Rodman, H., and Honig, B. (1985). On the mechanism of wavelength regulation in visual pigments. *Photochem. Photobiol.* **41,** 471–479.

Kalisky, O., Goldschmidt, C. R., and Ottolenghi, M. (1977). On the photocycle and light adaptation of dark-adapted bacteriorhodopsin. *Biophys. J.* **19,** 185–189.

Kalisky, O., Ottolenghi, M., Honig, B., and Korenstein, R. (1981). Environmental effects on formation and photoreaction of the M_{412} photoproduct of bacteriorhodopsin: Implications for the mechanism of proton pumping. *Biochemistry* **20,** 649–655.

Kamikubo, H., Kataoka, M., Váró, G., Oka, T., Tokunaga, F., Needleman, R., and Lanyi, J. K. (1996). Structure of the N intermediate of bacteriorhodopsin revealed by x-ray diffraction. *Proc. Natl. Acad. Sci. USA* **93,** 1386–1390.

Kamikubo, H., Oka, T., Imamoto, Y., Tokunaga, F., Lanyi, J. K., and Kataoka, M. (1997). The last phase of the reprotonation switch in bacteriorhodopsin: The transition between the M-type and the N-type protein conformation depends on hydration. *Biochemistry* **36,** 12282–12287.

Kandori, H., Yamazaki, Y., Sasaki, J., Needleman, R., Lanyi, J. K., and Maeda, A. (1995). Water-mediated proton transfer in proteins: An FTIR study of bacteriorhodopsin. *J. Am. Chem. Soc.* **117,** 2118–2119.

Kataoka, M., Kamikubo, H., Tokunaga, F., Brown, L. S., Yamazaki, Y., Maeda, A., Sheves, M., Needleman, R., and Lanyi, J. K. (1994). Energy coupling in an ion pump: The reprotonation switch of bacteriorhodopsin. *J. Mol. Biol.* **243,** 621–638.

Khorana, H. G. (1993). Two light-transducing membrane proteins: Bacteriorhodopsin and the mammalian rhodopsin. *Proc. Natl. Acad. Sci USA* **90,** 1166–1171.

Kimura, Y., Ikegami, A., and Stoeckenius, W. (1984). Salt and pH-dependent changes of the purple membrane absorption spectrum. *Photochem. Photobiol.* **40,** 641–646.

Kimura, Y., Vassylyev, D. G., Miyazawa, A., Kidera, A., Matsushima, M., Mitsuoka, K., Murata, K., Hirai, T., and Fujiyoshi, Y. (1997a). Surface of bacteriorhodopsin revealed by high-resolution electron crystallography. *Nature* **389,** 206–211.

Kimura, Y., Vassylyev, D. G., Miyazawa, A., Kidera, A., Matsushima, M., Mitsuoka, K., Murata, K., Hirai, T., and Fujiyoshi, Y. (1997b). High resolution structure of bacteriorhodopsin determined by electron crystallography. *Photochem. Photobiol.* **66,** 764–767.

Kouyama, T., Nasuda-Kouyama, A., Ikegami, A., Mathew, M. K., and Stoeckenius, W. (1988). Bacteriorhodopsin photoreaction: Identification of a long-lived intermediate N (P,R_{350}) at high pH and its M-like photoproduct. *Biochemistry* **27,** 5855–5863.

Kuschmitz, D., and Hess, B. (1982). *Trans-cis* isomerization of the retinal chromophore of bacteriorhodopsin during the photocycle. *FEBS Lett.* **138,** 137–140.

Landau, E. M., and Rosenbusch, J. P. (1996). Lipidic cubic phases: A novel concept for the crystallization of membrane proteins. *Proc. Natl. Acad. Sci. USA* **93,** 14532–14535.

Lanyi, J. K. (1986). Mechanism of base-catalyzed Schiff-base deprotonation in halorhodopsin. *Biochemistry* **25,** 6706–6711.

Lanyi, J. K. (1992). Proton transfer and energy coupling in the bacteriorhodopsin photocycle. *J. Bioenerg. Biomembr.* **24,** 169–179.

Lanyi, J. K. (1993). Proton translocation mechanism and energetics in the light-driven pump bacteriorhodopsin. *Biochim. Biophys. Acta Bio-Energetics* **1183,** 241–261.

Lanyi, J. K. (1994). The photochemical reaction cycle of bacteriorhodopsin. *In* "Molecular and Biomolecular Electronics" (R. R. Birge, ed.), pp. 491–510. *Am. Chem. Soc.*, Washington, DC.

Lanyi, J. K. (1995). Bacteriorhodopsin as a model for proton pumps. *Nature* **375,** 461–463.

Lanyi, J. K. (1997). Mechanism of ion transport across membranes. Bacteriorhodopsin as a prototype for proton pumps. *J. Biol. Chem.* **272,** 31209–31212.

Lanyi, J. K., and Váró, G. (1995). The photocycles of bacteriorhodopsin. *Israel J. Chem.* **35,** 365–386.

Lanyi, J. K., Tittor, J., Váró, G., Krippahl, G., and Oesterhelt, D. (1992). Influence of the size and protonation state of acidic residue 85 on the absorption spectrum and photoreaction of the bacteriorhodopsin chromophore. *Biochim. Biophys. Acta* **1099,** 102–110.

Le Coutre, J., Tittor, J., Oesterhelt, D., and Gerwert, K. (1995) Experimental evidence for hydrogen-bonded network proton transfer in bacteriorhodopsin shown by Fourier-transform infrared spectroscopy using azide as catalyst. *Proc. Natl. Acad. Sci. USA* **92,** 4962–4966.

Li, Q., Govindjee, R., and Ebrey, T. G. (1984). A correlation between proton pumping and the bacteriorhodopsin photocycle. *Proc. Natl. Acad. Sci. USA* **81,** 7079–7082.

Lin, S. W., and Mathies, R. A. (1989). Orientation of the protonated retinal Schiff base group in bacteriorhodopsin from absorption linear dichroism. *Biophys. J.* **56,** 653–660.

Lozier, R. H., Bogomolni, R. A., and Stoeckenius, W. (1975). Bacteriorhodopsin: A light-driven proton pump in *Halobacterium halobium. Biophys. J.* **15,** 955–963.

Lozier, R. H., Niederberger, W., Bogomolni, R. A., Hwang, S., and Stoeckenius, W. (1976). Kinetics and stoichiometry of light-induced proton release and uptake from purple membrane fragments, *Halobacterium halobium* cell envelopes, and phospholipid vesicles containing oriented purple membrane. *Biochim. Biophys. Acta* **440,** 545–556.

Lozier, R. H., Xie, A., Hofrichter, J., and Clore, G. M. (1992). Reversible steps in the bacteriorhodopsin photocycle. *Proc. Natl. Acad. Sci. USA* **89,** 3610–3614.

Maeda, A. (1995). Application of FTIR spectroscopy to the structural study on the function of bacteriorhodopsin. *Israel J. Chem.* **35,** 387–400.

Maeda, A., Iwasa, T., and Yoshizawa, T. (1977). Isomeric composition of retinal chromophore in dark-adapted bacteriorhodopsin. *J. Biochem. (Tokyo)* **82,** 1599–1604.

Maeda, A., Sasaki, J., Pfefferlé, J.-M., Shichida, Y., and Yoshizawa, T. (1991). Fourier transform infrared spectral studies of the Schiff base mode of all-trans bacteriorhodopsin and its photointermediates, K and L. *Photochem. Photobiol.* **54,** 911–921.

Maeda, A., Sasaki, J., Shichida, Y., and Yoshizawa, T. (1992a). Water structural changes in the bacteriorhodopsin photocycle: Analysis by Fourier transform infrared spectroscopy. *Biochemistry* **31,** 462–467.

Maeda, A., Sasaki, J., Shichida, Y., Yoshizawa, T., Chang, M., Ni, B., Needleman, R., and Lanyi, J. K. (1992b). Structures of aspartic acid-96 in the L and N intermediates of bacteriorhodopsin: Analysis by Fourier transform infrared spectroscopy. *Biochemistry* **31,** 4684–4690.

Maeda, A., Sasaki, J., Yamazaki, Y., Needleman, R., and Lanyi, J. K. (1994). Interaction of aspartate 85 with a water molecule and the protonated Schiff base in the L intermediate of bacteriorhodopsin: A Fourier transform infrared spectroscopy study. *Biochemistry* **33,** 1713–1717.

Marinetti, T., and Mauzerall, D. (1983). Absolute quantum yields and proof of proton and nonproton transient release and uptake in photoexcited bacteriorhodopsin. *Proc. Natl. Acad. Sci. USA* **80,** 178–180.

Marti, T., Otto, H., Mogi, T., Rösselet, S. J., Heyn, M. P., and Khorana, H. G. (1991). Bacteriorhodopsin mutants containing single substitutions of serine or threonine residues are all active in proton translocation. *J. Biol. Chem.* **266,** 6919–6927.

Mathies, R. A., Lin, S. W., Ames, J. B., and Pollard, W. T. (1991). From femtoseconds to biology: Mechanism of bacteriorhodopsin's light-driven proton pump. *Annu. Rev. Biophys. Biophys. Chem.* **20,** 491–518.

Maurer, R., Vogel, J., and Schneider, S. (1987). Analysis of flash photolysis data by a global fit with multiexponentials—I. Determination of the minimal number of intermediates in the photocycle of bacteriorhodopsin by the "stability criterion." *Photochem. Photobiol.* **46,** 247–253.

Merz, H., and Zundel, G. (1981). Proton conduction in bacteriorhodopsin via a hydrogen-bonded chain with large proton polarizability. *Biochem. Biophys. Res. Commun.* **101,** 540–546.

Metz, G., Siebert, F., and Engelhard, M. (1992). Asp85 is the only internal aspartic acid that gets protonated in the M intermediate and the purple-to-blue transition of bacteriorhodopsin: A solid-state ^{13}C CP-MAS NMR investigation. *FEBS Lett.* **303,** 237–241.

Miercke, L. J. W., Betlach, M. C., Mitra, A. K., Shand, R. F., Fong, S. K., and Stroud, R. M. (1991). Wild-type and mutant bacteriorhodopsins D85N, D96N, and R82Q: Purification to homogeneity, pH dependence of pumping, and electron diffraction. *Biochemistry* **30,** 3088–3098.

Milder, S. J. (1991). Correlation between absorption maxima and thermal isomerization rates in bacteriorhodopsin. *Biophys. J.* **60,** 440–446.

Milder, S. J., Thorgeirsson, T. E., Miercke, L. J. W., Stroud, R. M., and Kliger, D. S. (1991). Effects of detergent environments on the photocycle of purified monomeric bacteriorhodopsin. *Biochemistry* **30,** 1751–1761.

Miller, A., and Oesterhelt, D. (1990). Kinetic optimization of bacteriorhodopsin by aspartic acid 96 as an internal proton donor. *Biochim. Biophys. Acta Bio-Energetics* **1020,** 57–64.

Mitchell, D., and Rayfield, G. W. (1986). Order of proton uptake and release by bacteriorhodopsin at low pH. *Biophys. J.* **49,** 563–566.

Mogi, T., Stern, L. J., Hackett, N. R., and Khorana, H. G. (1987). Bacteriorhodopsin mutants containing single tyrosine to phenylalanine substitutions are all active in proton translocation. *Proc. Natl. Acad. Sci. USA* **84,** 5595–5599.

Mogi, T., Stern, L. J., Marti, T., Chao, B. H., and Khorana, H. G. (1988). Aspartic acid substitutions affect proton translocation by bacteriorhodopsin. *Proc. Natl. Acad. Sci. USA* **85,** 4148–4152.

Moltke, S., Heyn, M. P., Krebs, M. P., Mollaaghababa, R., and Khorana, H. G. (1992). Low pH photovoltage kinetics of bacteriorhodopsin with replacements of Asp-96, -85, -212 and Arg-82. *In* "Structures and Functions of Retinal Proteins" (J. L. Rigaud, ed.), pp. 201–204. Libbey Eurotex, Montrouge.

Moore, T. A., Edgerton, M. E., Parr, G., Greenwood, C., and Perham, R. N. (1978). Studies of an acid-induced species of purple membrane from *Halobacterium halobium. Biochem. J.* **171,** 469–476.

Mowery, P. C., Lozier, R. H., Chae, Q., Tseng, Y. W., Taylor, M., and Stoeckenius, W. (1979). Effect of acid pH on the absorption spectra and photoreactions of bacteriorhodopsin. *Biochemistry* **18,** 4100–4107.

Müller, K.-H., Butt, H. J., Bamberg, E., Fendler, K., Hess, B., Siebert, F., and Engelhard, M. (1991). The reaction cycle of bacteriorhodopsin: An analysis using visible absorption, photocurrent and infrared techniques. *Eur. Biophys. J.* **19,** 241–251.

Nagel, G., Kelety, B., Möckel, B., Büldt, G., and Bamberg, E. (1998). Voltage dependence of proton pumping by bacteriorhodopsin is regulated by the voltage sensitive ratio of M_1 to M_2. *Biophys. J.* **74,** 403–412.

Nagle, J. F. (1991). Solving complex photocycle kinetics. Theory and direct method. *Biophys. J.* **59,** 476–487.

Nagle, J. F., and Mille, M. (1981). Molecular models of proton pumps. *J. Chem. Phys.* **74,** 1367–1372.

Nagle, J. F., and Tristram-Nagle, S. (1983). Hydrogen bonded chain mechanisms for proton conduction and proton pumping. *J. Membr. Biol.* **74,** 1–14.

Nagle, J. F., Parodi, L. A., and Lozier, R. H. (1982). Procedure for testing kinetic models of the photocycle of bacteriorhodopsin. *Biophys. J.* **38,** 161–174.

Nagle, J. F., Zimányi, L., and Lanyi, J. K. (1995). Testing BR photocycle kinetics. *Biophys. J.* **68,** 1490–1499.

Nakanishi, K., Balogh-Nair, V., Arnaboldi, M., Tsujimoto, K., and Honig, B. (1980). An external point-charge model for bacteriorhodopsin to account for its purple colour. *J. Am. Chem. Soc.* **102,** 7945–7947.

Nakasako, M., Kataoka, M., Amemiya, Y., and Tokunaga, F. (1991). Crystallographic characterization by X-ray diffraction of the M-intermediate from the photocycle of bacteriorhodopsin at room temperature. *FEBS Lett.* **292,** 73–75.

Needleman, R., Chang, M., Ni, B., Váró, G., Fornes, J., White, S. H., and Lanyi, J. K. (1991). Properties of asp212-asn bacteriorhodopsin suggest that asp212 and asp85 both participate in a counterion and proton acceptor complex near the Schiff base. *J. Biol. Chem.* **266,** 11478–11484.

Nonella, M., Windemuth, A., and Schulten, K. (1991). *Photochem. Photobiol.* **54,** 937–948.

Oesterhelt, D., Tittor, J., and Bamberg, E. (1992). A unifying concept for ion translocation by retinal proteins. *J. Bioenerg. Biomembr.* **24,** 181–191.

Ohno, K., Takeuchi, Y., and Yoshida, M. (1977). Effect of light-adaptation on the photoreaction of bacteriorhodopsin from *Halobacterium halobium. Biochim. Biophys. Acta* **462,** 575–582.

Oka, T., Kamikubo, H., Tokunaga, F., Lanyi, J. K., Needleman, R., and Kataoka, M. (1997). X-ray diffraction studies of bacteriorhodopsin. Determination of the positions of mercury label at several engineered cysteine residues. *Photochem. Photobiol.* **66,** 768–773.

Orlandi, G., and Schulten, K. (1979). Coupling of stereochemistry and proton donor-acceptor properties of a Schiff base. A model of a light-driven proton pump. *Chem. Phys. Lett.* **64,** 370–374.

Ormos, P. (1991). Infrared spectroscopic demonstration of a conformational change in bacteriorhodopsin involved in proton pumping. *Proc. Natl. Acad. Sci. USA* **88,** 473–477.

Ormos, P., Chu, K., and Mourant, J. (1992). Infrared study of the L, M, and N intermediates of bacteriorhodopsin using the photoreaction of M. *Biochemistry* **31,** 6933–6937.

Ort, D. R., and Parson, W. W. (1979). Enthalpy changes during the photochemical cycle of bacteriorhodopsin. *Biophys. J.* **25,** 355–364.

Otto, H., Marti, T., Holz, M., Mogi, T., Lindau, M., Khorana, H. G., and Heyn, M. P. (1989). Aspartic acid-96 is the internal proton donor in the reprotonation of the Schiff base of bacteriorhodopsin. *Proc. Natl. Acad. Sci. USA* **86,** 9228–9232.

Otto, H., Marti, T., Holz, M., Mogi, T., Stern, L.J., Engel, F., Khorana, H. G., and Heyn, M. P. (1990). Substitution of amino acids Asp-85, Asp-212, and Arg-82 in bacteriorhodopsin affects the proton release phase of the pump and the pK of the Schiff base. *Proc. Natl. Acad. Sci. USA* **87,** 1018–1022.

Papadopoulos, G., Dencher, N. A., Zaccai, G., and Büldt, G. (1990). Water molecules and exchangeable hydrogen ions at the active centre of bacteriorhodopsin localized by neutron diffraction. Elements of the proton pathway. *J. Mol. Biol.* **214,** 15–19.

Pebay-Peyroula, E., Rummel, G., Rosenbusch, J. P., and Landau, E. M. (1997). X-ray structure of bacteriorhodopsin at 2.5 angstroms from microcrystals grown in lipidic cubic phases. *Science* **277,** 1676–1681.

Perkins, G. A., Liu, E., Burkard, F., Berry, E. A., and Glaeser, R. M. (1992). Characterization of the conformational change in the M_1 and M_2 substates of bacteriorhodopsin by the combined use of visible and infrared spectroscopy. *J. Struct. Biol.* **109,** 142–151.

Pfefferlé, J.-M., Maeda, A., Sasaki, J., and Yoshizawa, T. (1991). Fourier transform infrared study of the N intermediate of bacteriorhodopsin. *Biochemistry* **30,** 6548–6556.

Pusch, C., Diller, R., Eisfeld, W., Lohrmann, R., and Stockburger, M. (1992). The light-induced proton-pump of bacteriorhodopsin studied by resonance Raman and optical transient spectroscopy. *In* "Structures and Functions of Retinal Proteins" (J. L. Rigaud, ed.), pp. 143–146. Libbey Eurotext, Montrouge.

Rath, P., Marti, T., Sonar, S., Gobind Khorana, H., and Rothschild, K. J. (1993). Hydrogen bonding interactions with the Schiff base of bacteriorhodopsin. Resonance Raman spectroscopy of the mutants D85N and D85A. *J. Biol. Chem.* **268,** 17742–17749.

Richter, H. T., Brown, L. S., Needleman, R., and Lanyi, J. K. (1996a). A linkage of the pK_a's of asp-85 and glu-204 forms part of the reprotonation switch of bacteriorhodopsin. *Biochemistry* **35,** 4054–4062.

Richter, H. T., Needleman, R., and Lanyi, J. K. (1996b). Perturbed interaction between residues 85 and 204 in tyr-185-phe and asp-85-glu bacteriorhodopsins. *Biophys. J.* **71,** 3392–3398.

Riesle, J., Oesterhelt, D., Dencher, N. A., and Heberle, J. (1996). D38 is an essential part of the proton translocation pathway in bacteriorhodopsin. *Biochemistry* **35,** 6635–6643.

Rohr, M., Schulenberg, P., Gärtner, W., and Braslavsky, S. E. (1992). Detection of conformational changes during the photocycle of bacteriorhodopsin by laser-induced optoacoustic spectroscopy (LIOAS). *In* "Structures and Functions of Retinal Proteins" (J. L. Rigaud, ed.), pp. 151–154. Libbey Eurotext, Montrouge.

Rothschild, K. J. (1992). FTIR difference spectroscopy of bacteriorhodopsin: Toward a molecular model. *J. Bioenerg. Biomembr.* **24,** 147–167.

Rothschild, K. J., Roepe, P., Ahl, P. L., Earnest, T. N., Bogomolni, R. A., Das Gupta, S. K., Mulliken, C. M., and Herzfeld, J. (1986). Evidence for a tyrosine protonation change during the primary phototransition of bacteriorhodopsin at low temperature. *Proc. Natl. Acad. Sci. USA* **83,** 347–351.

Rothschild, K. J., Braiman, M. S., He, Y.-W., Marti, T., and Khorana, H. G. (1990). Vibrational spectroscopy of bacteriorhodopsin mutants. Evidence for the interaction of aspartic acid 212 with tyrosine 185 and possible role in the proton pump mechanism. *J. Biol. Chem.* **265,** 16985–16991.

Sampogna, R. V., and Honig, B. (1996). Electrostatic coupling between retinal isomerization and the ionization state of Glu-204: A general mechanism for proton release in bacteriorhodopsin. *Biophys. J.* **71,** 1165–1171.

Sasaki, J., Shichida, Y., Lanyi, J. K., and Maeda, A. (1992). Protein changes associated with reprotonation of the Schiff base in the photocycle of asp96-asn bacteriorhodopsin. The M_N intermediate with unprotonated Schiff base but N-like protein structure. *J. Biol. Chem.* **267,** 20782–20786.

Sasaki, J., Lanyi, J. K., Needleman, R., Yoshizawa, T., and Maeda, A. (1994). Complete identification of C—O stretching vibrational bands of protonated aspartic acid residues in the difference infrared spectra of M and N intermediates versus bacteriorhodopsin. *Biochemistry* **33,** 3178–3184.

Sass, H. J., Schachowa, I. W., Rapp, G., Koch, M. H. J., Oesterhelt, D., Dencher, N. A., and Büldt, G. (1997). The tertiary structural changes in bacteriorhodopsin occur between M states; X-ray diffraction and Fourier transform infrared spectroscopy. *EMBO J.* **16,** 1484–1491.

Scheiner, S., and Duan, X. (1991). Effect of intermolecular orientation upon proton transfer within a polarizable medium. *Biophys. J.* **60,** 874–883.

Scheiner, S., and Hillenbrand, E. A. (1985). Modification of pK values caused by change in H-bond geometry. *Proc. Natl. Acad. Sci. USA* **82,** 2741–2745.

Scherrer, P., Mathew, M. K., Sperling, W., and Stoeckenius, W. (1989). Retinal isomer ratio in dark-adapted purple membrane and bacteriorhodopsin monomers. *Biochemsitry* **28,** 829–834.

Scherrer, P., Alexiev, U., Otto, H., Heyn, M. P., Marti, T., and Khorana, H. G. (1992). Proton movement and surface charge in bacteriorhodopsin detected by selectively attached pH-indicators. *In* "Structures and Functions of Retinal Proteins" (J. L. Rigaud, ed.), pp. 205–211. Libbey Eurotext, Montrouge.

Scherrer, P., Alexiev, U., Marti, T., Khorana, H. G., and Heyn, M. P. (1994). Covalently bound pH-indicator dyes at selected extracellular or cytoplasmic sites in bacteriorhodopsin. 1. Proton migration along the surface of bacteriorhodopsin micelles and its delayed transfer from surface to bulk. *Biochemistry* **33,** 13684–13692.

Schulten, K., and Tavan, P. (1978). A mechanism for the light-driven proton pump of Halobacterium halobium. *Nature* **272,** 85–86.

Schulten, K., Schulten, Z., and Tavan, P. (1984). An isomerization model for the pump cycle of bacteriorhodopsin. *In* "Information and Energy Transduction in Biological Membranes." (A. Bolis, H. Helmreich, and H. Passow, eds.), pp. 113–131. A. R. Liss, New York.

Schulten, Z., and Schulten, K. (1986). Proton conduction through proteins: An overview of theoretical principles and applications. *Methods Enzymol.* **127,** 419–438.

Sherman, W. V., Eicke, R. R., Stafford, S. R., and Wasacz, F. M. (1979). Branching in the bacteriorhodopsin photochemical cycle. *Photochem. Photobiol.* **30,** 727–729.

Sheves, M., Albeck, A., Friedman, N., and Ottolenghi, M. (1986). Controlling the pK_a of the bacteriorhodopsin Schiff base by use of artificial retinal analogues. *Proc. Natl. Acad. Sci. USA* **83,** 3262–3266.

Shichida, Y., Matuoka, S., Hidaka, Y., and Yoshizawa, T. (1983). Absorption spectra of intermediates of bacteriorhodopsin measured by laser photolysis at room temperatures. *Biochim. Biophys. Acta* **723,** 240–246.

Smith, S. O., Pardoen, J. A., Mulder, P. P. J., Curry, B., Lugtenburg, J., and Mathies, R. A. (1983). Chromophore structure in bacteriorhodopsin's O_{640} photointermediate. *Biochemistry* **22,** 6141–6148.

Smith, S. O., Myers, A. B., Pardoen, J. A., Winkel, C., Mulder, P. P. J., Lugtenburg, J., and Mathies, R. A. (1984). Determination of retinal Schiff base configuration in bacteriorhodopsin. *Proc. Natl. Acad. Sci. USA* **81,** 2055–2059.

Smith, S. O., Braiman, M. S., Myers, A. B., Pardoen, J. A., Courtin, J., Winkel, C., Lugtenburg, J., and Mathies, R. A. (1987). Vibrational analysis of the all-*trans*-retinal chromophore in light-adapted bacteriorhodopsin. *J. Am. Chem. Soc.* **109,** 3108–3125.

Souvignier, G., and Gerwert, K. (1992). Proton uptake mechanism of bacteriorhodopsin as determined by time-resolved stroboscopic FTIR spectroscopy. *Biophys. J.* **63,** 1393–1405.

Sperling, W., Carl, P., Rafferty, C. N., and Dencher, N. A. (1977). Photochemistry and dark equilibrium of retinal isomers and bacteriorhodopsin isomers. *Biophys. Struct. Mech.* **3,** 79–94.

Sperling, W., Rafferty, C. N., Kohl, K. D., and Dencher, N. A. (1979). Isomeric composition of bacteriorhodopsin under different environmental conditions. *FEBS Lett.* **97,** 129–132.

Spudich, J. L., and Bogomolni, R. A. (1988). Sensory rhodopsins of halobacteria. *Annu. Rev. Biophys. Biophys. Chem.* **17,** 193–215.

Stern, L. J., Ahl, P. L., Marti, T., Mogi, T., Dun;tı;ach, M., Berkovitz, S., Rothschild, K. J., and Khorana, H. G. (1989). Substitution of membrane-embedded aspartic acids in bacteriorhodopsin causes specific changes in different steps of the photochemical cycle. *Biochemistry* **28,** 10035–10042.

Subramaniam, S., Marti, T., and Khorana, H. G. (1990). Protonation state of Asp (Glu)-85 regulates the purple-to-blue transition in bacteriorhodopsin mutants Arg-82 → Ala and Asp-85 → Glu: The blue form is inactive in proton translocation. *Proc. Natl. Acad. Sci. USA* **87,** 1013–1017.

Subramaniam, S., Greenhalgh, D. A., Rath, P., Rothschild, K. J., and Khorana, H. G. (1991). Replacement of leucine-93 by alanine or threonine slows down the decay of the N and O intermediates in the photocycle of bacteriorhodopsin: Implications for proton uptake and 13-cis-retinal → all-*trans*-retinal reisomerization. *Proc. Natl. Acad. Sci. USA* **88,** 6873–6877.

Subramaniam, S., Greenhalgh, D. A., and Khorana, H. G. (1992). Aspartic acid 85 in bacteriorhodopsin functions both as proton acceptor and negative counterion to the Schiff base. *J. Biol. Chem.* **267,** 25730–25733.

Subramaniam, S., Gerstein, M., Oesterhelt, D., and Henderson, R. (1993). Electron diffraction analysis of structural changes in the photocycle of bacteriorhodopsin. *EMBO J.* **12,** 1–8.

Subramaniam, S., Faruqi, A. R., Oesterhelt, D., and Henderson, R. (1997). Electron diffraction studies of light-induced conformational changes in the Leu-93 → Ala bacteriorhodopsin mutant. *Proc. Natl. Acad. Sci. USA* **94,** 1767–1772.

Száraz, S., Oesterhelt, D., and Ormos, P. (1994). pH-induced structural changes in bacteriorhodopsin studied by Fourier transform infrared spectroscopy. *Biophys. J.* **67,** 1706–1712.

Takeuchi, Y., Ohno, K., Yoshida, M., and Nagano, K. (1981). Light-induced proton dissociation and association in bacteriorhodopsin. *Photochem. Photobiol.* **33,** 587–592.

Tanford, C. (1982). Simple model for the chemical potential change of a transported ion in active transport. *Proc. Natl. Acad. Sci. USA* **79,** 2882–2884.

Tavan, P., Schulten, K., and Oesterhelt, D. (1985). The effect of protonation and electrical interactions on the stereochemistry of retinal Schiff bases. *Biophys. J.* **47,** 415–430.

Terner, J., Hsieh, C. L., Burns, A. R., and El-Sayed, M. A. (1979a). Time-resolved resonance Raman characterization of the bO_{640} intermediate of bacteriorhodopsin. Reprotonation of the Schiff base. *Biochemistry* **18,** 3629–3634.

Terner, J., Hsieh, C. L., and El-Sayed, M. A. (1979b). Time-resolved resonance Raman characterization of the bL_{550} intermediate and the two dark-adapted bRDA/560 forms of bacteriorhodopsin. *Biophys. J.* **26,** 527–541.

Thorgeirsson, T. E., Milder, S. J., Miercke, L. J. W., Betlach, M. C., Shand, R. F., Stroud, R. M., and Kliger, D. S. (1991). Effects of Asp-96 → Asn, Asp-85 → Asn, and Arg-82 → Gln single-site substitutions on the photocycle of bacteriorhodopsin. *Biochemistry* **30,** 9133–9142.

Thorgeirsson, T. E., Xiao, W., Brown, L. S., Needleman, R., Lanyi, J. K., and Shin, Y.-K. (1997). Opening of the cytoplasmic proton channel in bacteriorhodopsin. *J. Mol. Biol.* **273,** 951–957.

Tittor, J., Soell, C., Oesterhelt, D., Butt, H.-J., and Bamberg, E. (1989). A defective proton pump, point-mutated bacteriorhodopsin Asp96 → Asn is fully reactivated by azide. *EMBO J.* **8,** 3477–3482.

Tittor, J., Schweiger, U., Oesterhelt, D., and Bamberg, E. (1994a). Inversion of proton translocation in bacteriorhodopsin mutants D85N, D85T and D85, D96N. *Biophys. J.* **67,** 1682–1690.

Tittor, J., Wahl, M., Schweiger, U., and Oesterhelt, D. (1994b). Specific acceleration of de- and reprotonation steps by azide in mutated bacteriorhodopsins. *Biochim. Biophys. Acta Bio-Energetics* **1187,** 191–197.

Tittor, J., Haupts, U., Haupts, C., Oesterhelt, D., Becker, A., and Bamberg, E. (1997). Chloride and proton transport in bacteriorhodopsin mutant D85T: Different modes of ion translocation in a retinal protein. *J. Mol. Biol.* **271,** 405–416.

Turner, G. J., Miercke, L. J. W., Thorgeirsson, T. E., Kliger, D. S., Betlach, M. C., and Stroud, R. M. (1993). Bacteriorhodopsin D85N: Three spectroscopic species in equilibrium. *Biochemistry* **32,** 1332–1337.

Váró, G., and Lanyi, J. K. (1989). Photoreactions of bacteriorhodopsin at acid pH. *Biophys. J.* **56,** 1143–1151.

Váró, G., and Lanyi, J. K. (1990a). Pathways of the rise and decay of the M photointermediate of bacteriorhodopsin. *Biochemistry* **29,** 2241–2250.

Váró, G., and Lanyi, J. K. (1990b). Protonation and deprotonation of the M, N, and O intermediates during the bacteriorhodopsin photocycle. *Biochemistry* **29,** 6858–6865.

Váró, G., and Lanyi, J. K. (1991a). Distortions in the photocycle of bacteriorhodopsin at moderate dehydration. *Biophys. J.* **59,** 313–322.

Váró, G., and Lanyi, J. K. (1991b). Kinetic and spectroscopic evidence for an irreversible step between deprotonation and reprotonation of the Schiff base in the bacteriorhodopsin photocycle. *Biochemistry* **30,** 5008–5015.

Váró, G., and Lanyi, J. K. (1991c). Thermodynamics and energy coupling in the bacteriorhodopsin photocycle. *Biochemistry* **30,** 5016–5022.

Váró, G., and Lanyi, J. K. (1991d). Effects of the crystalline structure of purple membrane on the kinetics and energetics of the bacteriorhodopsin photocycle. *Biochemistry* **30,** 7165–7171.

Váró, G., and Lanyi, J. K. (1995). Effects of hydrostatic pressure on the kinetics reveal a volume increase during the bacteriorhodopsin photocycle. *Biochemistry* **34,** 12161–12169.

Váró, G., Duschl, A., and Lanyi, J. K. (1990). Interconversions of the M, N, and O intermediates of the bacteriorhodopsin photocycle. *Biochemistry* **29,** 3798–3804.

Váró, G., Zimányi, L., Chang, M., Ni, B., Needleman, R., and Lanyi, J. K. (1992). A residue substitution near the b-ionone ring of the retinal affects the M substates of bacteriorhodopsin. *Biophys. J.* **61,** 820–826.

Váró, G., Needleman, R., and Lanyi, J. K. (1996). Protein structural change at the cytoplasmic surface as the cause of cooperativity in the bacteriorhodopsin photocycle. *Biophys. J.* **70,** 461–467.

Vonck, J. (1996). A three-dimensional difference map of the N intermediate in the bacteriorhodopsin photocycle: Part of the F helix tilts in the M to N transition. *Biochemistry* **35,** 5870–5878.

Warshel, A. (1978). Charge stabilization mechanism in the visual and purple membrane pigments. *Proc. Natl. Acad. Sci. USA* **75,** 2558–2562.

Warshel, A., and Ottolenghi, M. (1979). Kinetic and spectroscopic effects of protein–chromophore electrostatic interactions in bacteriorhodopsin. *Photochem. Photobiol.* **30,** 291–293.

Xie, A. H., Nagle, J. F., and Lozier, R. H. (1987). Flash spectroscopy of purple membrane. *Biophys. J.* **51,** 627–635.

Yamazaki, Y., Hatanaka, M., Kandori, H., Sasaki, J., Karstens, W. F. J., Raap, J., Lugtenburg, J., Bizounok, M., Herzfeld, J., Needleman, R., Lanyi, J. K., and Maeda, A. (1995a). Water structural changes at the proton uptake site (the Thr46-Asp96 domain) in the L intermediate of bacteriorhodopsin. *Biochemistry* **34,** 7088–7093.

Yamazaki, Y., Sasaki, J., Hatanaka, M., Maeda, A., Kandori, H., Needleman, R., Shinada, T., Yoshihara, K., Brown, L. S., and Lanyi, J. K. (1995b). Interaction of tryptophan 182 with the 9-methyl group of the retinal in the L intermediate of bacteriorhodopsin. *Biochemistry* **34,** 577–582.

Yamazaki, Y., Tuzi, S., Saitô, H., Kandori, H., Needleman, R., Lanyi, J. K., and Maeda, A. (1996). Hydrogen bonds of water and C—O groups coordinate long-range structural in the L photointermediate of bacteriorhodopsin. *Biochemistry* **35,** 4063–4068.

Zhou, F., Windemuth, A., and Schulten, K. (1993). Molecular dynamics study of the proton pump cycle of bacteriorhodopsin. *Biochemistry* **32,** 2291–2306.

Zimányi, L., and Lanyi, J. K. (1993). Deriving the intermediate spectra and photocycle kinetics from time-resolved difference spectra of bacteriorhodopsin. The simpler case of the recombinant D96N protein. *Biophys. J.* **64,** 240–251.

Zimányi, L., Keszthelyi, L., and Lanyi, J. K. (1989). Transient spectroscopy of bacterial rhodopsins with optical multichannel analyser. 1. Comparison of the photocycles of bacteriorhodopsin and halorhodopsin. *Biochemistry* **28,** 5165–5172.

Zimányi, L., Cao, Y., Chang, M., Ni, B., Needleman, R., and Lanyi, J. K. (1992a). The two consecutive M substates in the photocycle of bacteriorhodopsin are affected specifically by the D85N and D96N residue replacements. *Photochem. Photobiol.* **56,** 1049–1055.

Zimányi, L., Váró, G., Chang, M., Ni, B., Needleman, R., and Lanyi, J. K. (1992b). Pathways of proton release in the bacteriorhodopsin photocycle. *Biochemistry* **31,** 8535–8543.

Zimányi, L., Cao, Y., Needleman, R., Ottolenghi, M., and Lanyi, J. K. (1993). Pathway of proton uptake in the bacteriorhodopsin photocycle. *Biochemistry* **32,** 7669–7678.

Zscherp, C., and Heberle, J. (1997). Infrared difference spectra of the intermediates L, M, N, and O of the bacteriorhodopsin photoreaction obtained by time-resolved attenuated total reflection spectroscopy. *J. Phys. Chem. B.* **101,** 10542–10547.

The Regulation of Apoptosis by Microbial Pathogens

Jeremy E. Moss, Antonios O. Aliprantis, and Arturo Zychlinsky
The Skirball Institute of Biomolecular Medicine, New York University Medical Center, New York City, New York 10016

In the past few years, there has been remarkable progress unraveling the mechanism and significance of eukaryotic programmed cell death (PCD), or apoptosis. Not surprisingly, it has been discovered that numerous, unrelated microbial pathogens engage or circumvent the host's apoptotic program. In this chapter, we briefly summarize apoptosis, emphasizing those studies which assist the reader in understanding the subsequent discussion on PCD and pathogens. We then examine the relationship between virulent bacteria and apoptosis. This section is organized to reflect both common and diverse mechanisms employed by bacteria to induce PCD. A short discussion of parasites and fungi is followed by a detailed description of the interaction of viral pathogens with the apoptotic machinery. Throughout the review, apoptosis is considered within the broader contexts of pathogenesis, virulence, and host defense. Our goals are to update the reader on this rapidly expanding field and identify topics in the current literature which demand further investigation.

KEY WORDS: Apoptosis, Bacteria, Infections, Virus, Necrosis.

I. Introduction

Cells die by either of two mechanisms: apoptosis or necrosis. Apoptotic cells can be identified by morphologic and biochemical changes, including chromatin condensation, nuclear segmentation, cytoplasmic vacuolization, cell shrinkage, membrane blebbing, and DNA fragmentation (Arends and Wyllie, 1991; Ellis and Horvitz, 1986). Necrotic cells demonstrate organelle destruction, chromatin flocculation, swelling, and release of intracellular contents. Whereas apoptosis results when the cell's intrinsic suicide program is activated, necrosis is usually the outcome of overwhelming mechanical or chemical injury. Apoptosis, also known as programmed cell death (PCD), has been reviewed extensively in the past few years. In this introduction we refer only to general reviews on each topic.

Apoptosis is essential to the development and homeostasis of multicellular organisms. Studies in the nematode *Caenorhabditis elegans* demonstrate that a genetic program for cell death exists. In this worm, 1090 somatic cells are generated during development. Of these, 131 must die by apoptosis to form a mature animal. Classical studies, using *C. elegans* genetics, identified a number of genes required for the regulation of PCD. Interestingly, some of these genes have mammalian homologs also involved in apoptosis. Two *C. elegans* cell death (*ced*) genes are crucial for the induction of apoptosis, *ced-3* and *ced-4.* A third gene, *ced-9,* is an inhibitor of apoptosis (Ellis and Horvitz, 1986; Horvitz *et al.,* 1994).

Ced-3 is the prototype of a family of cytoplasmic cysteine proteases that all cleave after aspartate residues (caspases) but with slightly different sequence specificity. At least 10 caspase homologs have been identified in humans. When overexpressed in tissue culture cells, all those tested induce apoptosis. Caspases are synthesized as zymogens which are activated through limited proteolysis by other caspases or autocatalytically. Although numerous caspases are activated during an apoptotic event, it is still unclear whether they act within hierarchical, or parallel, proteolytic cascades. In addition to other caspases, several caspase substrates have recently been identified. It is apparent that some of these molecules are important in executing the apoptotic changes outlined above, although the requirement of individual substrates for PCD remains to be established (Nicholson and Thornberry, 1997; Yuan, 1997).

Ced-4 is essential for the induction of apoptosis in *C. elegans* and acts upstream of Ced-3. Ced-4 binds both caspases and members of the Bcl-2 family of proteins. A mammalian gene with homology to *ced-4* has been recently described (Vaux, 1997) and its role in mammalian cell death is emerging.

Ced-9 is the worm prototype of a family of proteins involved in the induction and inhibition of cell death. Proapoptotic family members include Bad, Bax, Bik, and Bak. Bcl-2 and Bcl-xL, on the other hand, inhibit apoptosis. Members of this family form homodimers or heterodimers. Changes in the relative concentrations of pro- and anti-apoptotic proteins either promote or inhibit PCD. The direct function of these proteins is still unclear. Their localization to the mitochondria and structural similarity to bacterial pore-forming proteins indicates that they might act as channels which regulate cytochrome c release into the cytosol. Cytochrome c has recently been identified as a key endogenous mediator of caspase activation. A more direct role in caspase regulation for Bcl-2, through a Ced-4-like molecule, has also been suggested (Hawkins and Vaux, 1997; Merry and Korsmeyer, 1997; Rudin and Thompson, 1997; Reed, 1997).

Microbial pathogens regulate apoptosis within their hosts. In many cases, this regulation is clearly important for pathogenesis. In other instances, the role of programmed cell death is less well-defined. Here, we review the mechanisms by which bacterial, parasitic, fungal, and viral pathogens induce or inhibit apoptosis.

II. Bacteria

In this section, we review the numerous mechanisms that extracellular and intracellular pathogenic bacteria employ to induce apoptosis in host cells. The relationship between apoptotic events and pathogenesis is also explored.

A. Pore-Forming Toxins

1. *Staphylococcus aureus*

Staphylococcus aureus is a gram-positive coccus which causes a variety of diseases, including skin lesions, food poisoning, toxic shock syndrome, endocarditis, and osteomyelitis. A number of virulence factors, including several exotoxins, have been implicated in the pathogenesis of infections with this bacterium. One of these, alpha toxin, induces apoptosis in eukaryotic cells.

Alpha toxin is a pore-forming molecule which, at low doses, causes apoptosis of T lymphocytes (Jonas *et al.,* 1994). Alpha toxin forms pores in many eukaryotic cell membranes (Ikigae and Nakae, 1987; Menestrina, 1986; Suttorp *et al.,* 1987; Bhakdi *et al.,* 1988; Ahnert-Hilger *et al.,* 1985). When present in low concentrations, the toxin binds to specific, but unidentified, receptors on the cell surface (Hildebrand *et al.,* 1991) and produces small holes which selectively facilitate the release of monovalent ions (Jonas *et al.,* 1994). At higher doses, the toxin attaches nonspecifically to membranes (Hildebrand *et al.,* 1991) and opens large pores, which result in massive cell injury and a quick, necrotic death (Jonas *et al.,* 1994). Hameed *et al.* (1989) were the first to demonstrate that alpha toxin, and other pore-forming agents, can induce DNA fragmentation in cell lines. Though their actual data may reflect the presence of a contaminating nuclease (Duke *et al.,* 1994; Jonas *et al.,* 1994), it has since been established that alpha toxin can indeed induce apoptosis (Jonas *et al.,* 1994).

The mechanism of alpha toxin-induced apoptosis probably involves entry of a toxic molecule, or exit of a necessary substance, through pores created by the toxin. A prime candidate, a priori, is calcium ions, whose entry into cells has been correlated with apoptosis (Wertz and Hanley, 1996). Alpha toxin-induced apoptosis, however, is dependent on the movement of sodium ions across the cell membrane, not calcium fluxes (Jonas *et al.*, 1994). Cells cultured in sodium-free media do not fragment their DNA when exposed to alpha toxin. A large influx of sodium may directly induce apoptosis. Alternatively, sodium movement may disrupt the ionic balance of cells, leading to the depletion or buildup of other ions (including, perhaps, calcium) which trigger apoptosis.

2. *Actinobacillus actinomycetemcomitans*

Actinobacillus actinomycetemcomitans is a gram-negative coccobacillus associated with periodontitis as well as meningitis and endocarditis. A virulence factor of this bacterium, leukotoxin, generates membrane pores and kills several lymphoid cell types (Taichman *et al.*, 1980; Tsai *et al.*, 1979) but spares fibroblasts, platelets, and endothelial and epithelial cells (Tsai *et al.*, 1979). Leukotoxin induces apoptosis, as well as necrosis, in lymphocytes and natural killer cells (Mangan *et al.*, 1991b; Shenker *et al.*, 1994). The mechanism of death depends on the concentration of leukotoxin employed. As with staphylococcal alpha toxin, beyond a threshold concentration, leukotoxin inflicts overwhelming damage and necrosis predominates (Mangan *et al.*, 1991b). Below this concentration, however, more subtle cellular changes occur which activate the PCD pathway.

Actinobacillus can also induce apoptosis independently of leukotoxin (Kato *et al.*, 1995). Strains lacking leukotoxin activate a PCD pathway in J774 cells, a murine macrophage cell line. This effect depends on the bacterium entering the macrophage because cytochalasin D prevents apoptosis. Cytochalasin D is an inhibitor of actin polymerization and therefore phagocytosis. Apoptosis depends on a protein kinase C pathway, whereas cAMP-dependent protein kinases are not involved (Kato *et al.*, 1995). CD14 molecules on the surface of the macrophage are important for actinobacillus internalization and the induction of PCD (Muro *et al.*, 1997). CD14 is a receptor for lipopolysaccharide, an outer membrane glycolipid of gram-negative bacteria. The relationship between the chronic inflammation of periodontitis and the ability of actinobacillus to induce cell death remains to be explored.

3. *Escherichia coli*

Escherichia coli hemolysin (HlyA) is another bacterial pore-forming molecule which has been linked to apoptosis. *Escherichia coli* pathogenicity

(Cavalieri *et al.,* 1984) correlates with production of this virulence factor. Unlike staphylococcal alpha toxin, HlyA is not thought to bind to a specific cellular receptor. The toxin forms pores by directly inserting into membranes through a calcium-dependent mechanism (Welch, 1991).

Jonas *et al.,* (1993) showed that HlyA induces DNA fragmentation in phytohemagluttinin-stimulated T lymphocytes. Interestingly, unactivated lymphocytes do not fragment their DNA when exposed to HlyA. This suggests that during stimulation T cells upregulate apoptosis competence factors and/or downregulate inhibitors. The authors question whether the fragmentation they observed is an epiphenomenon because they did not observe the decline in cellular ATP expected to accompany PCD. It remains to be determined whether HlyA induces DNA fragmentation outside the context of apoptosis.

4. *Listeria monocytogenes*

Listeria monocytogenes is an etiological agent of meningitis and sepsis in infants and immunosuppressed individuals. This gram-positive rod is acquired by ingesting contaminated food. Listeria invades the gastrointestinal tract and spreads systemically. When the bacterium contacts target cells, it is internalized into a phagolysosome. Listeria escapes from this compartment into the cytosol, where the bacterium moves by vectorially polymerizing actin filaments and subsequently spreads directly from cell to cell (Portnoy *et al.,* 1992; Tilney and Portnoy, 1989).

Listeria induces apoptosis in dendritic cells (Guzman *et al.,* 1996) and hepatocytes (Rogers *et al.,* 1996) but not macrophages (Zychlinsky *et al.,* 1992; Barsig and Kaufmann, 1997). Both a dendritic cell line and primary explanted bone marrow dendritic cells undergo PCD after infection with listeria (Guzman *et al.,* 1996). The authors conclude that the *L. monocytogenes* virulence factor listeriolysin mediates dendritic cell cytotoxicity because (i) mutations in the structural gene (*hly2*) or a regulator of *hly2* (*prfA*) abrogate the killing activity and (ii) purified listeriolysin was sufficient to trigger PCD. Mutations in genes involved in actin polymerization (*actA*) and cell to cell spread (*plcB*) do not affect cytotoxicity. Apoptosis is also independent of bacterial internalization since cytochalasin D does not inhibit the bacteria from killing cells. Listeriolysin is a pore-forming protein which, like the others discussed previously, induces apoptosis at sublytic concentrations and necrosis at higher doses. Mutations which decrease the hemolytic activity of listeriolysin also attenuate its ability to induce apoptosis in dendritic cells (Guzman *et al.,* 1996). Dendritic cells are the key antigen presenting cells in lymphoid tissue throughout the body. Guzman *et al.,* suggest that by killing dendritic cells, listeria could impair the acquired immune response and create a more favorable environment within the host.

The liver receives the venous drainage of the GI tract and is the first systemic organ colonized by listeria after infection. In addition to dendritic cells, listeria kills infected hepatocytes by apoptosis. This effect is observed *in vitro* (primary cells and cell lines) and *in vivo* (Rogers *et al.,* 1996). Tumor necrosis factor-α (TNF-α), Interferon-γ (IFN-γ), nitric oxide, and polymorphonuclear leukocytes (PMNs) are not important for the induction of PCD *in vivo,* suggesting that the apoptotic stimulus is not merely inflammation. In an experimental model of listeriosis, hepatocyte apoptosis was shown to be crucial for bacterial clearance (Rogers *et al.,* 1996). It is well established that neutrophils kill listeria and are the principal mediators of innate resistance to this organism. In fact, mice depleted of PMN prior to infection with listeria demonstrate increased liver damage (Conlan and North, 1991, 1994; Czuprynski *et al.,* 1994; Rogers and Unanue, 1993). Interestingly, listeria-infected hepatocytes release a powerful PMN chemoattractant and efficiently bind these cells (Rogers *et al.,* 1996). These data indicate that upon infection with listeria, hepatocyte apoptosis serves as a mechanism to recruit neutrophils which clear the invading agent.

B. Protein Synthesis Inhibitors

1. *Corynebacterium diphtheria*

Corynebacterium diphtheria is a gram-positive, non-spore-forming rod that causes diphtheria in humans. This disease is characterized by sore throat, fever, and the formation of a gray, fibrinous pseudomembrane on the posterior pharynx composed of bacteria and inflammatory cells. Airway obstruction, recurrent laryngeal nerve palsy, and cardiac manifestations, such as arrhythmia and circulatory failure, are complications of diphtheria.

Corynebacterium diphtheria is an extracellular pathogen which secretes a classic bacterial AB toxin, called diphtheria toxin (DTX). DTX is encoded by a temperate bacteriophage and is synthesized in response to low iron. The B subunit of AB toxins mediates attachment to cells and facilitates delivery of the catalytic A subunit into the cytosol. The DTX B subunit binds to an extracellular glycoprotein and the complex is endocytosed. In response to acidification, the T domain of the toxin is exposed and an aqueous pore is generated through which the A domain exits the endosome (Oh *et al.,* 1996). The A domain then ADP ribosylates elongation factor-2 (EF-2), a necessary component of the protein synthesis machinery (Choe *et al.,* 1992). ADP ribosylation inactivates EF-2 and inhibits protein synthesis.

DTX causes apoptosis in several epithelial and myeloid cell lines (Chang *et al.,* 1989a; Chang and Wisnieski, 1990; Sandvig and van Deurs, 1992; Morimoto and Bonavida, 1992; Kochi and Collier, 1993). PCD requires acidification of the toxin in the endosome and host energy expenditure (Chang *et al.,* 1989a; Chang and Wisnieski, 1990). The following evidence supports protein synthesis inhibition as the mechanism of DTX-induced cell death: (i) Levels of protein synthesis inhibition by DTX correlate with cytotoxicity; (ii) analysis of DTX mutants indicates that its ADP ribosylating activity, and hence ability to inhibit protein synthesis, is necessary for cytotoxicity (Morimoto and Bonavida, 1992); (iii) cell lines with mutant EF-2s, which are not susceptible to ADP ribosylation, cannot be killed by DTX (Kohno and Uchida, 1987; Phan *et al.,* 1993); and (iv) compounds which block ADP ribosylation of EF-2 by DTX inhibit apoptosis (Morimoto and Bonavida, 1992). Although necessary, inhibition of protein synthesis is not sufficient for DTX to kill cells. Cell lines exist in which DTX effectively blocks protein synthesis without inducing apoptosis (Chang *et al.,* 1989a; Morimoto and Bonavida, 1992; Kochi and Collier, 1993). Perhaps these cells contain an inhibitor of apoptosis or lack a factor needed to transmit a proapoptotic signal after attenuation of protein synthesis.

If protein synthesis inhibition is insufficient to cause apoptosis, perhaps DTX possesses other undefined functions. Interestingly, the A subunit of DTX was reported to have nuclease activity (Chang *et al.,* 1989b; Lessnick *et al.,* 1992). This proposed nuclease cleaves at internucleosomal junctions and creates a prototypical apoptotic DNA ladder. This is an intriguing but disputed finding (Bodley, 1990; Johnson, 1990; Wilson *et al.,* 1990). Several laboratories claim that the DTX nuclease activity represents a DNAse contaminant, but all of the criticisms of the original work have subsequently been addressed (Lessnick *et al.,* 1990, 1992; Morimoto and Bonavida, 1992). Perhaps, in addition to shutting off protein synthesis, DTX damages DNA. Without the ability to synthesize new proteins, cells might not be able to amend the DNA damage inflicted by the proposed DTX nuclease. DNA damage is a well-documented apoptotic stimulus. Additional experiments are necessary to document the nuclease activity and its relationship, if any, to cell death.

Recently, a gene has been cloned that, when expressed in an antisense fashion, confers resistance to cells normally sensitive to DTX's lethal effect. This gene, called cellular apoptosis susceptibility (*CAS*), also protects from pseudomonas exotoxin A and TNF-α induced cell death (Brinkmann *et al.,* 1995a). Although expression of antisense *CAS* abrogates DTX-induced apoptosis, it does not prevent protein synthesis inhibition. This finding supports the hypothesis that protein synthesis inhibition is insufficient for DTX-induced apoptosis (Brinkmann *et al.,* 1996). The CAS protein is homologous to the yeast protein CSE1, which is important for cell division and β cyclin degradation (Brinkmann *et al.,* 1995b; Scherf *et al.,* 1996). CAS is highly expressed in tissues with dividing cells and in cells exposed to proliferative stimuli (Brinkmann *et al.,* 1995b). Future work on CAS should help define its role in the apoptotic pathway and diphtheria infections.

The function of DTX in pathogenesis is unresolved. DTX may allow the bacterium to escape host defenses by killing immune cells. Additionally, apoptosis induction may be a mechanism for bacterium to acquire necessary environmental nutrients. In fact, low iron concentrations stimulate DTX expression. To address these possibilities, a good *in vivo* model of *C. diphtheria*-induced apoptosis is required.

2. *Pseudomonas aeruginosa*

Pseudomonas aeruginosa, a gram-negative rod, is found in soil and water and occasionally in the human gut. It causes sepsis, urinary tract infections, and pneumonia, particularly in immunocompromised patients. Pseudomonas is a troublesome nosocomial pathogen because it can grow in water and is resistant to many antibiotics and disinfectants.

Exotoxin A (ExoA) is a key virulence factor of psuedomonas. Although they are not homologous, ExoA, like DTX, ADP ribosylates EF-2 and inhibits protein synthesis. ExoA induces apoptosis in a human monoblastoid cell line (Morimoto and Bonavida, 1992; Kochi and Collier, 1993). Cells expressing mutant EF-2 molecules are resistant to ExoA, indicating that ADP ribosylation is critical for its cytotoxic activity (Kohno and Uchida, 1987). Additionally, ExoA reportedly contains the same nuclease activity as DTX (Chang *et al.,* 1989b). Notably, ExoA mutants are avirulent, suggesting that the ability to induce apoptosis is a critical feature of pseudomonas pathogenesis (Rahme *et al.,* 1995).

3. *Shigella dysenteriae* and Enterohemorrhagic *E. coli*

Shigella dysenteriae and enterohemorrhagic *E. coli* (EHEC) cause dysenteric syndromes with sequelae, including kidney and central nervous system manifestations. These bacteria produce similar toxins called shiga and shiga-like toxin (SLT), respectively. Shiga toxin is an AB toxin composed of one A and five B subunits. The A subunit cleaves eukaryotic rRNA, thereby disrupting ribosome function and protein synthesis (Tesh and O'Brien, 1991). The B subunits mediate binding of the toxin to globotriaosylceramide (Gb3) receptors on the cell surface (Tesh and O'Brien, 1991; Mangeney *et al.,* 1993). SLT, also known as Vero toxin, functions similarly.

Shiga and SLT induce apoptosis in several systems. Employing a rabbit ileal loop model of intestinal inflammation, Keenan *et al.,* (1986) demonstrated that both toxins cause apoptosis of mature, differentiated, absorptive villus epithelial cells but spare crypt cells. *In vitro,* shiga (Sandvig and van Deurs, 1992) and SLT (Inward *et al.,* 1995) kill vero cells by apoptosis.

Binding of the B subunit to its receptor, Gb3, is sufficient to trigger apoptosis (Mangeney *et al.*, 1993). The holotoxin, however, is more potent, suggesting that protein synthesis inhibition and Gb3 binding synergize to kill cells. A possible explanation for the renal sequelae caused by *S. dysenteriae* and EHEC is that kidney cells are rich in Gb3 receptors (Tesh and O'Brien, 1991) and are likely to be highly susceptible to shiga and SLT.

C. Superantigens

Superantigens activate T cells without undergoing antigen processing by simultaneously binding the T cell receptor (TCR)/CD3 complex and MHC class II molecules on antigen presenting cells (Fleischer, 1994). Individual superantigens engage different variable regions on the β chain of the TCR (Vβ), and therefore stimulate only a subset of T cells. In addition to activating T cells to produce cytokines, superantigens induce apoptosis.

1. *Staphylococcus aureus*

Staphylococcus aureus expresses several exotoxins that act as superantigens and kill T lymphocytes by apoptosis. Staphylococcus exotoxin B (SEB) causes V$\beta 8^+$ thymocytes and mature T cells to undergo PCD *in vivo* (Lin *et al.*, 1992; Kawabe and Ochi, 1991; D'Adamio *et al.*, 1993) and *in vitro* (Jenkinson *et al.*, 1989; D'Adamio *et al.*, 1993). SEA, SED, and SEE also trigger clonal T cell apoptosis (Kabelitz and Wesselborg, 1992). Of these, only SEA has been documented to induce cell death *in vivo* (Kuroda *et al.*, 1996).

The following is a brief outline of recent findings regarding the pathway of staphylococcal superantigen-induced apoptosis. SEB causes activation-induced cell death of peripheral T cells, whereas an antibody directed against Vβ8 results in passive, macrophage-dependent deletion (Gonzalo *et al.*, 1994). Proliferation (Gonzalo *et al.*, 1994; Renno *et al.*, 1995) and activation (Aroeira *et al.*, 1996) are necessary for SEB-induced T-cell apoptosis. SEB alters protein kinase C levels and activity in apoptotic thymocytes (Lin *et al.*, 1995). P2X1, and ATP-gated ion channel, mediates thymocyte apoptosis in response to SEB (Chvatchko *et al.*, 1996). IL-2 can block DNA fragmentation in SEA-reactive T cells *in vivo,* suggesting that IL-2 deprivation plays a significant role in the pathway (Kuroda *et al.*, 1996). Finally, RU-38486, a glucocorticoid antagonist, blocks SEB-induced V$\beta 8^+$ T-cell apoptosis (Gonzalo *et al.*, 1994). It is evident that, although the molecular mechanism of staphylococcal superantigen-mediated PCD has been well studied, a clear model has not emerged.

Recently, numerous studies have been published which either support or refute a role for Fas and Fas ligand (FasL) in superantigen-induced apoptosis. Fas is a TNF family receptor that triggers PCD when activated by FasL. T cells from lipopolysaccharide (LPS)-treated mice are resistant to both superantigen and Fas-induced PCD, implying these stimuli utilize a common apoptotic pathway (Vella *et al.,* 1995). Renno *et al* (1995) noticed a rapid induction of both Fas and FasL on Vβ8$^+$ T cells from mice injected with SEB. *In vitro,* SEB-induced apoptosis correlates with a rapid upregulation of FasL on CD4$^+$ cells, which constituitively express high levels of Fas. Moreover, soluble Fas–IgG blocks PCD in this system (Ettinger *et al.,* 1995). From results of experiments employing Fas and FasL mutant mice, however, the importance of these molecules in superantigen-induced apoptosis is questionable. While one group claims that T cells from these mice are less vulnerable than wild type (Ettinger *et al.,* 1995), another reports equal susceptibility (Kuroda *et al.,* 1996). It is hoped that future work will resolve this discrepancy.

Staphylococcal toxic shock syndrome toxin 1 (TSST1) is another superantigen produced by *S. aureus.* Bacteria with this toxin caused an outbreak of toxic shock among tampon-using women. When added to a culture of peripheral blood mononuclear cells (PMBCs), TSST1 causes B cells, but not T cells, to initiate an apoptotic program (Hofer *et al.,* 1996). Surprisingly, even at high doses TSST1 does not kill isolated B cells, implying that another cell type is required. T cells are prime candidates. TSST1 causes an upregulation of Fas on B cells and both Fas expression and apoptosis are inhibited by anti-IFN-γ antibodies (Hofer *et al.,* 1996). Therefore, a probable mechanism for B-cell PCD involves (i) activation of T cells by TSST, (ii) production of IFN-γ which stimulates Fas expression on B cells, and (iii) ligation of the proapoptotic receptor Fas by FasL expressed on T cells. (Hofer *et al.,* 1996).

2. *Streptococcus pyogenes*

Streptococcus pyogenes is a common cause of pharyngitis. A toxic shock-like syndrome has also been described in patients infected with strains of *S. pyogenes* which produce superantigens (Cone *et al.,* 1987). Apoptotic PBMCs, as visualized by DNA laddering, are observed in patients with this severe infection. A depletion of specific Vβ cells is also noted (Watanabe-Ohnishi *et al.,* 1995). It remains to be determined whether the apoptotic PBMCs represent the population of depleted Vβ cells.

Superantigens may have evolved as a bacterial survival mechanism to destroy host immune cells. Superantigen-induced cell death, however, could also benefit the host. Unregulated T-cell proliferation and activation is not only a waste of energy but also possibly pathologic. In fact, Mountz *et al.* (1995) found that Fas mutant mice are more susceptible to SEB-induced lethal shock, implying that superantigen-induced apoptosis retards exuberant immune responses.

D. Other Mechanisms

1. *Bordetella pertussis*

Bordetella pertussis causes an acute upper respiratory tract infection characterized by a cough with an inspiratory "whoop." This encapsulated gram-negative rod remains localized to the respiratory tract throughout the course of infection which can last up to 4 weeks. It is transmitted in airborne droplets and is highly contagious. Bordetella produces several important virulence factors. Fimbriae (FIM), filamentous hemagluttinin (FHA), and pertactin (PRN) mediate adhesion of the bacteria to respiratory epithelial cells (Relman *et al.,* 1990; Leninninger *et al.,* 1991). In a mouse model, however, bordetella is virulent without these adhesins. Only pertussis toxin (PTX) and adenylate cyclase hemolysin (Ac-Hly) are necessary for lethality (Weiss and Goodwin, 1989; Khelef *et al.,* 1992).

PTX is an AB family toxin. The A subunit ADP ribosylates, and thus inhibits, the G protein subunit responsible for regulating adenylate cyclase. This results in a stimulation of enzyme activity and an increase in intracellular cAMP (Katada, *et al.,* 1986). Although this phenomenon has been shown *in vitro,* that PTX increases intracellular cAMP *in vivo* has never been demonstrated. Bordetella's other toxin, Ac-Hly, definitely increases intracellular cAMP *in vivo* (Confer and Eaton, 1982; Rogel and Hanski, 1992; Weiss and Goodwin, 1989). The hemolysin portion of this secreted molecule is believed to create membrane pores through which the toxin transits into cells. Once inside, the adenylate cyclase subunit is activated by calmodulin and efficiently synthesizes cAMP (Hanski and Coote, 1991).

Bordatella pertussis kills macrophage cell lines, as well as murine alveolar macrophages, by apoptosis (Khelef *et al.,* 1993; Khelef and Guiso, 1995). Ac-Hly is necessary to cause apoptosis, whereas PRN, FHA, FIM, and, surprisingly, PTX are not (Khelef *et al.,* 1993). Each of these virulence factors has been purified. Only Ac-Hly induces PCD in macrophages, indicating that this molecule is also sufficient to kill cells (Khelef and Guiso, 1995). Given that purified Ac-Hly is cytotoxic, it is not surprising that bordetella does not need to enter macrophages to induce apoptosis. Intracellular bacteria, however, kill cells faster than extracellular bacteria (Khelef *et al.,* 1993). This kinetic difference probably reflects the increased time the toxin requires to enter the cytoplasm when the bacteria are outside the cell. The Hly subunit of Ac-Hly does not cause macrophage apoptosis (Khelef and Guiso, 1995). Therefore, it is unlikely that the holotoxin kills cells solely through pore formation. The relative contribution of each subunit to the induction of cell death remains to be determined.

Though PTX and Ac-Hly share a common ability to stimulate adenylate cyclase activity *in vitro,* only Ac-Hly appears to increase intracellular cAMP *in vivo.* Elevated cAMP levels have been linked, in a variety of circumstances, to apoptosis (Wertz and Hanley, 1996). With respect to pathogenesis, bordetella may use Ac-Hly to kill off the first line of cellular defense that it encounters, the alveolar macrophage. In addition, macrophages transport pathogens throughout the body. Perhaps, by killing these cells, bordetella ensures its localization to the upper respiratory tract from where it can be efficiently transmitted (Khelef *et al.,* 1993).

2. Shigella

Gram-negative bacteria of the genus Shigella cause a severe diarrheal syndrome called dysentery. Dysentery is characterized by blood and mucous in the stool, fever, abdominal pain, and tenesmus. It is endemic in developing countries where it is an important cause of infant mortality. In industrialized nations, localized outbreaks of shigellosis occur.

Shigellosis is transmitted through the fecal/oral route. After ingestion, the bacteria pass through the gut to the colon, where they invade the mucosa and induce an erosive inflammation. Shigella is a facultative intracellular pathogen which invades cells by a process described as pathogen-induced phagocytotsis. Since only the basolateral cell membrane is permissive for invasion, bacteria do not enter epithelial cells directly from the lumen of the colon (Mounier *et al.,* 1992). Shigellae overcome their colonic confinement by penetrating M cells, modified epithelial cells which transcytose enteric antigens to submucosal lymphoid follicles (Wassef *et al.,* 1989). After traversing M cells, shigellae encounter resident macrophages and are phagocytosed. The bacteria escape the phagosome and kill macrophages by inducing PCD (Zychlinsky *et al.,* 1992, 1996).

Macrophage apoptosis initiates an inflammatory response in the colon resulting in the dysenteric symptoms. Mature IL-1β, a key proinflammatory cytokine, is released as shigella-infected macrophages commit suicide (Zychlinsky *et al.,* 1994a). IL-1 plays a key role in an experimental model of shigellosis (Sansonetti *et al.,* 1995). IL-1β recruits neutrophils to the source of invading bacteria. These leukocytes are efficient killers of shigella and are not susceptible to shigella-induced apoptosis (Mandic-Mulec *et al.,* 1997). As the neutrophils migrate into the lumen of the intestine though, they break apart paracellular junctions and thereby disrupt the colonic epithelial barrier. This allows massive numbers of bacteria, previously restricted to the lumen of the intestine, free access to the colonic mucosa and the basolateral membranes of epithelial cells. Subsequent colonization of the epithelium and the intense inflammatory infiltrate result in profound mucosal damage and thus dysentery.

Shigella causes DNA fragmentation and the characteristic ultrastructural changes of apoptosis in macrophages. To trigger programmed cell death, shigella needs to enter macrophages (Zychlinsky *et al.,* 1992). *In vitro,* shigella kills macrophages rapidly; nearly 100% of cells are killed within 3 h of infection. Inhibition of host cell transcription and translation do not affect shigella's ability to induce apoptosis (Zychlinsky *et al.,* 1992).

The mechanism of shigella-induced apoptosis has been partially dissected. Shigella invasion and cytotoxicity are mediated by the secreted invasion plasmid antigen (Ipa) invasins. The Ipa operon consists of several genes important in cell invasion and escape from the phagosome, including *ipaB, ipaC, ipaD,* and *ipaA.* Escape from the phagosome is a prerequisite for shigella-induced cell death. Zychlinsky *et al.* 1994b), however, were able to separate the ability to lyse the phagosome from the induction of cell death and prove that *ipaB* is necessary for shigella-induced apoptosis. Subsequent work proved that IpaB is also sufficient to trigger PCD because microinjection of purified IpaB into mouse peritoneal macrophages resulted in death (Y. Chen, *et al.,* 1996).

The induction of macrophage apoptosis by IpaB is dependent on its interaction with IL-1β converting enzyme (ICE). ICE, or caspase 1, cleaves pro-IL-1β to its biologically active form. It is also capable of inducing apoptosis in cells when overexpressed (Miura *et al.,* 1993). IpaB is found in the cytoplasm of shigella-infected cells (Y. Chen *et al.,* 1996) and colocalizes with ICE (Thirumalai *et al.,* 1997). ICE was identified as an IpaB-binding protein in macrophage cell lysates. Furthermore, the specific ICE inhibitor, YVAD-CHO, attenuates shigella's ability to kill macrophages (Y. Chen *et al.,* 1996). ICE links macrophage apoptosis to the characteristic inflammation of dysentery. As previously described, IL-1 is released from shigella-infected macrophages and plays a key role the pathogenesis of shigellosis (Sansonetti *et al.,* 1995).

Activation of ICE by shigella seems to have the dual role of initiating the apoptotic cascade and activating an immunoregulatory cytokine, pro-IL-1β. IL-1β's lack of a classical signal sequence, lack of glycosylation, and diffuse cytoplasmic localization suggest that it is not released from cells through the endoplasmic reticulum–Golgi secretion pathway (Stevenson *et al.,* 1992; Rubartelli *et al.,* 1990). Perhaps apoptosis, induced by ICE, is the mechanism by which macrophages release this cytokine in its mature, active form.

3. Salmonella

The genus Salmonella includes several species of gram-negative pathogens which invade the gastrointestinal tract. After invasion, salmonella either remain localized to the gut or spread throughout the body. Diseases produced by these organisms include typhoid fever (*S. typhi*), colitis (several species), and sepsis (particularly *S. cholerasuis*).

Salmonella colonizes the intestinal tissue by passing through M cells (Jones *et al.,* 1994). In the mucosa, the bacteria localize to lymphoid follicles and interact with several cell types, including macrophages and lymphocytes. Secreted salmonella proteins cause cells to rearrange their cell membrane-associated cytoskeletons (Francis *et al.,* 1992, 1993) and phagocytose the bacterium. *In vitro,* salmonella survives inside macrophages (Buchmeier and Heffron, 1989; Abshire and Neidhardt, 1993; Alpuche-Aranda *et al.,* 1994). *In vivo,* however, evidence for survival is controversial (L. Chen *et al.,* 1996; Hsu, 1989). If salmonella does survive inside macrophages, then these cells may act as vehicles to transport bacteria from the intestine to distant systemic sites, such as the spleen, liver, and gall bladder. Salmonella, because it is encapsulated, is also able to live free in the bloodstream.

Salmonella not only survive inside macrophages but also kill these cells by inducing apoptosis (Monack *et al.,* 1996; Lindgren *et al.,* 1996; Arai *et al.,* 1995; L. Chen *et al.,* 1996). *Salmonella typhimurium* and *S. typhi* kill murine macrophage cell lines as well as primary cells (Monack *et al.,* 1996; Lindgren *et al.,* 1996; L. Chen *et al.,* 1996). Interestingly, *S. cholerasuis* induces apoptosis only in mouse peritoneal macrophages upon IL-10 blockade. Antagonizing IL-10 activity allows these macrophages to commit autocrine cell suicide through TNF-α (Arai *et al.,* 1995).

The necessity of bacterial internalization for salmonella to induce apoptosis is a controversial issue. Cytochalasin D was reported by one group to have no effect on salmonella-induced macrophage cell death (L. Chen *et al.,* 1996), whereas Monack *et al.* (1996) published the opposite result. This discrepancy could be accounted for by differences in the dose of cytochalasin D used by each group. It appears that activation of cytoskeletal changes by the bacteria correlate with the ability to induce apoptosis. Invasion of macrophages by *S. typhimurium* is known to alter second messenger levels (Bliska *et al.,* 1993; Pace *et al.,* 1993). Salmonella proteins which execute invasion must be functional for the bacteria to kill macrophages (Monack *et al.,* 1996). The type III secretion system, which discharges these proteins, is also required (L. Chen *et al.,* 1996). Several mutants defective in internalization demonstrate attenuated cytotoxicity: *invJ, spaO, sipB, sipC, sipD* (L. Chen *et al.,* 1996), and *nagA* (Lindgren *et al.,* 1996). Mutations of proteins secreted by the type III system, but not associated with invasion, do not alter salmonella's ability to kill macrophages (L. Chen *et al.,* 1996). Thus, something inherent to the bacterial entry pathway appears to activate the apoptotic cascade.

The salmonella protein SipB is homologous to IpaB of shigella. As stated earlier, IpaB is necessary and sufficient to induce apoptosis in macrophages. SipB is conserved across salmonella species and mutants are noncytotoxic (L. Chen *et al.,* 1996). It is tempting to speculate that these proteins function analogously. For shigella to induce apoptosis, IpaB must enter the cytoplasm (Zychlinsky *et al.,* 1994b). Salmonella, however, does not leave the phagolysosome. It remains to be determined whether SipB directly triggers PCD and, if so, whether it mediates apoptosis from the phagolysosome or escapes to the cytoplasm.

In an *in vivo* model of salmonellosis, type III secretion mutants have attenuated virulence when administered orally but are fully virulent when given by intraperitoneal injection. Perhaps the type III secretion system, and therefore apoptosis, has evolved as a mechanism for salmonella to break through the protective barrier of the gut. As in the model proposed for shigellosis, apoptosis may be a necessary proinflammatory event, which allows salmonella to establish an intestinal infection.

4. Yersinia

Yersinia, another gram-negative enteric pathogen, has recently been shown to cause apoptosis of murine and human macrophages (Monack *et al.,* 1997; Mills *et al.,* 1997; Ruckdeschel *et al.,* 1997). *Yersinia enterocolitica* and *Y. psuedotuberculosis* are acquired by ingesting contaminated food. Yersiniae cross the intestinal epithelium and replicate in Peyer's patches, aggregates of lymphoid tissue in the ileum (Hanski *et al.,* 1989). Unlike salmonella and shigella, yersinia does not enter macrophages, but instead exerts its cytotoxic effect from outside the cell (Mills *et al.,* 1997).

Several genes are associated with yersinia's ability to induce macrophage apoptosis. Mutants which cannot attach to host cells, via YadA or invasin, are not cytotoxic (Monack *et al.,* 1997). Yersina outer protein J (YopJ) and YopP, homologous proteins from different yersinia species with no other described function to date, are necessary for yersinia to kill macrophages (Mills *et al.,* 1997; Monack *et al.,* 1997). A salmonella gene, *avrA,* with homology to *yopJ* was recently cloned. Unlike *yopJ* mutants, *avrA* mutants are cytotoxic and virulent (Hardt and Galan, 1997). Whether YopP and YopJ are effectors of yersinia-induced apoptosis or are involved indirectly remains to be discerned.

Interestingly, YopB of yersinia is homologous to IpaB and SipB. YopB mutants are also noncytotoxic. It is unlikely, however, that YopB acts analogously to IpaB. YopB assists in translocating other Yops, including presumably YopP, into the cytosol of macrophages (Hakansson *et al.*, 1996). Abrogation of Yop translocation, and not a direct effect of YopB on macrophages, is a more likely explanation the *yopB* phenotype. In fact, other mutants with impaired Yop secretion do not kill macrophages (Mills *et al.*, 1997).

5. *Helicobacter pylori*

Helicobacter pylori is an important cause of gastritis, duodenal ulcers, gastric atrophy, and carcinoma. This gram-negative rod contains several proposed virulence factors but none has been demonstrated to induce PCD (Correa and Miller, 1995). However, several groups have documented that *H. pylori* kills gastric epithelial cells by apoptosis. (Moss *et al.*, 1996; Mannick *et al.*, 1996; Chen *et al.*, 1997; Peek *et al.*, 1997). In addition, apoptosis in gastric biopsy specimens is significantly associated with *H. pylori* infection (Moss *et al.*, 1996; Mannick *et al.*, 1996). Therapy aimed at eradicating bacteria reduces epithelial apoptosis to control levels. In one study, no correlation between inflammation and apoptosis was found (Moss *et al.*, 1996). Another study, though, reported that *H. pylori* infection is associated with the formation of reactive nitrogen species, and that these molecules correlate with levels of apoptosis (Mannick *et al.*, 1996). However, since helicobacter induces apoptosis of cultured epithelial cells, it is doubtful that inflammation is an absolute requirement for cytotoxicity (Chen *et al.*, 1997).

The ability of *H. pylori* to induce apoptosis may be the key to understanding the sequelae of chronic infection: gastric atrophy and neoplasia. Intuitively, atrophy could result from unrelenting cell death induction. It is more difficult, however, to explain how *H. pylori* promotes carcinogenesis. The chronic inflammation elicited by *H. pylori* may cause DNA damage (Correa and Miller, 1995; Moss *et al.*, 1996) and precancerous mutations. Consistent with this hypothesis is the finding that the antiapoptotic molecule Bcl-2 is expressed at higher levels in precancerous gastric lesions (Lauwers *et al.*, 1994). Additionally, gastric epithelial cell death acts as a signal for compensatory proliferation (Moss *et al.*, 1996). Excess proliferative stimuli could promote the survival and expansion of precancerous cells, eventually resulting in a malignancy.

6. *Legionella pneumophila*

The gram-negative pathogen *L. pneumophila* is the causative agent of Legionnaire's disease. This atypical pneumonia especially affects immunocompromised and elderly patients. Legionella is a facultative intracellular organism which invades macrophages and grows within a phagosome that never acidifies (Horwitz, 1983; Horwitz and Maxfield, 1984). Legionella produces several toxins whose role in pathogenesis is not established (Muller *et al.,* 1996).

Muller *et al.,* (1996) demonstrated that legionella induces PCD in differentiated HL60 cells, a human premonocytic leukemia cell line. Unlike the pathogens discussed previously, which kill cells within a few hours, legionella-induced apoptosis requires days to manifest. Perhaps chronic bacterial infection stimulates macrophages to produce autocrine apoptotic signals. From the experiments these authors present, it is unclear whether bacterial internalization is necessary for cell death induction. Bacterial viability is critical, however, indicating that macrophage apoptosis does not simply result from exposure to inflammatory stimuli. Mutations in two known virulence genes, *mip* and *msp,* have no effect on cytotoxicity (Muller *et al.,* 1996). Isolation of noncytotoxic mutants, however, will no doubt allow scientists to investigate the mechanism of legionella-induced apoptosis and its relationship to pathogenesis.

7. Lipopolysaccharide

LPS is a structural glycolipid component of the outer membrane of all gram-negative bacteria. It is one of the most potent biological compounds, capable of stimulating myeloid and nonmyeloid cells at picomolar concentrations. Also referred to as endotoxin, LPS is an important mediator of shock and inflammation associated with gram-negative bacterial infections (Henderson *et al.,* 1996). In addition to stimulating cells to produce cytokines and proliferate, this molecule can also influence the cell death pathway.

Intrathymic injection of *E. coli, Salmonella enteriditis,* or *Klebsiella pneumonia* LPS causes thymocytes to undergo apoptosis (Zhang *et al.,* 1993). This effect is blunted in adrenalectomized mice or mice pretreated with anti-TNF-α antibodies. Furthermore, *de novo* protein and RNA synthesis is necessary for gram-negative bacteria to induce apoptosis in thymocytes after intraperitoneal injection (S. Wang *et al.,* 1994). These findings, combined with the observation that LPS does not cause DNA fragmentation in thymocytes *in vitro,* indicate that LPS-induced thymocytes apoptosis requires the production of systemic, proapoptotic mediators (Zhang *et al.,* 1993).

Extrathymic lymphocytes are also susceptible to LPS-induced programmed cell death. This effect has been observed both *in vivo* (Norimatsu *et al.,* 1995a,b; Isogai *et al.,* 1996) and *in vitro* (Peng and Raveche, 1993). Peng and Ravache demonstrated LPS-induced apoptosis of malignant, B1 $CD5^+$ lymphocytes and suggested using this susceptibility as a therapy for chronic lymphocytic leukemia. *In vivo,* rises in cortisol and cytokine levels, especially TNF-α (Norimatsu *et al.,* 1995a,b; Isogai *et al.,* 1996, 1995b), correlate with lymphocyte programmed cell death after LPS administration.

Isogai *et al.*, speculate that LPS-induced lymphocyte apoptosis may represent a host mechanism to eliminate nonspecifically activated, and therefore potentially hazardous, lymphocytes.

Polymorphonuclear leukocytes (PMNs), or neutrophils, are phagocytic immune cells which, because of their potent antimicrobial activity, play a prominent role in acute inflammatory reactions. These cells have a short half-life in the peripheral circulation. Without appropriate stimuli, such as LPS, they commit PCD and are rapidly scavenged by macrophages (Savill *et al.*, 1989a,b). LPS increases survival of PMNs *in vitro* (Colotta *et al.*, 1992; Yamamoto *et al.*, 1993; Haslett *et al.*, 1991). Furthermore, peritoneal exudate neutrophils, exposed to LPS *in vivo,* disappear more slowly and are phagocytosed less than untreated cells by peritoneal macrophages (Yamamoto *et al.*, 1993). Prolonging the life of an important antimicrobial cell, at sites of infection with gram-negative bacteria, is an obvious teleological advantage of this system. Phagocytosis of gram-negative bacteria by PMNs has been reported to either cause (perhaps via oxygen radicals produced by the respiratory burst; Watson *et al.*, 1996) or inhibit apoptosis (Baran *et al.*, 1996). The experimental methods employed by these two groups are similar and the reason for their discrepancy is difficult to discern.

Monocytes and macrophages are the other principal antimicrobial phagocytes in mammals. In addition, these cells coordinate inflammatory responses by releasing a plethora of cytokines. *In vitro,* LPS induces apoptosis in macrophages but, in contrast, protects monocytes from PCD (Bingisser *et al.*, 1996; Adler *et al.*, 1995; Depraetere and Joniau, 1994; Mangan *et al.*, 1991a, 1992; Albina *et al.*, 1993). The cytotoxic effect of LPS on macrophages may limit inflammation and associated tissue damage. Inhibition of monocyte apoptosis, on the other hand, could aid in recruiting these cells to infected areas. Interestingly, cytokines modulate the cytotoxic effects of LPS on monocytes and macrophages (Bingisser *et al.*, 1996; Adler *et al.*, 1995; Depraetere and Joniau, 1994; Mangan *et al.*, 1991a, 1992a). Therefore, within the context of different inflammatory responses the effect of LPS on monocyte/macrophage PCD probably varies.

Laine *et al.*, (1996) found that intraperitoneal injection of LPS causes pancreatic tissue damage, indicating that endotoxin could play a causative role in acute pancreatitis. Histological analysis of the pancreas revealed a significant number of apoptotic cells (Laine *et al.*, 1996). Unfortunately, *in vitro* experiments were not performed to assess whether LPS is directly toxic to pancreatic cells.

8. *Clostridium dificile*

Here, we focus on *C. dificile,* a gram-positive organism that has not previously been discussed. Mechanisms of apoptosis induction by *S. aureus,*

S. pyogenes, L. monocytogenes, and *C. diphtheria,* all important gram-positive pathogens, were discussed previously.

The gram-positive anaerobic rod, *C. dificile,* is a normal floral constituent of the human colon. When antibiotics disrupt the balance of organisms living in the gut, *C. dificile* colonization can become pathologic. With fewer bacterial competitors, *C. difficile* grow to high levels and synthesize exotoxins A and B. This results in a diarrheal syndrome and a characteristic pseudomembrane on the colonic mucosa composed of fibrin, mucus, PMN, and dead epithelial cells.

The exotoxins of *C. dificile* are thought to be its key virulence factors. Although toxin B is more cytotoxic to cells *in vitro,* only purified toxin A induces inflammation in ileal and colonic models of infection (Mitchell *et al.,* 1986; Triadafilopolous *et al.,* 1987). Toxin A appears to allow toxin B access to the mucosa by disrupting the epithelial barrier (Lyerly *et al.,* 1985). *In vitro* studies indicate that toxin A causes epithelial cell apoptosis (Mahida *et al.,* 1996). The mechanism of PCD induction might involve the detachment of cells from their substratum. Mahida *et al.* claim that either chemical, via EDTA, or toxin A-dependent detachment of epithelial cells leads to apoptosis. Other studies (Frisch and Francis, 1994; Ruoslahti and Reed, 1994) corroborate this finding. Toxin A-mediated epithelial cell apoptosis was also demonstrated in cultured colonic biopsy specimens (Mahida *et al.,* 1996). Interestingly, toxin A-treated colonocytes release IL-8, a powerful PMN chemoattractant (Mahida *et al.,* 1996). Therefore, apoptotic death of detached enterocytes may account for both the damaged colonic mucosa and the intense inflammatory infiltrate observed in patients with *C. difficile* overgrowth.

9. Mycobacteria

Mycobacterium tuberculosis is an acid fast bacteria that causes the disease tuberculosis. Tuberculosis has reclaimed a position in the medical spotlight with the recent emergence of strains resistant to multiple drug therapies. *Mycobacterium tuberculosis* is transmitted via inhaled respiratory droplets. In the lung, the bacteria encounter alveolar macrophages, in which they reside during the course of infection (Keane *et al.,* 1997). Macrophages, however, are also responsible for coordinating the immune response against the bacteria, culminating in the hallmark pathologic lesion of tuberculosis: the caseating granuloma.

Mycobacterium avium intracellularae are a group of bacteria which cause diseases similar to tuberculosis. Although these organisms are prevalent in the environment, they primarily infect immunocompromised individuals. *Mycobacterium avium* intracellularae, like *M. tuberculosis,* live and replicate in macrophages. In addition to persisting in these cells, mycobacteria also induce PCD.

Mycobacterium tuberculosis triggers apoptosis in explanted human alveolar macrophages through a mechanism involving TNF-α (Keane *et al.,* 1997). Likewise, *M. avium* serovar 4 kills peripheral blood macrophages by apoptosis *in vitro. In vivo,* Cree *et al.* (1987) demonstrated apoptotic cells in tubercular granulomas as well as granulomas from another mycobacterial disease, leprosy. Keane *et al.* (1997) visualized apoptosis in active granulomas by terminal dUTP nick end labeling, a technique that identifies cells with fragmented DNA. In this study, significant apoptosis was not seen in hyalinized, inactive granulomas. It is unclear whether mycobacterial or host factors cause the apoptosis associated with granulomas.

In the setting of mycobacterial infections, macrophage apoptosis appears to benefit the host by depriving the organism of a place to live. Support for this hypothesis includes the following findings: (i) Virulent bacteria are less efficient at inducing apoptosis than avirulent strains (Keane *et al.,* 1997); (ii) macrophages infected with *M. bovis* which die by apoptosis yield fewer colony-forming units than necrotic cells (Molloy *et al.,* 1994); (iii) PCD of *M. avium*-infected macrophages prevents spreading of the bacteria and facilitates growth inhibition (Fratazzi *et al.,* 1997); (iv) transglutaminases, activated during apoptosis, are implicated in limiting bacterial spread; and (v) macrophage apoptosis induced by H_2O_2 decreases the number of viable, intracellular *M. avium* intracellularae (Laochumroonvorapong *et al.,* 1996). Since macrophages release H_2O_2 after phagocytosing mycobacteria, this might represent an actual bacterial containment mechanism (Gangadharam and Edwards, 1984; Gordon and Hart, 1994). In contrast, Fas-induced apoptosis does not affect *M. avium* intracellularae viability, suggesting that only certain apoptotic pathways hinder mycobacterial growth (Laochumroonvorapong *et al.,* 1996). A well-defined murine model of tuberculosis exists. Infection of mutant mice, with targeted insertions in genes involved in apoptosis, should help resolve the role of PCD in mycobacterial pathogenesis.

10. *Leptospira interrogans*

Leptospira interrogans serovar icterohemorrhagiae causes the most severe form of leptospirosis—a zoonotic disease characterized by fever, meningitis, liver disease, and renal failure. Virulent leptospires evade phagocytosis while avirulent organisms are cleared by the reticuloendothelial system (RES) (Faine, 1994). Although it is known that *L. interrogans* can invade certain cells (Merien *et al.,* 1997; Thomas and Higbie, 1990), its virulence factors are not well characterized.

Leptospira interrogans induces PCD in J774 cells, a murine macrophage cell line. A monkey fibroblast cell line was shown to be resistant. The induction of apoptosis depends on the organism invading the macrophage since cell death is inhibited by cytochalasin D. A saprophytic variant, *Leptospira biflexa* Patoc I, does not kill J774 cells (Merien *et al.,* 1997). The RES appears to plays a critical role in managing leptospire exposure (Faine, 1964, 1994). Perhaps virulent organisms escape innate immune surveillance by killing key RES effector cells.

III. Parasites and Fungi

Parasites and fungi, eukaryotic pathogens, can be added to the growing list of organisms which modulate apoptosis. Unlike bacteria though, relatively few investigators have focused on this aspect of parasitic and fungal pathogenesis. Therefore, the brevity of the following section reflects the infancy of this field and the need for further inquiry.

Infection with *Leishmania donovani,* a parasite that grows within macrophages, causes visceral leishmaniasis or kala-azar. Lipophysoglycan, a parasitic cell surface molecule, promotes the survival of growth factor-deprived macrophages by inducing the expression of antiapoptotic cytokines (Moore and Matlashewski, 1994). Leishmania's ability to prevent macrophage apoptosis might prolong the life span of infected cells and maintain a reservoir of cells for the parasite to infect.

As opposed to leishmania, several parasites have been identified as mediators of apoptosis. *Acanthamoeba castellani* causes potentially fatal brain infections. Intact parasites, as well as cell extracts, kill a variety of tumor cell lines by apoptosis (Alizadeh, *et al.,* 1994). In addition, a schizont-enriched extract of *Plasmodium falciparum,* a causative agent of malaria, induces PCD in human PBMCs *in vitro* (Balde *et al.,* 1996). Finally, *in vivo* infections with *P. falciparum* (Balde *et al.,* 1995, 1996), *Trypanosoma cruzi* (Lopes *et al.,* 1995), and *Toxoplasma gondii* (Khan *et al.,* 1996) are associated with enhanced lymphocyte apoptosis. The significance of these findings, with respect to parasite pathogenesis, has yet to be defined.

Pathogenic yeasts also interact with the PCD pathway. A variety of fungi produce an epipolythiodioxopiperzine metabolite, called gliotoxin, which induces apoptosis of numerous cells *in vitro* and lymphocytes *in vivo* (Sutton *et al.,* 1994, 1995; Waring *et al.,* 1988; Terrence, 1994). The mechanism of gliotoxin-induced cell death may involve changes in cAMP concentration (Sutton *et al.,* 1995) or NF-κB inhibition (Pahl *et al.,* 1996). The precise role of gliotoxin in yeast pathogenesis awaits further experimentation. Fumosins from *Fusarium moniliforme* and *Alternaria alternata lycopersisi* are mycotoxins that induce apoptosis *in vitro* (Wang *et al.,* 1996; Tolleson *et al.,* 1996) and *in vivo* (Tolleson *et al.,* 1996; Voss *et al.,* 1996). These toxins are sphingamine analogs. This has led investigators to postulate that fumosins trigger apoptosis by disrupting sphingolipid metabolism. Finally, *Candida albicans,* an extremely common opportunistic pathogen, reduces background levels of PCD and inhibits TNF-α-induced apoptosis of monocytes *in vitro* (Heidenreich *et al.,* 1996). Apoptosis is associated with a decreased ability of host cells to kill *Candida.* Therefore, in this circumstance, it is likely that the inhibition of apoptosis represents a fungal growth control mechanism.

IV. Viruses

Viruses are strict intracellular parasites. Hence, inhibiting host cell apoptosis is of obvious advantage to viruses as a mechanism to increase the life span of the cells in which they multiply. Inhibition of apoptosis, in fact, has been described for many viruses and is associated with oncogenesis, latency, and persistence. Some viruses also induce apoptosis. Here, we analyze the interaction of viruses with the PCD pathway, covering those which have been investigated in detail. In contrast to bacteria, much is known about the cellular targets involved in viral pathogenesis. Moreover, the use of viruses in experimental systems has been of critical value in elucidating the apoptotic program. This section, therefore, focuses more on molecular events than previous sections.

A. Adenoviruses

Adenovirus is a nonenveloped DNA virus which causes pharyngitis, atypical pneumonia, gastroenteritis, and conjunctivitis. This virus preferentially infects mucosal epithelial cells, where DNA replication and viral assembly occur in the nucleus. Subsequently, the target cell is lysed and progeny viruses are released. Adenovirus encodes numerous proteins which engage the host cell death machinery.

The E1A proteins of adenovirus trigger DNA synthesis and cell growth by binding to cellular retinoblastoma (Rb) and p300 proteins (Dyson and Harlow, 1992). E1A alone, however, does not transform cells but results in apoptosis (Rao *et al.,* 1992; White *et al.,* 1991, 1992). E1A-mediated apoptosis is probably the result of conflicting growth signals. E1A causes the accumulation of p53, which arrests cells at the G_1/S boundary (Lowe and Ruley, 1993; Querido *et al.,* 1997). Interestingly, E1A-mediated apoptosis requires either serum starvation or cell–cell contact, both antiproliferative stimuli (Mymryk *et al.,* 1994). This also suggests that conflicting growth signals are important for E1A to induce apoptosis.

E1A triggers PCD by p53-dependent and independent mechanisms. Cells with temperature-sensitive p53 mutations undergo E1A-mediated apoptosis only at the permissive temperature (Sabbatini *et al.,* 1995; Debbas and White, 1993). Furthermore, p53 transcription factor activity is required for E1A-induced apoptosis to proceed (Sabbatini *et al.,* 1995). Notably, p53 increases expression of Bax, a proapoptotic *bcl-2* family member (Miyashita and Reed, 1995), and proteins involved in the generation of, or response to, oxidative stress (Polyak *et al.,* 1997). Bax, and/or redox-related proteins, could execute E1A's cytotoxic effect.

E1A can also cause p53-independent apoptosis (Teodoro *et al.,* 1995; Subramanian *et al.,* 1995). Adenovirus 5 codes for two E1A proteins of 243 and 289 amino acid residues. The domain corresponding to this 46-amino acid difference, CR3, is a central sequence in the 289R protein. CR3 activates the expression of cellular and viral early genes (Shenk and Flint, 1991). The 243R protein induces apoptosis only in cell lines expressing p53. In contrast, the 289R protein kills p53-deficient cells (Teodoro *et al.,* 1995; Subramanian *et al.,* 1995). Transactivation of the adenoviral E4 gene, by CR3, seems to be a downstream event in E1A-mediated, p53-independent cell death (Marcellus *et al.,* 1996).

The adenoviral E1B-55K protein inhibits E1A-induced apoptosis by binding and inactivating p53, without displacing it from its DNA binding site (Rao *et al.,* 1992; Debbas and White, 1993; Yew *et al.,* 1994). Only E1A p53-dependent apoptosis can be inhibited by E1B-55K; p53-independent apoptosis is resistant (Marcellus *et al.,* 1996). In addition, coexpression of E1A and E1B-55K results in transformation. This phenomenon is dependent on the ability of E1B-55K to bind p53 (Yew and Berk, 1992).

It has been known for some time that adenoviruses with mutations in E1B-19K produce a striking phenotype in infected cells, *cyt/deg,* characterized by cytolysis and DNA degradation (Takemori *et al.,* 1984; White *et al.,* 1984). *cyt/deg* is now recognized as apoptosis. E1B-19K, a member of the *bcl-2* family, inhibits adenoviral cytotoxicity (White *et al.,* 1991). Aside from protecting cells from E1A-mediated cell death, E1B-19K also confers resistance to TNF, Fas, and other proapoptotic stimuli (Huang *et al.,* 1997; Hashimoto *et al.,* 1991; White *et al.,* 1992). Resistance to these apoptotic stimuli may allow infected cells to escape immune surveillance. Bcl-2 can substitute for E1B-19K in protecting cells from both p53-dependent and independent E1A-induced apoptosis (Rao *et al.,* 1992; Teodoro *et al.,* 1995; Chiou *et al.,* 1994a, b). Bcl-2 and E1B-19K abrogate p53 transcriptional repression without affecting its ability to activate gene expression (Shen and Shenk, 1994). Whether these proteins overcome adenovirus-induced, p53-dependent apoptosis through this mechanism remains to be determined.

Bcl-2 and E1B-19K share only limited sequence homology. However, the domains important for transformation and regulation of apoptosis are conserved between these two proteins (Chiou *et al.,* 1994a). Moreover, these analogs bind a common set of cellular proteins important in the suppression of cell death, including Nip 1-3 (Boyd *et al.,* 1994). Bax, Bik, and Bak are proapoptotic Bcl-2 family members that interact with both E1B-19K and Bcl-2 (Farrow *et al.,* 1995; Boyd *et al.,* 1995; G. Chen *et al.,* 1996; Han *et al.,* 1996a,b). It has been suggested that E1B-19K and Bcl-2 inhibit apoptosis by antagonizing these death promoting molecules (G. Chen *et al.,* 1996; Han *et al.,* 1996a).

The E4orf6 protein of adenovirus also protects cells from apoptosis. Similar to E1B-55K, this protein interacts with p53, albeit at a distinct binding site, and blocks its ability to activate transcription (Dobner *et al.,* 1996). E4orf6 blocks apoptosis in cells expressing p53 but does not protect p53-deficient cells (Moore *et al.,* 1996). E4orf6 cooperates with other adenoviral oncogenes in transforming cells.

In addition to E1B-19K, adenoviruses encode several other proteins which protect cells from the host immune response. E3-14.7K and E3-10.4K/14.5K inhibit TNF-α, Fas, and cytotoxic T lymphocyte-induced apoptosis (Gooding *et al.,* 1988, 1991; Gooding, 1992). These molecules appear to inhibit TNF-α-mediated cytolysis by blocking activation of cytosolic phospholipase A2 (cPLA2) (Krajcsi *et al.,* 1996). Recently, FIP-1 was cloned as a candidate E14.7K-binding partner. Fip-1 is a low-molecular-weight GTP-binding protein and has been implicated in the TNF-α cell death pathway (Li *et al.,* 1997). How the interaction of E14.7K and Fip-1 blocks apoptotic signals remains to be determined.

E3-11.6K adenoviral death protein (ADP) is abundantly expressed at late stages of cellular infection. ADP, unrelated by homology to any known cell death protein, is necessary for efficient lysis of host cells and the release of infectious virions (Tollefson *et al.,* 1996). Lysis of cells by E3-11.3K may be apoptosis because DNA fragmentation has been demonstrated in infected cells.

B. Papillomaviruses

Papillomaviruses are the causative agents of warts, benign squamous cell tumors. Several subtypes of papillomaviruses, however, cause malignant tumors, especially cervical carcinomas. Papillomaviruses are double-stranded DNA viruses with a nonenveloped, icosahedral nucleocapsid.

These viruses preferentially replicate in the nucleus of differentiated squamous epithelial cells.

It is well established that papillomaviruses transform cells. Two early viral gene products, E6 and E7, are responsible for this activity. E7 induces DNA synthesis by binding to the Rb protein and liberating the transcription factor, E2F (Nevins, 1992). Furthermore, E7 overexpression allows cells to bypass p53-mediated growth arrest (Hickman *et al.,* 1997) and drives cells into the cell cycle. This results in both p53-dependent and -independent apoptosis (Pan and Griep, 1994, 1995; Howes *et al.,* 1994). These functions of E7 require an intact Rb-binding domain. Failure of cyclin A to inactivate E2F, after the completion of DNA synthesis, may cause p53-induced apoptosis in cells overexpressing E7. In fact, cell death in this experimental context correlates with reduced cyclin A kinase activity (Hickman *et al.,* 1997).

E6 proteins of carcinogenic human papillomavirus (HPV) subtypes inhibit apoptosis by stimulating p53 degradation. These E6 proteins bind to a cellular ubiquitin ligase, E6-associated protein (E6-AP), and this complex interacts with p53 (Scheffner *et al.,* 1993). The formation of this trimeric complex (E6/E6–Ap/p53) results in a rapid, ubiquitin-dependent degradation of p53 (Scheffner *et al.,* 1990, 1993). Expression of E6 inhibits G_1 arrest and apoptosis in response to DNA damage (Kessiss *et al.,* 1993; Thomas *et al.,* 1996). Furthermore, E6 inhibits E7-induced apoptosis (Pan and Griep, 1994, 1995). Coexpression of E6 and E7 transforms primary cells *in vitro* (Hawley-Nelson *et al.,* 1989) and causes tumor formation in mice (Pan and Griep, 1994, 1995). Thus, the interaction of E6 and E7 proteins with p53 and Rb is likely crucial for oncogenesis.

Recently, the E2 protein of HPV was reported to trigger apoptosis when expressed in HeLa cells. E2 activates p53-dependent transcription in part by downregulating E6 expression. Although p53 is probably necessary for E2-mediated apoptosis (p53-deficient SAOS cells are not susceptible to E2-mediated apoptosis), its transcriptional activity is dispensable (Desaintes *et al.,* 1997). The E2 ORF is usually disrupted upon viral integration into host DNA (Berumen *et al.,* 1994). Therefore, because E2 promotes apoptosis and downregulates transcription of E6 and E7, loss of its activity is likely required for HPV to transform cells (Desaintes *et al.,* 1997). As discussed earlier, the E6 and E7 proteins are critical for papillomavirus-mediated transformation. In the case of productive cellular infections, E2-mediated apoptosis may facilitate the release of progeny virus, enabling the pathogen to spread to other cells or hosts (Desaintes *et al.,* 1997).

C. Poxviruses

The poxvirus family includes the cowpox, myxoma, vaccinia, smallpox, and molluscum contagiousum viruses. These are large, double-stranded DNA viruses that, as opposed to other DNA viruses, replicate in cytoplasm. Disease manifestations of poxviruses include fever, rashes, and skin lesions which, in severe cases, can be fatal.

A cowpox viral gene product, CrmA, binds and inactivates cysteine proteases, including granzyme B, ICE, and Cpp32 (Komiyama *et al.*, 1994; Miura *et al.*, 1993; Tewari *et al.*, 1995; Quan *et al.*, 1995). CrmA was initially identified because mutants caused hemorrhagic lesions. This suggests that the inhibition of apoptosis attenuates poxvirus pathology. The sequence of CrmA is similar to those of plasma serine protease inhibitors (serpins) (Pickup *et al.*, 1986). Programmed cell death induction by many disparate stimuli is inhibited by CrmA (Uren and Vaux, 1997). A vaccinia virus gene product, SPI-2 (Dobblestein and Shenk, 1996), is a serpin which displays high homology to CrmA. SPI-2 inhibits Fas and TNF-induced apoptosis. It too is likely a caspase inhibitor.

Poxviruses encode other inhibitors of apoptosis. Infection of HeLa cells with a mutant vaccinia virus carrying a deletion of the *E3L* gene results in apoptosis (Kibler *et al.*, 1997). E3L is a dsRNA-binding protein, which probably prevents the activation of a dsRNA responsive host enzyme, perhaps protein kinase PKR. Activated PKR is known to cause cells to undergo apoptosis (Lee and Esteban, 1994). The vaccinia virus CHOhr protein prevents cell death in infected CHO cells by an unknown mechanism (Ink *et al.*, 1995). Lymphocytes infected with myxoma viruses lacking functional M-T2 and M11L undergo apoptosis, whereas cells infected with the wild-type virus do not (Mace *et al.*, 1996). The mechanism by which M11L, a transmembrane protein, inhibits apoptosis is unknown. M-T2 is a TNF receptor homolog which binds extracellular TNF and inhibits programmed cell death through two distinct domains (Schrieber *et al.*, 1997). Both M11L and M-T2 mutant myxoma viruses demonstrate attenuated virulence phenotypes in rabbits (Upton *et al.*, 1991; Opgenorth *et al.*, 1992). Finally, molluscum contagiousum virus and herpesvirus family members (herpesvirus saimiri, HHV-8, bovine herpesvirus-4, and equine herpesvirus-2) were recently shown to produce, or code for, putative FLICE inhibitory proteins (FLIPs). These proteins contain death effector domains which inhibit FLICE activation and interfere with apoptosis triggered through death receptors (Thome, *et al.*, 1997; Bertin *et al.*, 1997). The role of these proteins in pathogenesis has yet to be determined but could involve evasion of antiviral immune responses.

D. Baculoviruses

Baculoviruses are DNA viruses which infect insect cells and replicate in the nucleus. Some insect cells undergo apoptosis when infected with a baculovirus strain carrying a mutation in the *p35* gene (Clem and Miller, 1993). This effect can be reversed by complementation with wild-type *p35* (Hershberger *et al.,* 1992). P35 is an early viral protein which localizes to the cytosol (Hershberger *et al.,* 1994) and suppresses apoptosis by inhibiting caspases. When P35 is cleaved by a caspase it remains irreversibly attached to the enzyme (Bump *et al.,* 1995; Xue and Horvitz, 1995). Apoptosis in response to a variety of stimuli (Uren and Vaux, 1997) is hindered by *p35.* Production of virus from infected cells, as well as *in vivo* viral virulence, is impaired in *p35* mutants, suggesting that the inhibition of apoptosis is important for pathogenesis (Clem and Miller, 1993).

Another baculovirus gene, *iap* (inhibitor of apoptosis protein), also inhibits apoptosis. IAP proteins inhibits apoptosis in a variety of circumstances (Hawkins *et al.,* 1996; Dorstyn and Kumar, 1997). The mechanism may involve transcriptional regulation since a zinc finger motif is conserved throughout this family of proteins (Crook *et al.,* 1993). Several homologs of *iap* exist. African swine fever virus contains an IAP homolog (Chacon *et al.,* 1995). In addition, a variety of mammalian and fly homologues of IAPs have also been described that inhibit apoptosis (Dorstyn and Kumar, 1997).

E. Herpesviruses

Herpes viruses, like most DNA viruses, replicate in the nucleus of infected cells. Family members include the herpes simplex viruses (HSVs), Epstein–Barr virus (EBV), cytomegalovirus (CMV), and varicella–zoster virus (VSZ). A characteristic feature of this group is the ability to establish long-term latent infections which reactivate if the host becomes immunocompromised.

EBV is transmitted in salivary secretions and causes mononucleosis. Neoplastic diseases, such as Burkitt's lymphoma and nasopharyngeal carcinoma, are also associated with EBV infection. EBV infects primarily B lymphocytes, in which it either produces progeny or persists latently in the cytoplasm (Yao *et al.,* 1989; Gratama *et al.,* 1988). B cells normally have a limited life span due to apoptosis of aging cells (Gregory *et al.,* 1991). The ability of EBV to persist in B cells would be irrelevant if cells harboring the inactive virus died before reactivation. EBV overcomes this problem by inhibiting apoptosis through several mechanisms.

Eight proteins are expressed by EBV during the latent stage of infection (Gregory *et al.,* 1991). One of these, latent membrane protein 1 (LMP1), protects B cells from apoptosis (Gregory *et al.,* 1991). LMP-1 upregulates *bcl-2* expression (Henderson *et al.,* 1991) and induces the cellular antiapoptotic gene, *A20* (Fries *et al.,* 1996). Promotion of cell survival by LMP-1 may not only ensure viral persistence but also contribute to EBV-mediated oncogenesis (Takano *et al.,* 1997).

Apoptotic changes occur in cultured cells following lytic infection with EBV (Kawanishi, 1993; Taga *et al.,* 1994). The EBV gene product, BCRF-1, is homologous and functionally similar to the human cytokine IL-10 (Taga *et al.,* 1994). Both human and viral IL-10 (BCRF-1) inhibit EBV-induced programmed cell death. EBV may have incorporated the IL-10 gene to delay PCD of infected cells and thereby maximize viral production.

BHRF-1, an early lytic cycle viral protein, is an EBV Bcl-2 homolog which protects cells from a variety of apoptosis-inducing agents (Tarodi *et al.,* 1994; Henderson *et al.,* 1993). BHRF-1 has a similar subcellular localization pattern to Bcl-2 and colocalizes with its homolog in cotransfected cells. Despite weak sequence homology, it is likely that BHRF-1 and Bcl-2 control apoptosis by similar mechanisms. Other herpesviruses which encode Bcl-2 homologs include herpesvirus saimiri (Nava *et al.,* 1997) and Kaposi's sarcoma-associated virus (Cheng *et al.,* 1997).

Herpes simplex virus type-1 (HSV-1), which causes vesicular lesions (cold sores) and encephalitis, encodes proteins which inhibit apoptosis. The *g34.5* gene of HSV-1 is an antiapoptotic gene. HSV-1 *g34.5* mutants kill neuroblastoma (Chou and Roizman, 1992) and foreskin (Chou *et al.,* 1994) cells by apoptosis. In cells infected with this mutant, the translation initiation factor, eIF-2, becomes phosphorylated. This results in protein synthesis inhibition, which triggers PCD (Chou and Roizman, 1995). Although *g34.5* is not required for viral replication in cell culture (Chou *et al.,* 1990), it facilitates protein synthesis and suppresses apoptosis in HSV-1-infected neurons (Chou and Roizman, 1992). As mentioned earlier, encephalitis is a significant complication of herpes virus infections. Moreover, in a mouse model of herpes encephalitis, neurovirulence is greatly enhanced by *g34.5* (Chou *et al.,* 1990). Taken together, the works of Chou *et al.* indicate that the pathogenesis of encephalitis likely involves suppression of apoptosis by *g34.5.*

HSV-1 encodes another PCD inhibitor, ICP4. ICP4 inhibits apoptosis induced by HSV-1 or by hyperthermia (Leopoldi and Roizman, 1996). Koyama and Miwa (1997) report that an HSV-1 *g34.5* mutant suppresses sorbitol-induced apoptosis. They suspect this effect is mediated by ICP4. Cell suicide is probably a host defense reaction to viral infection because inhibition of PCD enhances viral replication in HSV-1-infected cells. The ability to control apoptosis is certainly a key feature of pathogenic HSV-1.

Cytomegalovirus causes a variety of diseases which predominantly afflict the immunocompromised and neonates. As with other herpesviruses, CMV is associated with pro- and antiapoptotic phenotypes. The CMV gene products IE1 and IE2 inhibit apoptosis *in vitro.* These immediate early (IE) proteins are among the first expressed during a CMV infection. IE1 and IE2 inhibit apoptosis induced by TNF-α or adenovirus E1-19kDa mutant virus infection, but not UV light (Zhu *et al.,* 1995). IE1 and IE2 are transcriptional regulators (Lukac *et al.,* 1994; Sambucetti *et al.,* 1989). IE2 also alters Rb activity (Hagemeier *et al.,* 1994) and interferes with *p53*-mediated transcription (Spier, *et al.,* 1994). Since IE1 and IE2 localize to the nucleus (Zhu *et al.,* 1995), they are probably neither Bcl-2 homologs nor caspase inhibitors. A detailed mechanism regarding IE1- and IE2-mediated PCD suppression remains to be determined.

Apoptosis induction is also associated with herpesvirus infections. The thymocytes of mice infected with CMV are highly sensitive to apoptosis triggered by anti-CD3 antibodies (Koga *et al.,* 1994). In this model, CMV also induces PCD of T cells and hematopoetic progenitors (Yoshida *et al.,* 1995; Mori *et al.,* 1997). HSV-1 (Ito *et al.,* 1997a, b; Tropea *et al.,* 1995), human herpesvirus 6 (Inoue *et al.,* 1997), and VZV have also been shown to induce apoptosis (Sadzot-Delvaux *et al.,* 1995). The significance of cell death induction in herpesvirus infections is not yet understood.

F. Human Immunodeficiency Virus

The human immunodeficiency virus (HIV) is the etiologic agent of acquired immunodeficiency syndrome (AIDS). AIDS is the consequence of chronic immune system deficiency and dysregulation. AIDS patients are extremely susceptible to a range of common and opportunistic pathogens. CD4 T-cell depletion is a hallmark of the disease. Mechanistic explanations for T-cell depletion include (i) cytotoxic T-lymphocyte responses against infected cells, (ii) complement-mediated lysis following antibody recognition of viral components expressed on the surface of infected cells, (iii) syncitia formation (Sodroski *et al.,* 1986; Lifson *et al.,* 1986), and (iv) direct HIV toxicity due to budding of vast quantities of virus from the cell surface. In 1991, Amiesen and Capron first proposed apoptosis as a mechanism of HIV-induced lymphocyte deletion. Since then, much information has been amassed regarding apoptosis in HIV infection. This subject has been reviewed in detail by several authors (Amiesen and Capron, 1991; Amiesen, 1992; Finkel and Banda, 1994; Amiesen *et al.,* 1994; Oyaizu and Pahwa, 1997; Famularo *et al.,* 1997). Here, we highlight the major findings in this field.

HIV is a diploid, single-stranded, positive RNA retrovirus. A required cellular receptor is the CD4 molecule. Thus, HIV preferentially infects T-helper lymphocytes and macrophages. Glycoprotein 120 (gp120), a product of the HIV *env* gene, binds to CD4 and promotes the fusion of viral and host cell membranes. Recently, coreceptors for HIV adsorption and entry have been described. These molecules further specify cell tropism.

Integration of reverse-transcribed HIV DNA into the host cell genome is a mandatory step in the retroviral life cycle.

In vitro, HIV kills lymphocytes by apoptosis (Terai *et al.,* 1991; Laurent-Crawford *et al.,* 1991). Lymphocytes derived from HIV-infected patients show accelerated apoptosis *in vitro* compared to those isolated from normal volunteers (Meyaard *et al.,* 1992, 1994; Groux *et al.,* 1992; Oyaizu *et al.,* 1993). Furthermore, lymph nodes from HIV-infected patients demonstrate increased numbers of apoptotic cells (Fauci, 1994; Muro-Cacho *et al.,* 1995). Interestingly, less lymphocyte apoptosis is observed in individuals infected with HIV-2 than in those who have contracted HIV-1 (Jaleco *et al.,* 1994). The reduced clinical severity of HIV-2 infection correlates with this decreased level of apoptosis. In asymptomatic HIV-1 infections, however, the extent of T-cell apoptosis does not correlate with CD4 T-cell numbers or viral load (Meyaard and Miedema, 1997). It is not known whether HIV directly kills cells *in vivo* and, if so, how large a contribution this effect has on T-cell numbers.

Many researchers have suggested that CD4 T-cell depletion is the result of an increase in the rate of PCD in uninfected lymphocytes. This phenomenon, called the bystander effect, is supported by the following findings: (i) The amount of CD4 lymphocytes in HIV patients which undergo an abnormal apoptotic death following activation greatly exceeds the number of infected cells (Fauci, 1993; Groux *et al.,* 1992; Gougeon *et al.,* 1993; Oyaizu *et al.,* 1993); (ii) lymphocyte apoptosis outpaces lymphocyte infection in an HIV-infected immunodeficient mouse reconstituted with a human immune system (SCID-hu) (Bonyhadi *et al.,* 1993; Su *et al.,* 1995; Aldrovandi *et al.,* 1993); (iii) in a macaque model of HIV infection, apoptotic lymph node cells are rarely HIV infected and HIV-infected cells are seldom apoptotic (Finkel *et al.,* 1995); and (iv) CD8 T cells, which cannot be infected with HIV, show accelerated apoptosis in HIV patients (Meyaard *et al.,* 1992, 1994; Lewis *et al.,* 1994). Multiple mechanisms of the bystander effect have been proposed.

One theory regarding how HIV causes lymphocyte apoptosis is cross-linking of CD4 by gp120. Cross-linking of CD4 molecules with antibodies causes apoptosis when the T-cell receptor is stimulated (Newell *et al.,* 1990). Gp120 cross-links CD4 molecules in the presence of anti-gp120 antibody. This event primes T cells for activation-induced apoptosis (Banda *et al.,* 1992). Whether cross-linking of CD4 by gp120 accounts for enhanced programmed cell death *in vivo* is unclear. However, several models suggest that it plays a role. When transgenic mice expressing human CD4 are injected with gp120 and anti-gp120 antibodies, lymphocytes expressing the transgene are specifically destroyed (Z., Wang *et al.,* 1994a). Furthermore, transgenic mice expressing both human CD4 and gp120 demonstrate T-cell depletion after administration of anti-gp120 antibodies and contain apoptotic T cells in their spleens (Finco *et al.,* 1997). It will be interesting to determine whether knowledge of this mechanism can be exploited therapeutically to reduce T-cell apoptosis in HIV-infected patients.

As mentioned previously, HIV infection is not restricted to lymphocytes. Macrophages are also infected with the virus and are considered to be a reservoir of HIV in patients (Levy, 1993). Macrophage infection with HIV usually does not result in cytotoxicity (Levy, 1993; Greene, 1993). Macrophages may mediate T-cell apoptosis, however. Antigen presenting macrophages provide critical costimulatory signals to CD4 T cells (Bretscher and Cohn, 1970; Janeway, 1992; Janeway and Bottomly, 1994) which, if blocked, can promote lymphocyte apoptosis (Gribben *et al.,* 1995; Sprent, 1994). It has been proposed that infection of macrophages may be critical to HIV-induced T-cell apoptosis *in vivo* and AIDS pathogenesis (Amiesen and Capron, 1991). Evidence for this notion includes the following: (i) Although HIV is cytotoxic to chimpanzee lymphocytes *in vitro,* it cannot infect chimpanzee monocytes, and despite similar viral loads, only humans infected with HIV develop a T-cell depletion (Watanabe *et al.,* 1991; Johnson *et al.,* 1993; Schuitmaker *et al.,* 1993; Saksela *et al.,* 1993), and (ii) SCID-hu mice infected with HIV clones that infect macrophages, but are noncytotoxic to lymphocytes *in vitro,* develop a more dramatic CD4 lymphocyte depletion than those infected with a potently cytotoxic clone, which does not infect macrophages well (Mosier *et al.,* 1993). Perhaps, HIV infection of antigen presenting cells (APCs) alters their ability to deliver proper costimulatory signals to T cells, resulting in apoptosis.

Gp120-mediated CD4 cross-linking and lack of APC-derived costimulatory signals may be important mechanisms for killing CD4 lymphocytes. They do not, however, account for apoptosis of CD8 T cells observed in HIV infection. Other factors must be involved in mediating apoptosis of CD8, and possibly CD4, lymphocytes.

The cytokine profile of HIV-infected individuals is abnormal. Increased TNF-α, IFN-γ, IL-10, and IL-6 are documented along with a decrease in IL-2 (Lahdevirta *et al.,* 1988; Breen *et al.,* 1990; Emille *et al.,* 1990; Fan *et al.,* 1993). This may be a result of the interaction of gp120 with CD4, altering the T cell's cytokine expression profile (Wahl *et al.,* 1989; Clouse *et al.,* 1990). Perhaps alterations in relative cytokine levels promotes apoptosis of both CD4 and CD8 T cell populations. For example, TNF-α is a well-described proapoptotic molecule and has been associated with lymphocyte death (Clement and Stamenkovic, 1994; Z. Wang *et al.,* 1994a). IL-2, on the other hand, protects T cells from apoptosis (Leonardo, 1991).

Fas–FasL interactions may also be important for HIV-induced T-cell apoptosis. Cross-linking of CD4 causes T cells to produce TNF-α and IFN-γ, which in turn stimulate the expression of Fas on CD4 and CD8 cells (Oyaizu *et al.,* 1994). This event sensitizes these cells to Fas-induced cell death (Desbarats *et al.,* 1996). The drug vesarinone inhibits this cascade of events (Oyaizu *et al.,* 1996). In HIV-infected individuals, both CD4 and CD8 cells show an increase in Fas expression (Debatin *et al.,* 1994; Katsikis *et al.,* 1995). This observation is magnified as the disease progresses (Aries *et al.,* 1995). In addition, T lymphocytes from HIV-infected individuals are more susceptible to Fas-mediated killing (Katsikis *et al.,* 1995). Finally, Z. Wang *et al.* (1994b) report that normal T lymphocytes, but not lymphocytes derived from Fas-defective *lpr* mice, undergo apoptosis in response to anti-CD4 antibody injection. Taken together, these data indicate that the abnormal cytokine profile of HIV-infected patients causes their T cells to upregulate a molecule (Fas) which facilitates their ultimate demise.

In order for Fas to induce cell death *in vivo,* it must be engaged by membrane-bound or soluble FasL. The source of FasL in HIV infection may be macrophages. Macrophages infected with HIV, and monocytes after CD4 cross-linking, increase expression of cell surface FasL *in vitro* (Badley *et al.,* 1996, 1997; Oyaizu *et al.,* 1997a). FasL on macrophages can induce apoptosis of Fas-expressing T lymphocytes. This cytotoxic reaction can be inhibited by anti-Fas antibodies (Badley *et al.,* 1996, 1997; Oyaizu *et al.,* 1997a).

Other HIV gene products have been linked to apoptosis. Apoptosis is inhibited in cells expressing Tat possibly due to Bcl-2 upregulation (McCloskey *et al.,* 1997; Zauli *et al.,* 1993). Conversely, uninfected cells exposed to exogenous Tat undergo apoptosis (McCloskey *et al.,* 1997; Li *et al.,* 1995). Cell death induced by Tat may be mediated by Fas (Westendorf *et al.,* 1995). Nef causes endocytosis of cell surface CD4 molecules, decreasing the possibility of gp120-induced cross-linking (Aiken *et al.,* 1994). Intracellular gp120 and Vpu also decrease surface CD4 expression (Crise *et al.,* 1990; Willey *et al.,* 1992). Finally, Vpr was recently reported to kill HeLa cells, fibroblasts, and lymphoid cells by apoptosis (Stewart *et al.,* 1997). The significance of these findings in HIV-infected patients remains undetermined.

In addition to peripheral T cells, HIV depletes thymocytes. Thymocytes from SCID-hu mice infected with HIV are killed by apoptosis (Bonyhadi *et al.,* 1993). $CD4^+$ thymocytes are affected to a greater extent than $CD8^+$. The CD4 lymphocyte population in these mice is obliterated, implying that thymocyte apoptosis hinders the ability of the thymus to reconstitute peripheral CD4 cells. The role of thymocyte depletion in human HIV infections is unclear but it may be a contributing factor in the pathogenesis of the immunodeficient state.

It is clear that HIV-induced T-cell depletion is mediated, at least in part, through apoptosis. One particular mechanism, however, has not emerged that satisfactorily accounts for this clinically devastating phenomenon. On the contrary, numerous plausible explanations, backed by credible scientific data, have been proposed. It is very possible that each contributes to the pathogenesis of the immunodeficiency syndrome. It is hoped that further experimentation will resolve this issue and lead to novel therapeutic interventions in HIV infection.

Atrophy of the cerebral cortex along with neurologic and psychiatric symptoms are also disease manifestations of HIV infection. An increase in the number of apoptotic neurons has been observed in the brains of HIV-infected patients (Adie-Biassette *et al.,* 1995; Gelbard *et al.,* 1995; Petito and Roberts, 1995; Shi *et al.,* 1996). Patients suffering from AIDS dementia demonstrate more apoptotic cells than those whose central nervous system disease has not progressed to this severe stage (Shi *et al.,* 1996). Since neurons are not infected with HIV, indirect triggers are probably responsible for apoptosis. Studies implicate viral gp120 and Tat and host TNF-α, arachadonic acid metabolites, free radicals, nitric oxide, and platelet-activating factor as the cause of neuronal apoptosis in HIV patients (Charriaut-Marlangue *et al.,* 1996; Pulliam *et al.,* 1991; Gelbard *et al.,* 1994; Genis *et al.,* 1992; Magnuson *et al.,* 1995; Wesselingh *et al.,* 1993; New *et al.,* 1997). *In vitro* models, in which HIV-1 induces neuronal apoptosis, are available and will help elucidate the mechanisms of HIV neurotoxicity (Shi *et al.,* 1996; He *et al.,* 1997).

G. Sindbis Virus

Sindbis virus is a togavirus of the alphavirus genus. It is related to equine encephalitis viruses. It is a positive, ssRNA virus that replicates in the cytoplasm of infected cells. Mosquitoes transmit the virus to mice, in which it causes persistent infections in the nervous system (Levine *et al.,* 1991, 1994).

Sindbis virus kills cells by apoptosis. Overexpression of Bcl-2 protects cells and transforms a lytic infection into a persistent one (Levine *et al.,* 1993). Bcl-xL also protects cells from sindbis virus-induced apoptosis (Cheng *et al.,* 1996). The E2 protein is involved in alphavirus cytotoxicity and neurovirulence. A single-amino acid modification of this protein allows viruses to overcome the inhibitory effect of Bcl-2 (Ubol *et al.,* 1994). The Ras signaling pathway is important for sindbis-induced apoptosis. Dominant-negative Ras molecules delay the appearance of apoptosis in sindbis virus-infected PC12 cells, a rat pheochromocytoma cell line (Joe *et al.,* 1996). *In vivo,* alphavirus infection causes neuronal apoptosis and the extent of cell death correlates with neurovirulence (Lewis *et al.,* 1996). Moreover, mice infected with sindbis viruses that encode Bcl-2 show decreased viral replication, neuronal apoptosis, and mortality compared to those exposed to wild-type viruses (Levine *et al.,* 1996).

H. Chicken Anemia Virus

Chicken anemia virus (CAV) is a circular, ssDNA virus (Noteborn *et al.*, 1991). Three overlapping reading frames are transcribed from CAV mRNA (Noteborn *et al.*, 1992). This virus, which causes anemia and immunosuppression in chickens (Goryo *et al.*, 1985), is of considerable interest to poultry farmers.

CAV induces apoptosis *in vivo* and *in vitro* (Jeurissen *et al.*, 1992). Apoptin is a CAV protein that induces apoptosis in cultured cells (Noteborn *et al.*, 1994). Apoptosis induced by apoptin is *p53* independent (Zhuang *et al.*, 1995c) and not inhibited by *bcl-2* (Zhuang *et al.*, 1995b, c). Interestingly, apoptin only kills transformed cells; normal cells are resistant. Apoptin localizes to the cytoplasm of normal cells but resides in the nucleus of transformed cells, suggesting that nuclear localization is important for apoptin-induced PCD (Danen-Van Oorschot *et al.*, 1997). Since apoptin specifically kills tumor cells in a *p53*-independent, Bcl-2-insensitive fashion, it is a logical candidate for anticancer gene therapy.

I. Summary of Other Viruses Which Induce Apoptosis

A variety of other viruses also cause apoptosis of cells *in vitro* and *in vivo*. Other viruses which induce apoptosis, and their relevant gene products, are outlined in Table I.

V. Concluding Remarks

We have explored the mechanisms that microbes employ to engage the apoptotic machinery of their eukaryotic hosts. These include, among others, changes in membrane permeability (pore-forming toxins) and intracellular second messengers (Ac-Hly), protein synthesis inhibition (DTX), and the synthesis of caspase inhibitors (poxviruses). That the stimuli leading to a common endpoint are so diverse reflects the numerous pathways to PCD in eukaryotes. Since apoptosis plays a critical role in homeostasis and the generation of immune responses, it is likely that its regulation by pathogens is a critical virulence determinant.

Some examples suggest that there is no ubiquitous function for pathogen-regulated PCD. Rather, apoptosis has been linked to such diverse pathophysiologic processes as inflammation (shigella), bacterial clearance (mycobacteria), oncogenesis (*H. pylori* and EBV), and immunodeficiency (HIV). Therefore, the influence of each apoptotic event on disease is probably

TABLE I

Other Viruses That Induce Apoptosis

Virus	Gene product(s) involved	Reference
Arenavirus		Aronson *et al.* (1994)
Dengue virus		Marianneau *et al.* (1997), Despres *et al.* (1996)
Hepatitis C virus	Core protein	Mita *et al.* (1994), Ruggieri *et al.* (1997)
HTLV	Tax protein	Chlichila *et al.*) (1995, 1997), Yamada *et al.* (1994)
Bursal disease virus	VP2 protein	Vasconcelos and Lam (1994, 1995), Tham and Moon (1996), Ojeda *et al.* (1997), Lam (1997), Fernandez-Arias, *et al.* (1997)
Influenza virus		Mori *et al.* (1995), Saito *et al.* (1996), Takizawa *et al.* (1993), Hinshaw *et al.*) (1994)
Japanese encephalitis virus		Liao *et al.* (1997)
LaCrosse virus		Pekosz *et al.* (1996)
Measles virus		Esolen *et al.* (1995), Ito *et al.* (1996, 1997b)
Newcastle disease virus		Lam *et al.* (1995), Lam (1996)
Parainfluenza virus		Sieg *et al.* (1996)
Parvovirus B19		Morey *et al.* (1993)
Poliovirus		Tolskaya *et al.* (1995, 1996)
PRRS	ORF5	Suarez *et al.* (1996)
Reovirus		Tyler *et al.* (1995, 1996), Oberhaus *et al.* (1997), Rodgers *et al.* (1997)
Rhabdovirus		Bjorklund *et al.* (1997)
Semliki forest virus		Scallan *et al.* (1997), Glasgow *et al.* (1997)
SV40		McCarthy *et al.* (1994), Fromm *et al.* (1994)
Vesicular stomatitis virus		Koyama (1995)

context dependent. Only further research will help clarify the significance of the observations summarized in this chapter. In addition to examining the role of apoptosis in animal models of infection, it may eventually be possible to exploit this phenomenon as a therapeutic target.

References

Abshire, K., and Neidhardt, F. (1993). Growth rate paradox of *Salmonella typhimurium* within macrophages. *J. Bacteriol.* **175,** 3744–3748.

Adie-Biassette, H., Levy, Y., Colombel, M., Poron, F., Natchev, S., Keohane, C., and Gray, F. (1995). Neuronal apoptosis in HIV infection in adults. *Neuropathol. Appl. Neurobiol.* **21,** 218–227.

Adler, B., Adler, H., Jungi, T., and Peterhans, E. (1995). Interferon alpha primes macrophages for lipopolysaccharide induced apoptosis. *Biochem. Biophys. Res. Commun.* **215,** 921–927.

Ahnert-Hilger, G., Bhakdi, S., and Gratzl, M. (1985). Minimal requirements for exocytosis: A study using PC12 cell permeabilized with staphylococcal alpha toxin. *J. Biol. Chem.* **260,** 12730–12734.

Aiken, C., Konner, J., Landau, N., Lenburg, M., and Trono, D. (1994). Nef induces CD4 endocytosis: Requirement for a critical dileucine motif in the membrane-proximal $CD4^+$ cytoplasmic domain. *Cell* **76,** 853–864.

Albina, J., Cui, S., Mateo, R., and Reichner, J. (1993). Nitric oxide-mediated apoptosis in murine peritoneal macrophages. *J. Immunol.* **150,** 5080–5085.

Aldrovandi, G., Feuer, G., Gao, L., Jamieson, B., Kristeva, M., Chen, I., and Zack, J. (1993). The SCID-Hu mouse as a model for HIV-1 infection. *Nature* **363,** 732–736.

Alizadeh, H., Pidherney, M. S., McCulley, J. P., and Niederkorn, J. Y. (1994). Apoptosis as a mechanism of cytolysis of tumor cells by a pathogenic free-living amoeba. *Infect. Immun.* **62,** 1298–1303.

Alpuche-Aranda, C., Racoosin, M., Swanson, J., and Miller, S. (1994). Salmonella stimulate macrophage macropinocytosis and persist within spacious phagosomes. *J. Exp. Med.* **179,** 601–608.

Amiesen, J. (1992). Programmed cell death and AIDS: From hypothesis to experiment. *Immunol. Today* **13,** 388–391.

Amiesen, J., and Capron, A. (1991). Cell dysfunction and depletion in AIDS; The programmed cell death hypothesis. *Immunol. Today* **12,** 102–105.

Amiesen, J., Estaquier, J., and Idziorek, T. (1994). From AIDS to parasite infection: Pathogen mediated subversion of programmed cell death as a mechanism for immune dysregulation. *Immunol. Rev.* **142,** 9–51.

Andrieu, J., and Lu, W. (1995). "Cell Activation and Apoptosis in HIV Infection." Plenum, New York.

Arai, T., Hiromatsu, K., Nishimura, H., Kimura, Y., Kobayashi, N., Ishida, H., Nimura, Y., and Yoshikai, Y. (1995). Endogenous interleukin-10 prevents apoptosis in macrophages during salmonella infection. *Biochem. Biophys. Res. Commun.* **213,** 600–617.

Arends, M. J., and Wyllie, A. H. (1991). Apoptosis: Mechanisms and roles in pathology. *Int. Rev. Exp. Pathol.* **32,** 223–254.

Aries, S., Schaaf, C., Muller, C., Dennin, R., and Dalhoff, K. (1995). Fas expression on $CD4^+$ T Cells from HIV infected patients increases with disease progression. *J. Mol. Med.* **73,** 591–593.

Aroeira, L., Moreno, M., and Martinez, C. (1996). In vivo activation of Y cell induction into the primed phenotype and programmed cell death by staphylococcal enterotoxin B. *Scand. J. Immunol.* **43,** 545–550.

Aronson, J. F., Herzog, N. K., and Jerrells, T. R. (1994). Pathological and virological features of arenavirus disease in guinea pigs. Comparison of two Pichinde virus strains. *Am. J. Pathol.* **145,** 228–235.

Badley, A., McElhinny, J., Leibson, P., Lynch, D., Alderson, M., and Paya, C. (1996). Upregulation of Fas ligand expression by human immunodeficiency virus in human macrophages mediates apoptosis of uninfected T lymphocytes. *J. Virol.* **70,** 199–206.

Badley, A., Dockrell, D., Simpson, M., Schut, R., Lynch, D., Leibson, P., and Paya, C. (1997). Macrophage dependent apoptosis of $CD4^+$ T lymphocytes from HIV infected individuals is mediated by FasL and tumor necrosis factor. *J. Exp. Med.* **185,** 55–64.

Balde, A. T., Sarthou, J. L., and Roussilhon, C. (1995). Acute *Plasmodium falciparum* infection is associated with increased percentages of apoptotic cells. *Immunol. Lett.* **46,** 59–62.

Balde, A., Sarthou, J., Aribot, G., Michel, P., Trape, J., Rogier, C., and Roussilhon, C. (1996). *Plasmodium falciparum* induces apoptosis in human mononuclear cells. *Infect. Immun.* **64,** 744–750.

Banda, N., Bernier, J., Kurahara, D., Kurrle, R., Haigwood, N., Sekaly, R., and Finkel, T. (1992). Crosslinking CD4 by human immunodeficiency virus gp 120 primes T cells for activation induced apoptosis. *J. Exp. Med.* **176,** 1099–1106.

Baran, J., Guzic, K., Hryniewicz, W., Ernst, M., Flad, H., and Pryjma, J. (1996). Apoptosis of monocytes and prolonged survival of granulocytes as a result of phagocytosis of bacteria. *Infect. Immun.* **64,** 4242–4248.

Barsig, J., and Kaufmann, S. (1997). The mechanism of cell death in *Listeria monocytogenes* infected murine macrophages is distinct from apoptosis. *Infect. Immun.* **65,** 4075–4081.

Bertin, J., Armstrong, R., Ottlie, S., Martin, D., Wang, Y., Banks, S., Wang, G., Senkevich, T., Alenmeri, E., Moss, B., Leonardo, M., Tomaselli, K., and Cohen, J. (1997). Death effector domain-containing herpesvirus and poxvirus proteins inhibit both Fas- and TNFR1-induced apoptosis. *Proc. Natl. Acad. Sci. USA* **94,** 1172–1176.

Berumen, J., Casas, L., Segura, E., Amezuca, J., and Garcia-Carranca, A. (1994). Genome amplification of human papillomavirus types 16 and 18 in cervical carcinomas is related to the retention of E1/E2 genes. *Int. J. Cancer* **56,** 640–645.

Bhakdi, S., Muhly, M., Mannhardt, U., Hugo, F., Klapettek, K., Muller-Eckhardt, C., and Roka, L. (1988). Staphyloccoal alpha toxin promotes blood coagulation via attack on human platelets. *J. Exp. Med.*

Bingisser, R., Stey, C., Weller, M., Groscurth, P., Russi, E., and Frei, K. (1996). Apoptosis in human alveolar macrophages is induced by endotoxin and is modulated by cytokines. *Am. J. Resp. Cell Mol. Biol.* **15,** 64–70.

Bjorklund, H. V., Johansson, T. R., and Rinne, A. (1997). Rhabdovirus-induced apoptosis in a fish cell line is inhibited by a human endogenous acid cysteine proteinase inhibitor. *J. Virol.* **71,** 5658–5662.

Bliska, J., Galan, J., and Falkow, S. (1993). Signal transduction in the mammalian cell during bacterial attachment and entry. *Cell* **73,** 903–920.

Bodley, J. (1990). Does diphtheria toxin have nuclease activity? *Science* **250,** 832.

Bonyhadi, M., Rabin, L., Salimi, S., Brown, D., Kosek, J., McCune, J., and Kaneshima, H. (1993). HIV induces thymus depletion in vivo. *Nature* **363,** 728–732.

Boyd, J. M., Malstrom, S., Subramanian, T., Venkatesh, L. K., Schaeper, U., Elangovan, B., D'Sa-Eipper, C., and Chinnadurai, G. (1994). Adenovirus E1B 19 kDa and Bcl-2 proteins interact with a common set of cellular proteins (see comments; published erratum appears in *Cell* **79**(6), following 1120, 1994). *Cell* **79,** 341–351.

Boyd, J. M., Gallo, G. J., Elangovan, B., Houghton, A. B., Malstrom, S., Avery, B. J., Ebb, R. G., Subramanian, T., Chittenden, T., Lutz, R. J., *et al.* (1995). Bik, a novel death-inducing protein shares a distinct sequence motif with Bcl-2 family proteins and interacts with viral and cellular survival-promoting proteins. *Oncogene* **11,** 1921–1928.

Breen, E., Rezai, A., Nakajima, K., Beall, G., Mitsuyasu, R., Hirano, T., Kishimoto, T., and Martinez-Maza, O. (1990). Infection with HIV is associated with elevated IL-6 levels and production. *J. Immunol.* **144,** 480–484.

Bretscher, P., and Cohn, M. (1970). A theory of self-nonself discrimination. *Science* **169,** 1042–1049.

Brinkmann, U., Brinkmann, E., and Pastan, I. (1995a). Expression cloning of cDNAs that render cancer cells resistant to pseudomonas and diphtheria toxin and immunotoxins. *Mol. Med.* **1,** 206–216.

Brinkmann, U., Brinkmann, E., Gallo, M., and Pastan, I. (1995b). Cloning and characterization of a cellular apoptosis susceptibility gene, the human homologue to the yeast chromosome segregation gene CSE1. *Proc. Natl. Acad. Sci. USA* **92,** 10427–10431.

Brinkmann, U., Brinkmann, E., Gallo, M., Scherf, U., and Pastan, I. (1996). Role of CAS, a human homologue to the yeast chromosome segregation gene CSE1, in toxin and tumor necrosis factor mediated apoptosis. *Biochemistry* **35,** 6891–6899.

Buchmeier, N., and Heffron, F. (1989). Intracellular survival of wild type *Salmonella typhimurium* and macrophage sensitive mutants in diverse populations of macrophages. *Infect. Immun.* **57,** 1–7.

Bump, N., Hackett, M., Hugunin, M., Seshagiri, S., Brady, K., Chen, P., Ferenz, C., Franklin, S., Ghayur, T., Li, P., Licari, P., Mankovich, J., Shi, L., Greenberg, A., Miller, L., and Wong, W. (1995). Inhibition of ICE family proteases by baculovirus antiapoptotic protein p35. *Science* **269,** 1885–1888.

Cavalieri, S., Bohach, G., and Snyder, I. (1984). Escherichia coli alpha hemolysin: Characteristics and probable role in pathogenicity. *Microbiol. Rev.* **48,** 326–343.

Chacon, M., Almazan, F., Nogal, M., Vinuela, E., and Rodriguez, J. (1995). The African swine fever virus IAP homolog is a late structural protein. *Virology* **214,** 670–674.

Chang, M., and Wisnieski, B. (1990). Comparison of the intoxication pathways on tumor necrosis factor and diphtheria toxin. *Infect. Immun.* **58,** 2644–2650.

Chang, M., Bramhal, J., Graves, S., Bonavida, B., and Wisnieski, B. (1989a). Internucleosomal DNA cleavage precedes diphtheria toxin induced cytolysis. *J. Biol. Chem.* **264,** 15261–15267.

Chang, M., Baldwin, R., Bruce, C., and Wisnieski, B. (1989b). Second cytotoxic pathway of diphtheria toxin suggested by nuclease activity. *Science* **246,** 1165–1168.

Charriaut-Marlangue, C., Aggoun-Zouaoui, D., Represa, A., and Ben-Ari, Y. (1996). Apoptotic features of selective neuronal cell death in ischemia, epilepsy, and gp 120 toxicity. *Trends Neurosci.* **19,** 109–114.

Chen, G., Branton, P. E., Yang, E., Korsmeyer, S. J., and Shore, G. C. (1996). Adenovirus E1B 19-kDa death suppressor protein interacts with Bax but not with Bad. *J. Biol. Chem.* **271,** 24221–24225.

Chen, G., Sordillo, E., Ramey, W., Reidy, J., Holt, P., Krajerski, S., Reed, J., Blaser, M., and Moss, S. (1997). Apoptosis in gastric epithelial cells is induced by helicobacter pylori and accompanied by increased expression of BAK. *Biochem. Biophys. Res. Commun.* **239,** 626–632.

Chen, L. M., Kaniga, K., and Galan, J. E. (1996). *Salmonella* spp. are cytotoxic for cultured macrophages. *Mol. Microbiol.* **21,** 1101–1115.

Chen, Y., Smith, M. R., Thirumalai, K., and Zychlinsky, A. (1996). A bacterial invasin induces macrophage apoptosis by directly binding ICE. *EMBO J.* **15,** 3853–3860.

Cheng, E., Levine, B., Boise, L, Thompson, C., and Hardwick, J. (1996). Bax-independent inhibition of apoptosis by Bcl-XL. *Nature* **379,** 554–556.

Cheng, E., Nicholas, J., Bellows, D., Hayward, G., Guo, H., Reitz, M., and Hardwick, J. (1997). A Bcl-2 homolog encoded by Kaposi sarcoma-associated virus, human herpesvirus 8, inhibits apoptosis but does not heterodimerize with Bax or Bak. *Proc. Natl. Acad. Sci. USA* **94,** 690–694.

Chiou, S. K., Tseng, C. C., Rao, L., and White, E. (1994a). Functional complementation of the adenovirus E1B 19-kilodalton protein with Bcl-2 in the inhibition of apoptosis.

Chiou, S. K., Rao, L., and White, E. (1994b). Bcl-2 blocks p53-dependent apoptosis (published erratum appears in *Mol. Cell. Biol.* **14**(6), 4333, 1994). *Mol. Cell. Biol.* **14,** 2556–2563.

Chlichila, K., Moldenhauer, G., Daniel, P., Busslinger, M., Gazzalo, L., Schirrmacher, V., and Khazaie, K. (1995). Immediate effects of reversible HTLV-1 tax function: T-cell activation and apoptosis. *Oncogene* **10,** 269–277.

Chlichlia, K., Busslinger, M., Peter, M., Walczak, H., Krammer, P., Schirrmacher, V., and Khazaie, K. (1997). ICE-proteases mediate HTLV-I tax induced apoptotic T cell death. *Oncogene* **14,** 2265–2272.

Choe, S., Bennett, M., Fugii, G., Curmi, K., Kantardjieff, K., Collier, R., and Eisenberg, D. (1992). The crystal structure of diptheria toxin. *Nature* **357,** 216–222.

Chou, J., and Roizman, B. (1992). The 134.5 gene of herpes simplex virus 1 precludes neuroblastoma cells from triggering total shutoff of protein synthesis characteristic of programmed cell death in neuronal cells. *Proc. Natl. Acad. Sci. USA* **89,** 3266–3270.

Chou, J., and Roizman, B. (1995). Association of a Mr 90,000 phosphoprotein with protein kinase PKR in cells exhibiting enhanced phosphorylation of translation initiation factor eIF-2 and premature shutoff of protein synthesis after infection with 34.5- mutants of herpes simplex virus 1. *Proc. Natl. Acad. Sci. USA* **92,** 10516–10520.

Chou, J., Kern, E., Whitley, R., and Roizman, B. (1990). Mapping of herpes simplex virus-1 neurovirulence to 134.5, a gene nonessential for growth in culture. *Science* **250,** 1262–1266.

Chou, J., Poon, A., Johnson, J., and Roizman, B. (1994). Differential response of human cells to deletions and stop codons in the 34.5 of herpes simplex virus. *J. Virol.* **68,** 8304–8311.

Chvatchko, Y., Valera, S., Aubry, J., Renno, T., Buell, G., and Bonnefoy, J. (1996). The involvement of an ATP gated ion channel, P2X1, in thymocyte apoptosis. *Immunity* **5,** 275–283.

Clem, R., and Miller, L. (1993). Apoptosis reduces both the in vitro replication and the in vivo infectivity of a baculovirus. *J. Virol.* **67,** 3730–3738.

Clement, M., and Stamenkovic, I. (1994). Fas and tumor necrosis factor receptor mediated cell death: Similarities and distinctions. *J. Exp. Med.* **180,** 557–567.

Clouse, K., Cosentino, L., Weih, K., Pyle, S., Robbins, P., Hochstein, H., Natarajan, V., and Farrar, W. (1990). The HIV-1 gp120 envelope protein has the intrinsic capacity to stimulate monokine secretion. *J. Immunol.* **147,** 2892–2901.

Colotta, F., Re, F., Polentarutti, N., Sozzani, S., and Mantovani, A. (1992). Modulation of granulocyte survival and programmed cell death by cytokines and bacterial products. *Blood* **80,** 2012–2020.

Cone, L., Woodard, D., Schlievert, P., and Tomory, G. (1987). Clinical and bacteriological observations of a toxic shock like syndrome due to streptococcus pyogenes. *N. Engl. J. Med.* **317,** 146–149.

Confer, D., and Eaton, J. (1982). Phagocyte impotence caused by an invasive bacterial adenylate cyclase. *Science* **217,** 948–950.

Conlan, J., and North, R. (1991). Neutrophil mediated dissolution of infected host cells as a defense strategy against a facultative intracellular bacterium. *J. Exp. Med.* **174,** 741–744.

Conlan, J., and North, R. (1994). Neutrophils are essential for early anti-listeria defense in the liver, but not in the spleen or peritoneal cavity, as revealed by a granulocyte depleting monoclonal antibody. *J. Exp. Med.* **179,** 259–268.

Correa, P., and Miller, M. (1995). Helicobacter pylori and gastric atrophy-cancer paradoxes. *J. Natl. Cancer Inst.* **87,** 1731–1732.

Cree, I., Nurbhai, S., Milne, G., and Beck, J. (1987). Cell death in granulomata: The role of apoptosis. *J. Clin. Pathol.* **40,** 1314–1319.

Crise, B., Buonocore, L., and Rose, J. (1990). CD4 is retained in the endoplasmic reticulum by the human immunodeficiency type I glycoprotein precursor. *J. Virol.* **64,** 5585–5593.

Crook, N., Clem, R., and Miller, L. (1993). An apoptosis inhibiting baculovirus gene with a zinc finger like motif. *J. Virol.* **67,** 2168–2174.

Czuprynski, C., Brown, F., Maroushek, N., Wagner, R., and Steinberg, H. (1994). Administration of anti-granulocyte mAb RB6-8C5 impairs the resistance of mice to listeria monocytogenes infection. *J. Immunol.* **152,** 1836–1846.

D'Adamio, L., Awad, K., and Reinherz, E. (1993). Thymic and peripheral apoptosis of antigen-specific T cells might cooperate in establishing self tolerance. *Eur. J. Immunol.* **23,** 747–753.

Danen-Van Oorschot, A., Fischer, D., Grimbergen, J., Klein, B., Zhuang, S., Falkenburg, J., Backendorf, C., Quax, P., Van der Eb, A., and Noteborn, M. (1997). Apoptin induces apoptosis in human transformed and malignant cells, but not in normal cells. *Proc. Natl. Acad. Sci. USA* **94,** 5843–5847.

Debatin, M., Fahrig-Farssner, A., Enekel-Stoodt, S., Kreuz, W., Bennder, A., and Krammer, P. (1994). High expression of apo-1 (CD95) on T lymphocytes from human immunodeficiency virus-1 infected children. *Blood* **83,** 3101–3103.

Debbas, M., and White, E. (1993). Wild-type p53 mediates apoptosis by E1A, which is inhibited by E1B. *Genes Dev.* **7,** 546–554.

Depraetere, S., and Joniau, M. (1994). Polar agents with differentiation inducing capacity prime myelomonocytic cell lines to lipopolysaccharide induced cytolysis: The role of endogenous tumor necrosis factor. *Leukemia* **8,** 1951–1959.

Desaintes, C., Demeret, C., Goyat, S., Yaniv, M., and Thierry, F. (1997). Expression of papillomavirus E2 protein in HeLa cells leads to apoptosis. *EMBO J.* **16,** 504–514.

Desbarats, J., Freed, J., Campbell, P., and Newell, M. (1996). Fas expression and death-mediating function are induced by CD4 cross linking on CD4$^+$ T cells. *Proc. Natl. Acad. Sci. USA* **93,** 11014–11018.

Despres, P., Flamand, M., Ceccaldi, P. E., and Deubel, V. (1996). Human isolates of dengue type 1 virus induce apoptosis in mouse neuroblastoma cells. *J. Virol.* **70,** 4090–4096.

Dobblestein, M., and Shenk, T. (1996). Protection against apoptosis by the vaccinia virus SPI-2 gene product. *J. Virol.* **70,** 6479–6485.

Dobner, T., Horikoshi, N., Rubenwolf, S., and Shenk, T. (1996). Blockage by adenovirus E4orf6 of transcriptional activation by the p53 tumor suppressor. *Science* **272,** 1470–1473.

Dorstyn, L., and Kumar, S. (1997). Differential inhibitory effects of CrmA, P35, IAP, and three mammalian homolgues on apoptosis in NIH 3T3 cells following various death stimuli. *Cell Death Differ.* **4,** 570–579.

Duke, R., Witter, R., Nash, P., Young, J., and Ojcius, D. (1994). Cytolysis mediated by ionophores and pore forming agents: Role of intracellular calcium in apoptosis. *FASEB J.* **8,** 237–246.

Dyson, N., and Harlow, E. (1992). Adenovirus E1A targets key regulators of cell proliferation. *Cancer Survival* **12,** 161–195.

Ellis, H., and Horvitz, H. (1986). Genetic control of programmed cell death in the nematode Caenorhabditis elegans. *Cell* **44,** 817–829.

Emille, D., Permutter, M., Malliot, M., Brousse, N., Delfraissy, J., Dormont, J., and Galanaud, P. (1990). Production of interleukins in human immunodeficiency virus-1 replicating lymph nodes. *J. Clin. Invest.* **86,** 148–159.

Esolen, L. M., Park, S. W., Hardwick, J. M., and Griffin, D. E. (1995). Apoptosis as a cause of death in measles virus-infected cells. *J. Virol.* **69,** 3955–3958.

Ettinger, R., Panka, D., Wang, J., Stanger, B., Ju, S., and Rothstein, A. (1995). Fas ligand mediated cytotoxicity is directly responsible for apoptosis of normal CD4$^+$ T cells responding to a bacterial superantigen. *J. Immunol.* **154,** 4302–4308.

Faine, S. (1964). Reticuloendothelial phagocytosis of virulent leptospires. *Am. J. Vet. Res.* **25,** 830–835.

Faine, S. (1994). "*Leptospira* and Leptospirosis." CRC Press, Boca Raton, FL.

Famularo, G., De Simone, C., and Marcellini, S. (1997). Apoptosis: Mechanisms and relation to AIDS. *Med. Hypothesis* **48,** 423–429.

Fan, J., Bass, H., and Fahey, J. (1993). Elevated IFN-γ and decreased IL-2 expression are associated with HIV infection. *J. Immunol.* **151,** 5031–5040.

Farrow, S. N., White, J. H., Martinou, I., Raven, T., Pun, K. T., Grinham, C. J., Martinou, J. C., and Brown, R. (1995). Cloning of a bcl-2 homologue by interaction with adenovirus E1B 19K (published erratum appears in *Nature* **375**(6530), 431, 1995). *Nature* **374,** 731–733.

Fauci, A. (1993). Multifactorial nature of HIV disease: Implications for therapy. *Science* **262,** 1011–1018.

Fauci, A. (1994). Multifactorial and multiphasic components of the immunopathogenic mechanisms of HIV disease. *In* "Retroviruses of Human AIDS and Related Animal Diseases," No. 81-5. Fondation M. Merieux, Lyon. 81-5.

Fernandez-Arias, A., Martinez, S., and Rodriguez, J. F. (1997). The major antigenic protein of infectious bursal disease virus, VP2, is an apoptotic inducer. *J. Virol.* **71,** 8014–8018.

Finco, O., Nuti, S., De Magistris, M., Mangiavacchi, L., Aiuti, A., Forte, P., Fantoni, A., van der Putten, H., and Abrignani, S. (1997). Induction of $CD4^+$ T cell depletion in mice doubly transgenic for HIV gp120 and human CD4. *Eur. J. Immunol.* **27,** 1319–1324.

Finkel, T., and Banda, N. (1994). Indirect mechanisms of HIV pathogenesis: How does HIV kill T cells? *Curr. Opin. Immunol.* **6,** 605–615.

Finkel, T., Tudor-Williams, G., Banda, N., Cotton, M., Curiel, T., Monks, C., Baba, T., Ruorecht, R., and Kupfer, A. (1995). Apoptosis occurs predominantly in bystander cells and not in productively infected cells of HIV and SIV infected lymph nodes. *Nature Med.* **1,** 129–134.

Fleischer, B. (1994). Superantigens produced by infectious pathogens: Molecular mechanism of action and biological significance. *Int. J. Clin. Lab. Res.* **24,** 193–197.

Francis, C., Starnbuch, M., and Falkow, S. (1992). Morphological and cytoskeletal changes in epithelial cells occur immediately upon interaction with *Salmonella typhimurium* grown under low oxygen conditions. *Mol. Microbiol.* **6,** 3077–3087.

Francis, C., Ryan, T., Jones, B., Smith, S., and Falkow, S. (1993). Ruffles induced by salmonella and other stimuli direct macropinocytosis of bacteria. *Nature* **364,** 639–642.

Fratazzi, C., Arbeit, R., Carini, C., and Remold, H. (1997). Programmed cell death of *Mycobacterium avium* serovar 4-infected macrophages prevents the mycobacteria from spreading and induces mycobacteria growth inhibition by freshly added, uninfected macrophages. *J. Immunol.* **158,** 432–437.

Fries, K., Miller, W., and Raab-Traub, B. (1996). Epstein–Barr virus latent membrane protein 1 blocks p53-mediated apoptosis through the induction of the A20 gene. *J. Virol.* **70,** 8653–8659.

Frisch, S., and Francis, H. (1994). Disruption of epithelial cell–matrix interactions induces apoptosis. *J. Cell Biol.* **124,** 619–926.

Fromm, L., Shawlot, W., Gunning, K., Butel, J., and Overbeek, P. (1994). The retinoblastoma binding protein region of simian virus 40 large T antigen alters cell cycle regulation in the lenses of transgenic mice. *Mol. Cell. Biol.* **14,** 6743–6754.

Gangadharam, P., and Edwards, C., III (1984). Release of superoxide anion from resident and activated mouse peritoneal macrophages infected with *Mycobacterium intracellularae. Am. Rev. Respir. Dis.* **130,** 834–838.

Gelbard, H., Nottet, H., Swindells, S., Jett, M., Dzenko, P., Gennis, P., White, R., Wang, L., Choi, Y., and Zhang, D. (1994). Platelet activating factor: A candidate human immunodeficiency virus type 1 induced neurotoxin. *J. Virol.* **68,** 4628–4635.

Gelbard, H., James, H., Sharer, S., Perry, S., Saito, Y., Kazee, A., Blumberg, B., and Epstein, L. (1995). Apoptotic neurons in brains from pediatric patients with HIV-1 encephalitis and progressive encephalopathy. *Neuropathol. Appl. Neurobiol.* **21,** 208–217.

Genis, P., Jett, M., Bernton, E., Boyle, T., Gelbard, H., Dzenki, K., Keane, R., Resnick, L., Mizrachi, Y., Volsky, D., *et al.* (1992). Cytokines and arachidonic acid metabolites produced during human immunodeficiency virus infected macrophage–astroglia interactions: Implications for neuropathogenesis of HIV disease. *J. Exp. Med.* **176,** 1703–1718.

Glasgow, G. M., McGee, M. M., Sheahan, B. J., and Atkins, G. J. (1997). Death mechanisms in cultured cells infected by Semliki Forest virus. *J. Gen. Virol.* **78,** 1559–1563.

Gonzalo, J., Baixeras, E., Garcia, A., Chandy, A., Rooijen, N., Martinez-A., C. and Kroemer, G. (1994). Differential in vivo effects of a superantigen and an antibody targeted to the same T cell receptor. *J. Immunol.* **152,** 1597–1608.

Gooding, L. R. (1992). Virus proteins that counteract host immune defenses. *Cell* **71,** 5–7.

Gooding, L. R., Elmore, L. W., Tollefson, A. E., Brady, H. A., and Wold, W. S. (1988). A 14,700 MW protein from the E3 region of adenovirus inhibits cytolysis by tumor necrosis factor. *Cell* **53,** 341–346.

Gooding, L. R., Ranheim, T. S., Tollefson, A. E., Aquino, L., Duerksen-Hughes, P., Horton, T. M., and Wold, W. S. (1991). The 10,400 and 14,500-dalton proteins encoded by region E3 of adenovirus function together to protect many but not all mouse cell lines against lysis by tumor necrosis factor. *J. Virol.* **65,** 4114–4123.

Gordon, A., and Hart, P. (1994). Stimulation or inhibition of the respiratory burst in cultured macrophages in a mycobacterium model: Initial stimulation followed by inhibition after phagocytosis. *Infect. Immun.* **62,** 4650–4651.

Goryo, M., Sugimura, H., Matsumoto, S., Umemura, T., and Itakura, C. (1985). Isolation of an agent inducing chicken anemia. *Avian Pathol.* **18,** 329–343.

Gougeon, M., Garcia, S., Heeney, J., Tschopp, R., Lecoeur, H., Guetard, D., Rame, V., Dauguet, C., and Montagnier, L. (1993). Programmed cell death in AIDS-related HIV and SIV infections. *AIDS Res. Hum. Retroviruses* **9,** 553–563.

Gratama, J., Oosterveer, M., Zwaan, F., Lepoutre, J., Klein, G., and Ernberg, I. (1988). Eradication of Epstein–Barr virus by allogeneic bone marrow transplantation: Implications for sites of viral latency. *Proc. Natl. Acad. Sci. USA* **85,** 8693–8696.

Greene, W. (1993). AIDS and the immune system. *Sci. Am.* **269,** 98–105.

Gregory, C., Dive, C., Henderson, S., Smith, C., Williams, G., Gordon, J., and Rickinson, A. (1991). Activation of Epstein–Barr virus latent genes protects human B cells from death by apoptosis. *Nature* **349,** 612–614.

Gribben, J., Freeman, G., Bousiotiss, V., Rennert, P., Jellis, C., Greenfield, E., Barber, M., Restivo, V., Ke, X., Gray, G., and Nadler, L. (1995). CTLA4 mediates antigen specific apoptosis of human T cells. *Proc. Natl. Acad. Sci. USA* **92,** 811–815.

Groux, H., Torpier, G., Monte, D., Mounton, Y., Capon, A., and Amiesen, J. (1992). Activation induced death by apoptosis in $CD4^+$ T cells from human immunodeficiency-infected asymptomatic individuals. *J. Exp. Med.* **175,** 331–340.

Guzman, C., Domann, E., Rohde, M., Bruder, D., Darji, A., Weiss, S., Wehland, J., Chakraborty, T., and Timmis, K. (1996). Apoptosis of mouse denritic cells is triggered by listeriolysin, the major virulence determinant of Listeria monocytogenes. *Mol. Microbiol.* **20,** 119–126.

Hagemeier, C., Caswell, R., Hayhurst, G., Sinclair, J., and Kouzarides, T. (1994). Functional interaction between the HCMV IE2 transactivator and the retinoblastoma protein. *EMBO J.* **1994,** 2897–2903.

Hakansson, S., Schesser, K., Persson, C., Galyov, E., Rosqvist, R., Homble, H., and Wolf-Watz, H. (1996). The YopB protein of *Yersinia psuedotuberculosis* is essential for the translocation of Yop effector proteins across the target cell plasma membrane and displays a contact dependent membrane disrupting activity. *EMBO J.* **15,** 5812–823.

Hameed, A., Olsen, K., Lee, M., Lichtenheld, M., and Podack, E. (1989). Cytolysis by Ca permeable transmembrane channels. *J. Exp. Med.* **169,** 765–777.

Han, J., Sabbatini, P., and White, E. (1996a). Induction of apoptosis by human Nbk/Bik, a BH3-containing protein that interacts with E1B 19K. *Mol. Cell. Biol.* **16,** 5857–5864.

Han, J., Sabbatini, P., Perez, D., Rao, L., Modha, D., and White, E. (1996b). The E1B 19K protein blocks apoptosis by interacting with and inhibiting the p53-inducible and death-promoting Bax protein. *Genes Dev.* **10,** 461–477.

Hanski, E., and Coote, G. (1991). *Bordetella Pertussis* adenylate cyclase toxin. *In* "Sourcebook for Bacterial Toxins" (J. Alouf and J. Freer, Eds.), pp. 349–366. Academic Press, London.

Hanski, C., Kutscchka, H., Schmoranzer, H., Naumann, M., Stallmach, A., Hahn, H., Menge, H., and Riecken, E. (1989). Immunohistochemical and electron microscopic study of interaction of *Yersinia enterocolitica* serotype O8 with intestinal mucosa during experimental enteritis. *Infect. Immun.* **57,** 673–678.

Hardt, W., and Galan, J. (1997). A secreted salmonella protein with homology to an avirulence determinant of plant pat hogenic bacteria. *Proc. Natl. Acad. Sci. USA* **94,** 9887–9892.

Hashimoto, S., Ishii, A., and Yonehara, S. (1991). The E1b oncogene of adenovirus confers cellular resistance to cytotoxicity of tumor necrosis factor and monoclonal anti-Fas antibody. *Int. Immunol.* **3,** 343–351.

Haslett, C., Lee, A., Savill, J., Meagher, L., and Whyte, M. (1991). Apoptosis and functional changes in aging neutrophils. *Chest Suppl.,* 6S.

Hawkins, C., and Vaux, D. (1997). The role of the Bcl-2 family of apoptosis regulatory proteins in the immune system. *Sem. Immunol.* **9,** 25–33.

Hawkins, C., Uren, A., Hacker, G., Medcalf, R., and Vaux, D. (1996). Inhibition of interleukin 1β-converting enzyme mediated apoptosis of mammalian cells by baculovirus IAP. *Proc. Natl. Acad. Sci. USA* **93,** 13786–13790.

Hawley-Nelson, P., Vousden, K., Hubbert, N., Lowy, D., and Schiller, J. (1989). HPV-16 E6 and E7 proteins cooperate to immortalize human foreskin keratinocytes. *EMBO J.* **8,** 3905–3910.

He, J., deCastro, C., Vandenbark, G., Busciglio, J., and Gabuzda, D. (1997). Astrocyte apoptosis induced by HIV-1 transactivation of the c-kit protooncogene. *Proc. Natl. Acad. Sci. USA* **94,** 3954–3959.

Heidenreich, S., Otte, B., Lang, D., and Schmidt, M. (1996). Infection by Candida albicans inhibits apoptosis of human monocytes and monocytic U937 cells. *J. Leukocyte Biol.* **60,** 737–743.

Henderson, B., Poole, S., and Wilson, M. (1996). Bacterial modulins: A novel class of virulence factors which cause host tissue pathology by inducing cytokine synthesis. *Microbiol. Rev.* **60,** 316–341.

Henderson, S., Rowe, M., Gregory, C., Croom-Carter, D., Wang, F., Longnecker, R., Kieff, E., and Rickinson, A. (1991). Induction of *bcl*-2 expression by Epstein–Barr virus latent membrane protein 1 protects infected B cells from programmed cell death. *Cell* **65,** 1107–1115.

Henderson, S., Huen, D., Rowe, M., Dawson, C., Johnson, G., and Rickinson, A. (1993). Epstein–Barr virus coded BHRF-1 protein, a viral homologoue of Bcl-2, protects human B cells from programmed cell death. *Proc. Natl. Acad. Sci. USA* **90,** 8479–8483.

Hershberger, P., Dickson, J., and Friesen, P. (1992). Site specific mutagenesis of the 35 kilodalton protein encoded by *Autographa californica* nuclear polyhedrosis virus: Cell line specific effects on viral replication. *J. Virol.* **67,** 3730–3738.

Hershberger, P., Lacount, D., and Friesen, P. (1994). The apoptotic supressor p35 is required early during baculovirus replication and is targeted to the cyotosol of infected cells. *J. Virol.* **68,** 3467–3477.

Hickman, E., Bates, S, and Vousden, K. (1997). Petrubation of the p53 response by human papillomavirus type 16 E7. *J. Virol.* **71,** 3710–3718.

Hildebrand, A., Roth, M., and Bhakdi, S. (1991). Staphyloccous alpha toxin: Dual mechanisms of binding to target cells. *J. Biol. Chem.* **266,** 17195–17200.

Hinshaw, V. S., Olsen, C. W., Dybdahl-Sissoko, N., and Evans, D. (1994). Apoptosis: A mechanism of cell killing by influenza A and B viruses. *J. Virol.* **68,** 3667–3673.

Hofer, M., Newell, K., Duke, R., Schlievert, P., Freed, J., and Leung, D. (1996). Differential effects of staphylococcal toxic shock syndrome toxin-1 on B cell apoptosis. *Proc. Natl. Acad. Sci. USA* **93,** 5425–5430.

Horvitz, H., Shaham, S., and Hengartner, M. (1994). The genetics of programmed cell death in the nematode *Caenorhabditis elegans. Cold Spring Harbor Symp. Quant. Biol.* **59,** 377–385.

Horwitz, M. (1983). The legionnaire's bacterium inhibits phagosome–lysosome fusion in human monocytes. *J. Exp. Med.* **158,** 1319–1331.

Horwitz, M., and Maxfield, F. (1984). Legionella pneumophila inhibits acidification of its phagosome in human monocytes. *J. Cell Biol.* **99,** 105–114.

Howes, K., Ransom, N., Papermaster, D., Lasudry, J., Albert, D., and Windle, J. (1994). Apoptosis or retinoblastoma: Alternative fates of photoreceptors expressing the HPV 16 E7 genes in the presence or absence of p53. *Genes Dev.* **8,** 1300–1310.

Hsu, H. (1989). Pathogenesis and immunity in murine salmonellosis. *Microbiol. Rev.* **53,** 390–409.

Huang, D. C., Cory, S., and Strasser, A. (1997). Bcl-2, Bcl-XL and adenovirus protein E1B19kD are functionally equivalent in their ability to inhibit cell death. *Oncogene* **14,** 405–414.

Ikigae, H., and Nakae, T. (1987). Interaction of the alpha toxin of S aureus with the liposome membrane. *J. Biol. Chem.* **262,** 2150–2155.

Ink, B., Gilbert, C., and Evan, G. (1995). Delay of vaccinia virus induced apoptosis in nonpermissive Chinese hamster ovary cells by the cowpox virus CHOhr and adenovirus E1B 19K genes. *J. Virol.* **69,** 661–668.

Inoue, Y., Yasukawa, M., and Fujita, S. (1997). Induction of T cell apoptosis by human herpesvirus 6. *J. Virol.* **71,** 3751–3759.

Inward, C., Williams, J., Chant, I., Crocker, J., Milford, D., Rose, P., and Taylor, C. (1995). Verocytotoxin-1 induces apoptosis in vero cells. *J. Infect. Dis.* **30,** 213–218.

Isogai, E., Isogai H., Kimura, K., Fujii, N., Takagi, S., Hirose, K., and Hayashi, M. (1996). In vivo induction of apoptosis and immune responses in mice by administration of LPS. *Infect. Immun.* **64,** 1461–1466.

Ito, M., Yamamoto, T., Watanabe, M., Ihara, T., Kamiya, H., and Sakurai, M. (1996). Detection of measles virus-induced apoptosis of human monocytic cell line (THP-1) by DNA fragmentation ELISA. *FEMS Immunol. Med. Microbiol.* **15,** 115–122.

Ito, M., Koide, W., Watanabe, M., Kamiya, H., and Sakurai, M. (1997a). Apoptosis of cord blood T lymphocytes by herpes simplex virus type 1. *J. Gen. Virol.* **78,** 1971–1975.

Ito, M., Watanabe, M., Kamiya, H., and Sakurai, M. (1997b). Herpes simplex virus type 1 induces apoptosis in peripheral blood T lymphocytes. *J. Infect. Dis.* **175,** 1220–1224.

Ito, M., Watanabe, M., Ihara, T., Kamiya, H., and Sakurai, M. (1997c) Measles virus induces apoptotic cell death in lymphocytes activated with phorbol 12-myristate 13-acetate (PMA) plus calcium ionophore. *Clin. Exp. Immunol.* **108,** 266–271.

Jaleco, A., Covas, M., and Victorino, R. (1994). Analysis of lymphocyte cell death and apoptosis in HIV-2 infected patients. *Clin. Exp. Immunol.* **98,** 185–189.

Janeway, C. (1992). The immune system evolved to discriminate infectious nonself from non infectious self. *Immunol. Today* **13,** 11–16.

Janeway, C., and Bottomly, K. (1994). Signals and signs for lymphocyte responses. *Cell* **76,** 275–285.

Jenkinson, E., Kingston, R., Smith, C., Williams, G., and Owen, J. (1989). Antigen induced apoptosis in developing T cells: A mechanism for negative selection of the T cell repertoire. *Eur. J. Immunol.* **19,** 2175–2177.

Jeurissen, S., Wagenaar, F., Pol, J., van der Eb, A., and Noteborn, M. (1992). Chicken anemia virus causes apoptosis of thymocytes after in vivo infection and of cell lines after in vitro infection. *J. Virol.* **66,** 7383–7388.

Joe, A., Ferrari, G., Jiang, H., Liang, X., and Levine, B. (1996). Dominant inhibitory Ras delays Sindbis virus induced apoptosis in neuronal cells. *J. Virol.* **70,** 7744–7751.

Johnson, B., Stone, G., Godec, M., Asher, D., Gajdusek, D., and Gibbs, C. (1993). Long term observation of human immunodeficiency virus infected chimpanzees. *AIDS Res. Hum. Retroviruses* **9,** 375–378.

Johnson, V. (1990). Does diphtheria toxin have nuclease activity? *Science* **250,** 832–834.

Jonas, D., Schultheis, B., Klas, C., Krammer, P., and Bhakdi, S. (1993). Cytocidal effects of *Eschericia coli* hemolysin on human t lymphocytes. *Infect. Immun.* **61**1715–1721.

Jonas, D., Walev, I., Berger, T., Liebetrau, M., Palmer, M., and Bhakdi, S. (1994). Novel path to apoptosis: Small transmembrane pores created by staphylococcal alpha toxin in T lymphocytes evoke internucleosomal degradation. *Infect. Immun.* **62,** 1304–1312.

Jones, B., Ghori, N., and Falkow, S. (1994). *Salmonella typhimurium* initiates murine infection by penetrating and destroying the specialized epithelial M cells of Peyer's patches. *J. Exp. Med.* **180,** 15–23.

Kabelitz, D., and Wesselborg, S. (1992). Life and death of a superantigen reactive human CD4⁺T cell clone: Staphylococcal enterotoxins induce death by apoptosis but simultaneously trigger a proliferative response in the presence of HLA-DR⁺ antigen presenting cells. *Int. Immunol.* **4,** 1381–1388.

Katada, T., Oinuma, M., and Ui, M. (1986). Mechanisms for inhibition of the catalytic activity of adenylate cyclase by the guanine nucleotide binding proteins serving as the substrate of islet activating protein, pertussis toxin. *J. Biol. Chem.* **261,** 5215–5221.

Kato, S., Muro, M., Akifusa, S., Hanada, N., Semba, I., Fujii, T., Kowashi, Y., and Nishihara, T. (1995). Evidence for apoptosis of murine macrophages by actinobacillus actinomycetemcomitans infection. *Infect. Immun.* **63,** 3914–3919.

Katsikis, P., Wunderlich, E., Smith, C., and Herzenberg, L. (1995). Fas antigen stimulation induces marked apoptosis of T lymphocytes in human immunodeficiency virus infected individuals. *J. Exp. Med.* **181,** 2029–2036.

Kawabe, Y., and Ochi, A. (1991). Programmed cell death and extrathymic reduction of VB8⁺ CD4⁺ T cells in mice tolerant to Staphylococcus aureus enterotoxin B. *Nature* **349,** 245–248.

Kawanishi, M. (1993). Epstein–Barr virus induces fragmentation of chromosomal DNA during lytic infection. *J. Virol.* **67,** 7654–7658.

Keane, J., Sablinska, M., Remold, H., Chupp, G., Meek, B., Fenton, M., and Kornfeld, H. (1997). Infection by *Mycobacterium tuberculosis* promotes human alveolar macrophage apoptosis. *Infect. Immun.* **65,** 298–304.

Keenan, K., Dharpnack, D., Formal, S., and O'Brien, A. (1986). Morphologic evaluation of the effects of shiga toxin and E. coli shiga like toxin on the rabbit intestine. *Am. J. Pathol.* **125,** 69–80.

Kessiss, T., Slebos, R., Nelson, W., Kastan, M., Plunkett, B., Han, S., Lorincz, A., Hedrick, L., and Cho, K. (1993). Human papillomavirus 16 E6 expression disrupts the p53 mediated cellular response to DNA damage. *Proc. Natl. Acad. Sci. USA* **90,** 3988–3992.

Khan, I. A., Matsuura, T., and Kasper, L. H. (1996). Activation-mediated CD4⁺ T cell unresponsiveness during acute Toxoplasma gondii infection in mice. *Int. Immunol.* **8,** 887–896.

Khelef, N., and Guiso, N. (1995). Induction of macrophage apoptosis by *Bordetella pertussis* adenylate cyclase-hemolysin. *FEMS Microbiol. Lett.* **134,** 27–32.

Khelef, N., Sakamoto, H., and Guiso, N. (1992). Both adenylate cyclase and hemolytic activities are required by bordatella pertussis to initiate infection. *Microbiol. Pathogen* **12,** 227–235.

Khelef, N., Zychlinsky, A., and Guiso, N. (1993). *Bordetella pertussis* induces apoptosis in macrophages: Role of adenylate cyclase-hemolysin. *Infect. Immun.* **61,** 4064–4071.

Kibler, K., Shors, T., Perkins, K., Zeman, C., Banaszak, M., Biesterfeldt, J., Langland, J., and Jacobs, B. (1997). Double stranded RNA is a trigger for apoptosis in vaccinia virus infected cells. *J. Virol.* **71,** 1992–2003.

Kochi, S., and Collier, J. (1993). DNA fragmentation and cytolysis in U937 cells treated with diphtheria toxin or other inhibitors of protein synthesis. *Exp. Cell Res.* **208,** 296–302.

Koga, Y., Tanaka, K., Lu, Y., Oh-Tsu, M., Sasaki, M., Kimura, G., and Nomoto, K. (1994). Priming of immature thymocytes to CD3-mediated apoptosis by infection with murine cytomegalovirus. *J. Virol.* **68,** 4322–4328.

Kohno, K., and Uchida, T. (1987). Highly frequent single amino acid substitution in mammalian EF-2 results in expression of resistance to EF-2 ADP-ribosylating toxins. *J. Biol. Chem.* **262,** 12298–12305.

Komiyama, T., Ray, C., Pickup, D., Howard, A., Thornberry, N., Peterson, E., and Salvesen, G. (1994). Inhibition of interleukin-1 beta converting enzyme by the cowpox virus serpin CrmA: An example of cross class inhibition. *J. Biol. Chem.* **269,** 19331–19337.

Koyama, A. H. (1995). Induction of apoptotic DNA fragmentation by the infection of vesicular stomatitis virus. *Virus Res.* **37,** 285–290.

Koyama, A., and Miwa, Y. (1997). Suppression of apoptotic DNA fragmentation in herpes simplex virus type-1 infected cells. *J. Virol.* **71,** 2567–2571.

Krajcsi, P., Dimitrov, T., Hermiston, T. W., Tollefson, A. E., Ranheim, T. S., Vande Pol, S. B., Stephenson, A. H., and Wold, W. S. (1996). The adenovirus E3-14.7K protein and the E3-10.4K/14.5K complex of proteins, which independently inhibit tumor necrosis factor (TNF)-induced apoptosis, also independently inhibit TNF-induced release of arachidonic acid. *J. Virol.* **70,** 4904–4913.

Kuroda, K., Yagi, J., Imanishi, K., Yan, X., Li, X., Fujimaki, W., Kato, H., Akiyama, T., Kumazawa, Y., Abe, H., and Uchiyama, T. (1996). Implantation of IL-2 containing osmotic pump prolongs the survival of superantigen reactive T cells expanded in mice injected with bacterial superantigen. *J. Immunol.* **157,** 1422–1431.

Lahdevirta, J., Maury, C., Teppo, A., and Repo, H. (1988). Elevated levels of circulating cachetin/tumor necrosis factor in patients with acquired immunodeficiency syndrome. *Am. J. Med.* **85,** 289–291.

Laine, V., Nyman, K., Peuravuori, H., Henriksen, K., Parvinen, M., and Nevalainen, T. (1996). Lipopolysaccharide induced apoptosis of rat pancreatic acinar cells. *Gut* **36,** 747–752.

Lam, K. M. (1996). Newcastle disease virus-induced apoptosis in the peripheral blood mononuclear cells of chickens. *J. Comp. Pathol.* **114,** 63–71.

Lam, K. M. (1997). Morphological evidence of apoptosis in chickens infected with infectious bursal disease virus. *J. Comp. Pathol.* **116,** 367–377.

Lam, K. M., Vasconcelos, A. C., and Bickford, A. A. (1995). Apoptosis as a cause of death in chicken embryos inoculated with Newcastle disease virus. *Microbiol. Pathogen.* **19,** 169–174.

Laochumroonvorapong, P., Paul, S., Elkon, K., and Kaplan, G. (1996). H202 induces monocytes apoptosis and reduces viability of Mycobacterium avium–M. intracellularae within cultured human monocytes. *Infect. Immun.* **64,** 452–459.

Laurent-Crawford, A., Krust, B., Muller, S., Riviere, Y., Rey-Cuille, M., Bechet, J., Montagnier, L., and Hovanessian, A. (1991). The cytopathic effect of HIV is associated with apoptosis. *Virology* **185,** 829–839.

Lauwers, G., Scoot, G., and Hendricks, J. (1994). Immunohistochemical evidence of aberrant bcl-2 expression in gastric epithelial dysplasia. *Cancer* **73,** 2900–2904.

Lee, S., and Esteban, M. (1994). The interferon induced double stranded RNA activated protein kinase induces apoptosis. *Virology* **199,** 491–496.

Leninger, E., Roberts, M., Kenimer, J., Charles, I., Fairweather, N., Novotny, P., and Brennan, M. (1991). Pertactin, an Arg-Gly-Asp containing Bordetella pertussis surface protein that promotes adherence of mammalian cells. *Proc. Natl. Acad. Sci. USA* **88,** 345–349.

Leonardo, M. (1991). Interleukin-2 programs T lymphocytes for apoptosis. *Nature* **363,** 858–861.

Leopoldi, R., and Roizman, B. (1996). The herpes simplex virus major regulatory protein ICP4 blocks apoptosis induced by the virus or by hyperthemia. *Proc. Natl. Acad. Sci. USA* **93,** 9583–9587.

Lessnick, S., Bruce, C., Baldwin, R., Chang, M., Nakumura, L., and Wisnieski, B. (1990). Does diphtheria toxin have nuclease activity? *Science* **250,** 836–838. [Reply]

Lessnick, S., Lyczak, J., Bruce, C., Lewis, D., Kim, P., Stolowirz, M., Hood, L., and Wisnieski, B. (1992). Localization of diphtheria toxin nuclease activity to fragment A. *J. Bacteriol.* **174,** 2032–2038.

Levine, B., Hardwick, J., Trapp, B., Crawford, T., Bollinger, R., and Griffin, D. (1991). Antibody mediated clearance of alphavirus infection from neurons. *Science* **254,** 856–860.

Levine, B., Huang, Q., Isaacs, J., Reed, J., Griffin, D., and Hardwick, J. (1993). Conversion of lytic to persistent alphavirus infection by the bcl-2 oncogene. *Nature* **361,** 739–742.

Levine, B., Hardwick, J., and Griffin, D. (1994). Persistence of alphavirus in vertebrate hosts. *Trends Microbiol.* **2,** 25–28.

Levine, B., Goldman, J., Jiang, H., Griffin, D., and Hardwick, J. (1996). Bcl-2 protects mice against fatal alphavirus encephalitis. *Proc. Natl. Acad. Sci. USA* **93,** 4810–4815.

Levy, J. (1993). Pathogenesis of human immunodeficiency virus infection. *Microbiol. Rev.* **57,** 183–289.

Lewis, D., Ng, D., Adu-oppong, A., Schoberg, W., and Rodgers, J. (1994). Anergy and apoptosis in CD8⁺ T cells from HIV infected persons. *J. Immunol.* **153,** 412–420.

Lewis, J., Wesselingh, L., Griffin, D., and Hardwick, J. (1996). Alphavirus induced apoptosis in mouse brains correlates with neurovirulence. *J. Virol.* **70,** 1828–1835.

Li, C., Friedman, D., Wang, C., Metelev, V., and Pardee, A. (1995). Induction of apoptosis in uninfected lymphocytes by the HIV Tat protein. *Science* **268,** 429–431.

Li, Y., Kang, J., and Horwitz, M. S. (1997). Interaction of an adenovirus 14.7-kilodalton protein inhibitor of tumor necrosis factor alpha cytolysis with a new member of the GTPase superfamily of signal transducers. *J. Virol.* **71,** 1576–1582.

Liao, C., Lin, Y., Wang, J., Huang, Y., Yeh, C., Ma, S., and Chen, L. (1997). Effect of enforced expression of human bcl-2 on Japanese encephalitis virus induced apoptosis in cultured cells. *J. Virol.* **71,** 5963–5971.

Lifson, J., Feinberg, M., Reyes, G., Rabin, L., Banapour, B., Chakrabarti, S., Moss, B., Wong-Staal, F., Steimer, K., and Engleman, E. (1986). Induction of CD4 dependent cell fusion by the HTLVIII/LAV envelope protein. *Nature* **323,** 725–728.

Lin, Y., Lei, H., Low, T. L. K., Shen, C. L., Chou, L. J., and Jan, M. S. (1992). In vivo induction of apoptosis in immature thymocytes by staphylococcal enterotoxin B. *J. Immunol.* **149,** 1156–1163.

Lin, Y., Kao, S., Jan, M., Cheng, M., Wing, L., Chang, W., Lei, H., and Lin, M. (1995). Changes of protein kinase C subspecies in Staphylococcal enterotoxin B induced thymocyte apoptosis. *Biochem. Biophys. Res. Commun.* **213,** 1132–1139.

Lindgren, S. W., Stojilkovic, I., and Heffron, F. (1996). Macrophage killing is an essential virulence mechanism of Salmonella typhimurium. *Proc. Natl. Acad. Sci. USA* **93,** 4197–4201.

Lopes, M. F., da Veiga, V. F., Santos, A. R., Fonseca, M. E., and DosReis, G. A. (1995). Activation-induced CD4⁺ T cell death by apoptosis in experimental Chagas' disease. *J. Immunol.* **154,** 744–752.

Lowe, S. W., and Ruley, H. E. (1993). Stabilization of the p53 tumor suppressor is induced by adenovirus 5 E1A and accompanies apoptosis. *Genes Dev.* **7,** 535–545.

Lukac, D., Manuppello, J., and Alwine, J. (1994). Transcriptional activation by the human cytomegalovirus immediate early proteins: Requirements for simple promoter structure and interactions with multiple components of the transcription complex. *J. Virol.* **68,** 5184–5193.

Lyerly, D., Saum, K., MacDonald, D., and Wilkins, T. (1985). Effects of clostridium dificile toxins given intragastrically to animals. *Infect. Immun.* **47,** 349–352.

Mace, J., Graham, K., Lee, S., Schrieber, M., Boshkov, L., and McFadden, G. (1996). Expression of myxoma virus tumor necrosis factor receptor homologue and M11L genes is required to prevent virus induced apoptosis in rabbit T lymphocytes. *Virology* **218,** 232–237.

Magnuson, D., Knudsen, B., Geiger, J., Brownstone, R., and Nath, A. (1995). Human immunodeficiency virus type 1 Tat activates non-*N*-methyl-D-aspartate excitatory amino acid receptors and causes neurotoxicity. *Ann. Neurol.* **37,** 373–380.

Mahida, Y., Makh, S., Gray, T., and Borriello, S. (1996). Effect of Clostridium dificile toxin A on human intestinal epithelial cells: Induction of interleukin-8 production and apoptosis after cell detachment. *Gut* **38,** 337–347.

Mandic-Mulec, I., Weiss, J., and Zychlinsky, A. (1997). *Shigella flexneri* is trapped in polymorphonuclear leukocyte vacuoles and efficiently killed. *Infect. Immun.* **65,** 110–115.

Mangan, D., Welch, G., and Wahl, S. (1991a). Lipopolysaccharide, tumor necrosis factor alpha, and IL-1β prevent programmed cell death (apoptosis) in human peripheral blood monocytes. *J. Immunol.* **146,** 1541–1546.

Mangan, D., Taichman, N., Lally, E., and Wahl, S. (1991b). Lethal effects of Actinobacillus actinomycetemcomitans leukotoxin on human T lymphocytes. *Infect. Immun.* **59,** 3267–3272.

Mangan, D., Robertson, B., and Wahl, S. (1992). IL-4 enhances programmed cell death (apoptosis) in stimulated human monocytes. *J. Immunol.* **148,** 1812–1816.

Mangeney, M., Lingwood, C., Taga, S., Caillou, B., Tursz, T., and Wiels, J. (1993). Apoptosis induced in Burkitt's lymphoma cells via Gb3/CD77, a glcolipid antigen. *Cancer Res.* **53,** 5314–5319.

Mannick, E., Bravo, L., Zarama, G., Realpe, J., Zhang, X., Ruiz, B., Fontham, E., Mera, R., Miller, M., and Correa, P. (1996). Inducible nitric oxide synthase, nitrotyrosine, and apoptosis in Helicobacter pylori gastritis: Effect of antibiotics and antioxidants. *Cancer Res.* **56,** 3238–3243.

Marcellus, R. C., Teodoro, J. G., Wu, T., Brough, D. E., Ketner, G., Shore, G. C., and Branton, P. E. (1996). Adenovirus type 5 early region 4 is responsible for E1A-induced p53-independent apoptosis. *J. Virol.* **70,** 6207–6215.

Marianneau, P., Cardona, A., Edelman, L., Deubel, V., and Despres, P. (1997). Dengue virus replication in human hepatoma cells activates NF-kappaB which in turn induces apoptotic cell death. *J. Virol.* **71,** 3244–3249.

McCarthy, S., Symonds, H., and Dyke, T. (1994). Regulation of apoptosis in transgenic mice by simian virus 40 T antigen-mediated inactivation of p53. *Proc. Natl. Acad. Sci. USA* **91,** 3979–3983.

McCloskey, T., Ott, M., Tribble, E., Khan, S., Teichberg, S., Paul, M., Pahwa, S., Verdin, E., and Narendra, C. (1997). Dual role of HIV Tat in regulation of apoptosis in T cells. *J. Immunol.* **158,** 1014–1019.

Menestrina, G. (1986). Ionic channels formed by Staphylococcal aureus alpha toxin: Voltage dependent inhibition by divalent and trivalent cations. *J. Membrane Biol.* **19,** 177–190.

Merien, F., Baranton, G., and Perolat, P. (1997). Invasion of vero cells and induction of apoptosis in macrophages by pathogenic *Leptospira interrogans* are correlated with virulence. *Infect. Immun.* **65,** 729–738.

Merry, D., and Korsmeyer, S. (1997). Bcl-2 gene family in the nervous system. *Annu. Rev. Neurol.* **20,** 245–267.

Meyaard, L., and Miedema, F. (1997). Immune dysregulation and $CD4^+$ T cell loss in HIV-1 infection. *Springer Sem. Immunopathol.* **18,** 285–303.

Meyaard, L., Otto, S., Jonker, R., Mijnister, M., Keet, R., and Miedema, F. (1992). Programmed death of T cells in HIV-1 infection. *Science* **257,** 217–219.

Meyaard, L., Otto, S., Keet, I., Roos, M., and Miedema, F. (1994). Programmed death of T cells in human immunodeficiency virus infection. *J. Clin. Invest.* **93,** 982–988.

Mills, S., Boland, A., Sory, M., van der Smissen, P., Kerbourch, C., Finlay, B., and Cornelis, G. (1997). *Yersinia enterocolitica* induces apoptosis in macrophages by a process requiring functional type III secretion and translocation mechanisms involving YopP, presumably acting as an effector protein. *Proc. Natl. Acad. Sci. USA* **94,** 12638–12643.

Mita, E., Hayashi, N., Iio, S., Takehara, T., Hijioka, T., Kasahara, A., Fusamoto, H., and Kamada, T. (1994). Role of Fas ligand in apoptosis induced by hepatitis C virus infection. *Biochem. Biophys. Res. Commun.* **204,** 468–474.

Mitchell, T., Ketley, J., Haslam, S., Stephen, J., Burdon, D., and Candy, D. (1986). Effect of toxin A and B of Clostridium dificile on rabbit ileum and colon. *Gut* **27,** 78–85.

Miura, M., Zhu, H., Rotello, R., Hartwieg, E. A., and Yuan, J. (1993). Induction of apoptosis in fibroblasts by IL-1β converting enzyme, a mammalian homolog of the *C. elegans* cell death gene *ced-3. Cell* **75,** 653–660.

Miyashita, T., and Reed, J. (1995). Tumor suppressor p53 is a direct transcriptional activator of the human bax gene. *Cell* **80,** 293–299.

Molloy, A., Laochumroonvorapong, P., and Kaplan, G. (1994). Apoptosis, but not necrosis, of infected monocytes is coupled with killing of intracellular Bacillus calmette-guerin. *J. Exp. Med.* **180,** 1499–1509.

Monack, D. M., Raupach, B., Hromockyj, A. E., and Falkow, S. (1996). *Salmonella typhimurium* invasion induces apoptosis in infected macrophages. *Proc. Natl. Acad. Sci. USA* **93,** 9833–9838.

Monack, D., Mecsas, J., Ghori, N., and Falkow, S. (1997). Yersinia signals macrophages to undergo apoptosis and YopJ is necessary for cell death. *Proc. Natl. Acad. Sci. USA* **94,** 10385–10390.

Moore, K. J., and Matlashewski, G. (1994). Intracellular infection by Leishmania donovani inhibits macrophage apoptosis. *J. Immunol.* **152,** 2930–2937.

Moore, M., Horikoshi, N., and Shenk, T. (1996). Oncogenic potential of the adenovirus E4orf6 protein. *Proc. Natl. Acad. Sci. USA* **93,** 11295–11301.

Morey, A. L., Ferguson, D. J., and Fleming, K. A. (1993). Ultrastructural features of fetal erythroid precursors infected with parvovirus B19 in vitro: Evidence of cell death by apoptosis. *J. Pathol.* **169,** 213–220.

Mori, I., Komatsu, T., Takeuchi, K., Nakakuki, K., Sudo, M., and Kimura, Y. (1995). In vivo induction of apoptosis by influenza virus. *J. Gen. Virol.* **76,** 2869–2873.

Mori, T., Ando, K., Tanaka, K., Ikeda, Y., and Koga, Y. (1997). Fas-mediated apoptosis of the hematopoetic progenitor cells in mice infected with murine cytomegalovirus. *Blood* **89,** 3565–3573.

Morimoto, H., and Bonavida, B. (1992). Diphtheria toxin and pseudomonas A toxin mediated apoptosis. *J. Immunol.* **149,** 2089–2094.

Mosier, D., Gulizia, R., MacIsaac, P., Torbett, B., and Levy, J. (1993). Rapid loss of $CD4^+$ T cells in human-PBL-SCID mice by noncytopathic HIV isolates. *Science* **260,** 689–692.

Moss, S., Calam, J., Agarwal, B., Wang, S., and Holt, P. (1996). Induction of gastric epithelial apoptosis by Helicobacter pylori. *Gut* **38,** 498–501.

Mounier, J., Vasselon, T., Hellio, R., Lesourd, M., and Sansonetti, P. J. (1992). *Shigella flexneri* enters human colonic Caco-2 epithelial cells through the basolateral pole. *Infect. Immun.* **60,** 237–248.

Mountz, J., Baker, T., Borcherding, D., Bluethmann, H., Zhou, T., and Edwards, C. (1995). Increased susceptibility of fas mutant lpr/lpr mice to Staphylococcal enterotoxin B induced septic shock. *J. Immunol.* **155,** 4829–4837.

Muller, A., Hacker, J., and Brand, B. (1996). Evidence for apoptosis of human macrophage like HL-60 cells by Legionella pneumophila infection. *Infect. Immun.* **64,** 4900–4906.

Muro-Cacho, C., Pantaleo, G., and Fauci, A. (1995). Analysis of apoptosis in lymph nodes of HIV-infected persons. *J. Immunol.* **154,** 5555–5566.

Muro, M., Koseki, T., Akifusa, S., Kato, S., Kowashi, Y., Ohsaki, Y., Yamato, K., Nishijima, M., and Nishihara, T. (1997). Role of CD14 molecules in internalization of *Actinobacillus actinomycetemcomitans* by macrophages and subsequent induction of apoptosis. *Infect Immun.* **65,** 1147–1151.

Mymryk, J. S., Shire, K., and Bayley, S. T. (1994). Induction of apoptosis by adenovirus type 5 E1A in rat cells requires a proliferation block. *Oncogene* **9,** 1187–1193.

Nava, V., Cheng, E., Veliuona, M., Zou, S., Clem, R., Mayer, M., and Hardwick, J. (1997). Herpesvirus saimiri encodes a functional homolog of the human *bcl-2* oncogene. *J. Virol.* **71,** 4118–4122.

Nevins, J. (1992). A link between the Rb tumor suppressor protein and viral oncoproteins. *Science* **258,** 424–429.

New, D., Ma, M., Epstein, L., Nath, A., and Gelbard, H. (1997). Human immunodeficiency virus type 1 Tat protein induces cell death by apoptosis in primary human neuron cultures. *J. Neurovirol.* **3,** 168–173.

Newell, M., Haughn, L., Maroun, C., and Julius, M. (1990). Death of mature T cells by separate ligation of the CD4 and T cell receptor for antigen. *Nature* **347,** 286–289.

Nicholson, D., and Thornberry, N. (1997). Caspases: Killer proteases. *Trends Biol. Sci.* **22,** 299–306.

Norimatsu, M., Ono, T., Aoki, A., Ohishi, K., and Tamura, Y. (1995a). In vivo induction of apoptosis in murine lymphocytes by bacterial lipopolysaccharides. *J. Med. Microbiol.* **43,** 251–257.

Norimatsu, M., Ono, T., Aoki, A., Ohishi, K., Takahashi, T., Watanabe, G., Taya, K., Sasamoto, S., and Tamura, Y. (1995b). Lipopolysaccharide induced apoptosis in swine lymphocytes in vivo. *Infect. Immun.* **63,** 1122–1126.

Noteborn, M., de Boer, G., van Roozelaar, D., Karreman, C., Kranenburg, O., Vos, J., Jeurissen, S., Hoeben, R., Zantema, A., Koch, G., van Ormondt, H., and van der Eb, A. (1991). Characterization of cloned chicken anemia virus DNA that contains all elements for the infectious replication cycle. *J. Virol.* **65,** 3131–3139.

Noteborn, M., Kranenburg, O., Zantema, A., Koch, G., de Boer, G., and van der Eb, A. (1992). Transcription of the chicken anemia virus (CAV) genome and synthesis of its 52 kDa protein. *Gene* **118,** 267–271.

Noteborn, M., Todd, D., Verschueren, C., de Gauw, H., Curran, W., Veldkamp, S., Douglas, A., McNulty, M., van der Eb, A., and Koch, G. (1994). A single chicken anemia virus protein induces apoptosis. *J. Virol.* **68,** 346–351.

Oberhaus, S. M., Smith, R. L., Clayton, G. H., Dermody, T. S., and Tyler, K. L. (1997). Reovirus infection and tissue injury in the mouse central nervous system are associated with apoptosis. *J. Virol.* **71,** 2100–2106.

Oh, K., Zhan, H., Cui, C., Hideg, K., Collier, R., and Hubbel, W. (1996). Organization of diphtheria toxin T domain in bilayers: A site directed spin labeling study. *Science* **273,** 810–812.

Ojeda, F., Skardova, I., Guarda, M. I., Ulloa, J., and Folch, H. (1997). Proliferation and apoptosis in infection with infectious bursal disease virus: A flow cytometric study. *Avian Dis.* **41,** 312–316.

Opgenorth, A., Graham, K., Nation, N., Strayer, D., and McFadden, G. (1992). Deletion analysis of two tandemly arranged virulence genes in myxoma virus, M1 1L and myxoma growth factor. *J. Virol.* **66,** 4720–4731.

Oyaizu, N., and Pahwa, S. (1997). Role of apoptosis in HIV disease pathogenesis. *J. Clin. Invest.* **15,** 217–231.

Oyaizu, N., McCloskey, T., Coronesi, M., Chirmule, N., Kalyanaraman, V., and Pahwa, S. (1993). Accelerated apoptosis in peripheral blood mononuclear cells from human immunodeficiency type 1 infected patients and in CD4 cross linked PBMCs from normal individuals. *Blood* **82,** 3392–3400.

Oyaizu, N., McCloskey, T., Than, S., Hu, R., Kalyanaraman, V., and Pahwa, S. (1994). Cross-linking of CD4 molecules up-regulates Fas antigen expression in lymphocytes by inuducing interferon-γ and tumor necrosis factor-α secretion. *Blood* **84,** 2622–2631.

Oyaizu, N., McCloskey, T., Than, S., and Pahwa, S. (1996). Inhibition of CD4 cross linking induced lymphocytes apoptosis by vesarinone as a novel immunomodulating agent: Vesarinone inhibits Fas expression and apoptosis by blocking cytokine secretion. *Blood* **87,** 6.

Oyaizu, N., Adachi, Y., Hashimoto, F., McCloskey, T., Hosaka, N., Kayagaki, N., Yagita, H., and Pahwa, S. (1997). Monocytes express Fas ligand upon CD4 cross-linking and induce T cell apoptosis. *J. Immunol.* **158,** 2456–2463.

Pace, J., Hayman, M., and Galan, J. (1993). Signal transduction and invasion of epithelial cells by Salmonella typhimurium. *Cell* **72,** 505–514.

Pahl, H. L., Krauss, B., Schulze-Osthoff, K., Decker, T., Traenckner, E. B., Vogt, M., Myers, C., Parks, T., Warring, P., Muhlbacher, A., Czernilofsky, A. P., and Baeuerle, P. A. (1996). The immunosuppressive fungal metabolite gliotoxin specifically inhibits transcription factor NF-kappaB. *J. Exp. Med.* **183,** 1829–1840.

Pan, H., and Griep, A. (1994). Altered cell cycle regulation in the lens of HPV-16 E6 or E7 trangenic mice: Implications for tumor suppressor gene function in development. *Genes Dev.* **8,** 1285–1299.

Pan, H., and Griep, A. (1995). Temporally distinct patterns of p53 dependent and p53 independent apoptosis during mouse lens development. *Genes Dev.* **9,** 2157–2169.

Peek, R., Moss, S., Tham, K., Perez-Perez, G., Wang, S., Miller, G., Atherton, J., Holt, P., and Blaser, M. (1997). Helicobacter pylori cagA$^+$ strains and dissociation of gastric epithelial cell proliferation from apoptosis. *J. Natl. Cancer Inst.* **89,** 863–869.

Pekosz, A., Phillips, J., Pleasure, D., Merry, D., and Gonzalez-Scarano, F. (1996). Induction of apoptosis by La Crosse virus infection and role of neuronal differentiation and human bcl-2 expression in its prevention. *J. Virol.* **70,** 5329–5335.

Peng, B., and Raveche, E. (1993). Apoptosis induction in CD5$^+$ malignant B cells. *Leukemia* **7,** 789–794.

Petito, C., and Roberts, V. (1995). Evidence of apoptotic cell death in HIV encephalitis. *Am. J. Pathol.* **146,** 1121–1130.

Phan, L., Perentesis, J., and Bodley, J. (1993). Saccharomyces cerevisiae elongation factor 2. *J. Biol. Chem.* **268,** 8665–8668.

Pickup, D., Ink, B., Hu, W., Ray, C., and Joklik, W. (1986). Hemorrhage in lesions caused by a cowpox virus is induced by a viral protein that is related to plasma protein inhibitors of serine proteases. *Proc. Natl. Acad. Sci. USA* **83,** 7698–7702.

Polyak, K., Xia, Y., Zweier, J., Kinzler, K., and Vogelstein, B. (1997). A model for p53 induced apoptosis. *Nature* **389,** 300–305.

Portnoy, D., Chakraborty, T., Goebel, W., and Cossart, P. (1992). Molecular determinants of Listeria monocytogenes pathogenesis. *Infect. Immun.* **60,** 1263–1267.

Pulliam, L., Herndier, B., Tang, N., and McGrath, M. (1991). Human immunodeficiency virus infected macrophages produce soluble factors that cause histological and neurochemical alterations in cultured human brains. *J. Clin. Invest.* **87,** 503–512.

Quan, L., Caputo, A., Bleackley, R., Pickup, D., and Salvesen, G. (1995). Granzyme B is inhibited by the cowpox virus serpin cytokine response modifier A. *J. Biol. Chem.* **270,** 10377–10379.

Querido, E., Teodoro, J., and Branton, P. (1997). Accumulation of p53 induced by the adenoviral E1A protein requires regions involved in the stimulation of DNA synthesis. *J. Virol.* **71,** 3526–3533.

Rahme, L. G., Stevens, E. J., Wolfort, S. F., Shao, J., Tompkins, R. G., and Ausubel, F. M. (1995). Common virulence factors for bacterial pathogenicity in plants and animals. *Science* **268,** 1899–1902.

Rao, L., Debbas, M., Sabbatini, P., Hockenbery, D., Korsmeyer, S., and White, E. (1992). The adenovirus E1A proteins induce apoptosis, which is inhibited by the E1B 19-kDa and Bcl-2 proteins (published erratum appears in *Proc. Natl. Acad. Sci. USA* **89**(20), 9974, 1992). *Proc. Natl. Acad. Sci. USA* **89,** 7742–7746.

Reed, J. (1997). Double identity for proteins of the Bcl-2 family. *Nature* **387,** 773–776.

Relman, D., Tuomanen, E., Falkow, S., Golenbock, D., Saukkonen, K., and Wright, S. (1990). Recognition of a bacterial adhesin by an integrin. Macrophage CR3 binds filamentous hemaglglutin of Bordetella Pertussis. *Cell* **61,** 1375–1382.

Renno, T., Hahne, M., and MacDonald, H. (1995). Proliferation is a prerequisite for bacterial superantigen induced T cell apoptosis in vivo. *J. Exp. Med.* **181,** 2283–2287.

Rodgers, S. E., Barton, E. S., Oberhaus, S. M., Pike, B., Gibson, C. A., Tyler, K. L., and Dermody, T. S. (1997). Reovirus-induced apoptosis of MDCK cells is not linked to viral yield and is blocked by Bcl-2. *J. Virol.* **71,** 2540–2546.

Rogel, A., and Hanski, E. (1992). Distinct steps in the penetration of adenylate cyclase toxin of Bordatella pertussis into sheep eryhtrocytes. *J. Biol. Chem.* **267,** 22599–22605.

Rogers, H., and Unanue, E. (1993). Neutrophils are involved in acute, nonspecific resistance to Listeria monocytogenes in mice. *Infect. Immun.* **61,** 5090–5096.

Rogers, H., Callery, M., Deck, B., and Unanue, E. (1996). Listeria monocytogenes induces apoptosis of infected hepatocytes. *J. Immunol.* **156,** 679–684.

Rubartelli, A., Cozzolino, F., Talio, M., and Sitia, R. (1990). A novel secretory pathway for interleukin-1β, a protein lacking a signal sequence. *EMBO J.* **9,** 1503–1510.

Ruckdeschel, K., Roggenkamp, A., Lafont, V., Mangeat, P., Heesemann, J., and Roulot, B. (1997). Interaction of Yersinia enterocolitica with macrophages leads to macrophage cell death through apoptosis. *Infect. Immun.* **65,** 4813–4821.

Rudin, C., and Thompson, C. (1997). Apoptosis and disease: Regulation and clinical relevance of programmed cell death. *Annu. Rev. Med.* **48,** 267–281.

Ruggieri, A., Harada, T., Matsuura, Y., and Miyamura, T. (1997). Sensitization to Fas-mediated apoptosis by hepatitis C virus core protein. *Virology* **229,** 68–76.

Ruoslahti, E., and Reed, J. (1994). Anchorage dependence, integrins and apoptosis. *Cell* **77,** 477–478.

Sabbatini, P., Lin, J., Levine, A. J., and White, E. (1995). Essential role for p53-mediated transcription in E1A-induced apoptosis. *Genes Dev.* **9,** 2184–2192.

Sadzot-Delvaux, C., Thonard, P., Schoonbroodt, S., Piette, J., and Rentier, B. (1995). Varicella zoster induces apoptosis in cell culture. *J. Gen. Virol.* **76,** 2875–2879.

Saito, T., Tanaka, M., and Yamaguchi, I. (1996). Effect of brefeldin A on influenza A virus-induced apoptosis in vitro. *J. Vet. Med. Sci.* **58,** 1137–1139.

Saksela, K., Muchmore, E., Girard, M., Fultz, P., and Baltimore, D. (1993). High viral load in lymph nodes and latent human immunodeficiency virus in peripheral blood cells of HIV-1 infected chimpanzees. *J. Virol.* **67,** 7423–7427.

Sambucetti, L., Cherrington, J., Wilkinson, G., and Mocarski, M. (1989). NF-kappa B activation of the cytomegalovirus enhancer is mediated by a viral trans-activator and by T-cell stimulation. *EMBO J.* **8,** 4251–4258.

Sandvig, K., and van Deurs, B. (1992). Toxin induced cell lysis: Protection by 3-methyladenine and cycloheximide. *Exp. Cell Res.* **200,** 253–262.

Sansonetti, P. J., Arondel, J., Cavaillon, J.-M., and Huerre, M. (1995). Role of IL-1 in the pathogenesis of experimental shigellosis. *J. Clin. Invest.* **96,** 884–892.

Savill, J., Wyllie, A., Henson, J., Walport, M., Henson, P. and Haslett, C. (1989a). Macrophage phagocytosis of aging neutrophils in inflammation. *J. Clin. Invest.* **83,** 865–875.

Savill, J., Henson, P., and Haslett, C. (1989b). Phagocytosis of aged neutrophils is mediated by a novel "charge sensitive" recognition mechanism. *J. Clin. Invest.* **84,** 1518–1527.

Scallan, M. F., Allsopp, T. E., and Fazakerley, J. K. (1997). bcl-2 acts early to restrict Semliki Forest virus replication and delays virus-induced programmed cell death. *J. Virol.* **71,** 1583–1590.

Scheffner, M., Werness, B., Huibregtse, J., Levine, A., and Howley, P. (1990). The E6 oncoprotein encoded by human papillomavirus types 16 and 18 promotes the degradation of p53. *Cell* **63,** 1129–1136.

Scheffner, M., Huibregtse, J., Viestra, R., and Howley, P. (1993). The HPV-16 E6 and E6-AP complex functions as a ubiquitin-protein ligase in the ubiquitination of p53. *Cell* **75,** 495–505.

Scherf, U., Pastan, I., Willingham, M. and Brinkmann, U. (1996). The human CAS protein which is homologous to the CSE1 yeast chromosome segregation gene product is associated with microtubules and mitotic spindle. *Proc. Natl. Acad. Sci. USA* **93,** 2670–2674.

Schrieber, M., Sedger, L., and McFadden, G. (1997). Distinct domains of M-T2, the myxoma virus tumor necrosis factor (TNF) receptor homolog, mediate extracellular TNF binding and intracellular apoptosis induction. *J. Virol.* **71,** 2171–2181.

Schuitmaker, H., Meyaard, L., Koostra, N., Dubbes, R., Otto, S., Termette, M., Heeney, J., and Miedema, F. (1993). Lack of T cell dysfunction and programmed cell death in human immunodeficiency type 1 infected chimpanzees correlates with absence of monocytotropic variants. *J. Infect. Dis.* **168,** 1140–1147.

Shen, Y., and Shenk, T. (1994). Relief of p53-mediated transcriptional repression by the adenovirus E1B 19-kDa protein or the cellular Bcl-2 protein. *Proc. Natl. Acad. Sci. USA* **91,** 8940–8944.

Shenk, T., and Flint, J. (1991). Transcriptional and transforming activities of the adenovirus E1A proteins. *Adv. Cancer Res.* **57,** 47–85.

Shenker, B., Vitale, L., Keiba, I., Harriso, G., Berthold, P., Golub, E., and Lally, E. (1994). Flow cytometric analysis of the cytotoxic effects of Actinobacillus actinomycetemcomitans leukotoxin on human natural killer cells. *J. Leukocyte Biol.* **55,** 153–160.

Sieg, S., King, C., Huang, Y., and Kaplan, D. (1996). The role of interleukin-10 in the inhibition of T-cell proliferation and apoptosis mediated by parainfluenza virus type 3. *J. Virol.* **70,** 4845–4848.

Shi, B., Girolami, U., Jianglin, H., Wang, S., Lorenzo, A., Busciglio, J., and Gabuzda, D. (1996). Apoptosis induced by HIV-1 infection of the central nervous system. *J. Clin. Invest.* **98,** 1979–1990.

Sodroski, J., Goh, W., Rosen, C., Campbell, K., and Haseltine, W. (1986). Role of the HTL VIII/LAV envelope in syncitium formation and cytopathicity. *Nature* **322,** 470–474.

Spier, E., Modali, R., Huang, E., Leon, M., Shawl, F., Finkel, T., and Epstein, S. (1994). Potential role of human cytomegalovirus and p53 interaction in coronary restenosis. *Science* **265,** 391–394.

Sprent, J. (1994). T and B memory cells. *Cell* **76,** 315–322.

Stevenson, F., Torrano, F., Locksley, R., and Lovett, D. (1992). Interleukin 1: The patterns of translation and intracellular distribution support alternative secretory mechanisms. *J. Cell. Physiol.* **152,** 223–231.

Stewart, S., Poon, B., Jowett, J., and Chen, I. (1997). Human immunodeficiency virus type 1 Vpr induces apoptosis following cell cycle arrest. *J. Virol.* **71,** 5579–5592.

Su, L., Kanesima, H., Bonyhadi, M., Salimi, S., Kraft, D., Rabin, L., and McCune, J. (1995). HIV-1 induced thymocyte depletion is associated with indirect cytopathicity and infection of progenitor cells in vivo. *Immunity* **2,** 25–36.

Suarez, P., Diaz-Guerra, M., Prieto, C., Esteban, M., Castro, J. M., Nieto, A., and Ortin, J. (1996). Open reading frame 5 of porcine reproductive and respiratory syndrome virus as a cause of virus-induced apoptosis. *J. Virol.* **70,** 2876–2882.

Subramanian, T., Tarodi, B., and Chinnadurai, G. (1995). p53-independent apoptotic and necrotic cell deaths induced by adenovirus infection: Suppression by E1B 19K and Bcl-2 proteins. *Cell Growth Differ.* **6,** 131–137.

Sutton, P., Newcombe, N. R., Waring, P., and Mullbacher, A. (1994). In vivo immunosuppressive activity of gliotoxin, a metabolite produced by human pathogenic fungi. *Infect. Immun.* **62,** 1192–1198.

Sutton, P., Beaver, J., and Waring, P. (1995). Evidence that gliotoxin enhances lymphocyte activation and induces apoptosis by effects on cyclic AMP levels. *Biochem. Pharmacol.* **50,** 2009–2014.

Suttorp, N., Seeger, W., Dewein, E., Bhakdi, S., and Roka, L. (1985). Staphylococcal alpha toxin stimulates synthesis of prostacyclin by cultured endothelial cells from pig pulmonary arteries. *Am. J. Physiol.* **248,** C127–C135.

Suttorp, N., Seeger, W., Zucker-Reinmann, J., Roka, L., and Bhakdi, S. (1987). Mechanisms of leukotriene generation in polymorphonuclear leukocytes by staphylococcal alpha toxin. *Infect. Immun.* **55,** 104–109.

Taga, K., Chretien, J., Cherney, B., Diaz, L., Brown, M., and Tosato, G. (1994). Interleukin-10 inhibits apoptotic cell death in infectious mononucleosis T cells. *J. Clin. Invest.* **94,** 251–260.

Taichman, N., Dean, R., and Sanderson, C. (1980). Biochemical and morphological characterization of the killing of human monocytes by a leukotoxin derived from Actinobacillus actinomycetemcomitans. *Infect. Immun.* **28,** 258–268.

Takano, Y., Saegusa, M., Masuda, M., Mikami, T., and Okayasu, I. (1997). Apoptosis, proliferative activity, and Bcl-2 expression in Epstein–Barr virus positive non-Hodgkin's lymphomas. *J. Cancer Res. Clin. Oncol.* **123,** 395–401.

Takemori, N., Cladaras, C., Bhat, B., Conley, A. J., and Wold, W. S. (1984). cyt gene of adenoviruses 2 and 5 is an oncogene for transforming function in early region E1B and encodes the E1B 19,000-molecular-weight polypeptide. *J. Virol.* **52,** 793–805.

Takizawa, T., Matsukawa, S., Higuchi, Y., Nakamura, S., Nakanishi, Y., and Fukuda, R. (1993). Induction of programmed cell death (apoptosis) by influenza virus infection in tissue culture cells. *J. Gen. Virol.* **74,** 2347–2355.

Tarodi, B., Subramanian, T., and Chinnadurai, G. (1994). Epstein–Barr virus BHRF1 protein protects against cell death induced by DNA-damaging agents and heterologous viral infection. *Virology* **201,** 404–407.

Teodoro, J. G., Shore, G. C., and Branton, P. E. (1995). Adenovirus E1A proteins induce apoptosis by both p53-dependent and p53-independent mechanisms. *Oncogene* **11,** 467–474.

Terai, C., Kornbluth, R., Pauza, C., Richman, D., and Carson, D. (1991). Apoptosis as a mechanism of cell death in cultured T lymphoblasts acutely infected with HIV-1. *J. Clin. Invest.* **87,** 1710–1715.

Terrence, J. (1994). Gliotoxin induces apoptosis in mouse L929 fibroblast cells. *Biochem. Mol. Biol. Int.* **33,** 411–419.

Tesh, V., and O'Brien, A. (1991). The pathogenic mechanisms of shiga toxin and the shiga like toxins. *Mol. Microbiol.* **5,** 1817–1822.

Tewari, M., Quan, L. T., O'Rourke, K., Desnoyers, S., Zeng, Z., Beidler, D. R., Poirier, G. G., Salvesen, G. S., and Dixit, V. M. (1995). Yama/CPP32, a mammalian homolog of CED-3, is a CrmA-inhibitable protease that cleaves the death substrate poly(ADP-ribose) polumerase. *Cell* **81,** 801–809.

Tham, K. M., and Moon, C. D. (1996). Apoptosis in cell cultures induced by infectious bursal disease virus following in vitro infection. *Avian Dis.* **40,** 109–113.

Thirumalai, K., Kim, K., and Zychlinsky, A. (1997). Ipa B, a *Shigella flexneri* invasin, colocalizes with interleukin-1B-converting enzyme in the cytoplasm of macrophages. *Infect. Immun.* **65,** 787–793.

Thomas, D., and Higbie, L. (1990). In vitro association of leptospires with host cells. *Infect. Immun.* **58,** 581–585.

Thomas, M., Matlashewski, G., Pim, D., and Banks, L. (1996). Induction of apoptosis by p53 is independent of its oligomeric state and can be abolished by HPV 18 E6 through ubiquitin mediated degradation. *Oncogene* **8,** 195–202.

Thome, M., Schneider, P., Hoffman, K., Fickenscher, H., Meinl, E., Neipel, F., Mattman, C., Burns, K., Bodmer, J., Schroter, M., Scaffaldi, C., Krammer, P., Peter, M., and Tschopp, J. (1997). Viral FLICE-inhibitory proteins (FLIPs) prevent apoptosis induced by death receptors. *Nature* **386,** 517–521.

Tilney, L., and Portnoy, D. (1989). Actin filaments and the growth, movement, and spread of the intracellular parasite, Listeria monocytogenes. *J. Cell Biol.* **109,** 1597–1608.

Tollefson, A. E., Scaria, A., Hermiston, T. W., Ryerse, J. S., Wold, L. J., and Wold, W. S. (1996). The adenovirus death protein (E3-11.6K) is required at very late stages of infection for efficient cell lysis and release of adenovirus from infected cells. *J. Virol.* **70,** 2296–2306.

Tolleson, W. H., Dooley, K. L., Sheldon, W. G., Thurman, J. D., Bucci, T. J., and Howard, P. C. (1996). The mycotoxin fumonisin induces apoptosis in cultured human cells and in livers and kidneys of rats. *Adv. Exp. Med. Biol.* **392,** 237–250.

Tolskaya, E. A., Romanova, L. I., Kolesnikova, M. S., Ivannikova, T. A., Smirnova, E. A., Raikhlin, N. T., and Agol, V. I. (1995). Apoptosis-inducing and apoptosis-preventing functions of poliovirus. *J. Virol.* **69,** 1181–1189.

Tolskaya, E. A., Romanova, L. I., Kolesnikova, M. S., Ivannikova, T. A., and Agol, V. I. (1996). Final checkpoint in the drug-promoted and poliovirus-promoted apoptosis is under posttranslational control by growth factors. *J. Cell. Biochem.* **63,** 422–431.

Triadafilopolous, G., Pothoulakis, C., O'Brien, M., and LaMont, T. (1987). Differential effect of Clostridium dificile toxins A and B on rabbit ileum. *Gastroenterology* **93,** 273–279.

Tropea, F., Troiano, L., Monti, D., Lovato, E., Malorni, W., Rainaldi, G., Mattana, P., Viscomi, G., Ingletti, M., Portolani, M., Cermelli, C., Cossarizza, A., and Franceschi, C. (1995). Sendai virus and herpes virus type 1 induce apoptosis in human peripheral blood mononuclear cells. *Exp. Cell Res.* **218,** 63–70.

Tsai, C., McArthur, W., Baehni, P., Hammond, B., and Taichman, N. (1979). Extraction and partial characterization of a leukotoxin from a plaque derived gram negative microorganism. *Infect. Immun.* **25,** 427–439.

Tyler, K. L., Squier, M. K., Rodgers, S. E., Schneider, B. E., Oberhaus, S. M., Grdina, T. A., Cohen, J. J., and Dermody, T. S. (1995). Differences in the capacity of reovirus strains to induce apoptosis are determined by the viral attachment protein sigma 1. *J. Virol.* **69,** 6972–6979.

Tyler, K. L., Squier, M. K., Brown, A. L., Pike, B., Willis, D., Oberhaus, S. M., Dermody, T. S., and Cohen, J. J. (1996). Linkage between reovirus-induced apoptosis and inhibition of cellular DNA synthesis: Role of the S1 and M2 genes. *J. Virol.* **70,** 7984–7991.

Ubol, S., Tucker, P., Griffin, D., and Hardwick, J. (1994). Neurovirulent strains of *Alphavirus* induce apoptosis in *bcl-2* expressing cells: Role of a single amino acid change in the E2 glycoprotein. *Proc. Natl. Acad. Sci. USA* **91,** 5202–5206.

Upton, C., Macen, J., Schrieber, M., and McFadden, G. (1991). Myxoma virus expresses a secreted protein with homology to the tumor necrosis factor receptor gene family that contributes to viral virulence. *Virology* **184,** 370–382.

Uren, A., and Vaux, D. (1997). Viral inhibitors of apoptosis. *Vitamins Horm.* **53,** 175–193.

Vasconcelos, A. C., and Lam, K. M. (1994). Apoptosis induced by infectious bursal disease virus. *J. Gen. Virol.* **75,** 1803–1806.

Vasconcelos, A. C., and Lam, K. M. (1995). Apoptosis in chicken embryos induced by the infectious bursal disease virus. *J. Comp. Pathol.* **112,** 327–338.

Vaux, D. (1997). CED-4 the third horseman of apoptosis. *Cell* **90,** 389–390.

Vella, A., McCormack, J., Linsley, P., Kappler, J., and Marrack, P. (1995). Lipopolysaccharide interferes with the induction of peripheral T cell death. *Immunity* **2,** 261–270.

Voss, K. A., Riley, R. T., Bacon, C. W., Chamberlain, W. J., and Norred, W. P. (1996). Subchronic toxic effects of Fusarium moniliforme and fumonisin B1 in rats and mice. *Nat. Toxins* **4,** 16–23.

Wahl, L., Corcoran, M., Pyle, S., Arthur, L., Harel-Bellan, A., and Farrar, W. (1989). Human immunodeficiency virus glycoprotein (gp120) induction of monocyte arachidonic acid metabolites and interleukin 1. *Proc. Natl. Acad. Sci. USA* **86,** 621–625.

Wang, J., Stolman, S., and Dennert, G. (1994). TCR cross linking induces CTL death via internal action of TNF. *J. Immunol.* **152,** 3824–3832.

Wang, S., Huang, K., Lin, Y., and Lei, H. (1994). Sepsis induced apoptosis of thymocytes in mice. *J. Immunol.* **152,** 5014–5021.

Wang, W., Jones, C., Ciacci-Zanella, J., Holt, T., Gilchrist, D. G., and Dickman, M. B. (1996). Fumonisins and alternaria alternata lycopersici toxins: Sphinganine analog mycotoxins induce apoptosis in monkey kidney cells. *Proc. Natl. Acad. Sci. USA* **93,** 3461–3465.

Wang, Z., Orlikowsky, T., Dudhane, A., Clarke, K., Li, X., Darzynkeiwicz, Z., and Hoffman, M. (1994a). Deletion of T lymphocytes in human CD4 transgenic mice induced by HIV gp120 and gp120 specific antibodies from AIDS patients. *Eur. J. Immunol.* **24,** 1553–1557.

Wang, Z., Dudhane, A., Orlikowsky, T., Clarke, K., Li, X., Darzynkeiwicz, Z., and Hoffman, M. (1994b). CD4 engagement induces Fas antigen dependent apoptosis in vivo. *Eur. J. Immunol.* **24,** 1553–1557.

Waring, P., Eichner, R. D., Mullbacher, A., and Sjaarda, A. (1988). Gliotoxin induces apoptosis in macrophages unrelated to its antiphagocytic properties. *J. Biol. Chem.* **263,** 18493–18499.

Wassef, J. S. Keren, D. F., and Mailloux, J. L. (1989). Role of M cells in initial antigen uptake and in ulcer formation in rabbit intestinal loop model of shigellosis. *Infect. Immun.* **57,** 858–863.

Watanabe, M., Ringler, D., Fultz, P., MacKey, J., Boyson, J., Levine, C., and Letvin, N. (1991). A chimpanzee passaged human immunodeficiency isolate is cytopathic for chimpanzee cells, but does not induce disease. *J. Virol.* **65,** 3344–3348.

Watanabe-Ohnishi, R., Low, D., McGeer, A., Stevens, D., Schlievert, P., Newton, D., Schwartz, B., Kreiswirth, B., Project, O. S. S., and Kotb, M. (1995). Selective depletion of VB-bearing T cells in patients with severe invasive group A streptococcal infections and streptococcal toxic shock syndrome. *J. Infect. Dis.* **171,** 74–84.

Watson, R. W., Redmond, H. P., Wang, J. H., Condron, C., and Bouchier-Hayes, D. (1996). Neutrophils undergo apoptosis following ingestion of *Escherichia coli. J. Immunol.* **156,** 3986–3992.

Weiss, A., and Goodwin, M. (1989). Lethal infection by Bordatella pertussis mutants in the infant mouse model. *Infect. Immun.* **57,** 3757–3764.

Welch, R. A. (1991). Pore-forming cytolysins of gram-negative bacteria. *Mol. Microbiol.* **5,** 521–528.

Wertz, I., and Hanley, M. (1996). Diverse molecular provocation of programmed cell death. *TIBS* **21,** 359–364.

Wesselingh, S., Power, C., Glass, J., Tyor, W., McArthur, J., Farber, J., Griffin, J., and Griffin, D. (1993). Intracerebral cytokine messenger RNA expression in AIDS dementia. *Ann. Neurol.* **33,** 576–582.

Westendorf, M., Frank, O., Ochsenbauer, C., Stricker, K., Dhein, J., Waiczak, H., Debatin, K., and Krammer, P. (1995). Sensitization of T cells to CD95-mediated apoptosis by HIV-1 Tat and gp120. *Nature* **375,** 497.

White, E., Grodzicker, T., and Stillman, B. W. (1984). Mutations in the gene encoding the adenovirus early region 1B 19,000- molecular-weight tumor antigen cause the degradation of chromosomal DNA. *J. Virol.* **52,** 410–419.

White, E., Cipriani, R., Sabbatini, P., and Denton, A. (1991). Adenovirus E1B 19-kilodalton protein overcomes the cytotoxicity of E1A proteins. *J. Virol.* **65,** 2968–2978.

White, E., Sabbatini, P., Debbas, M., Wold, W. S., Kusher, D. I., and Gooding, L. R. (1992). The 19-kilodalton adenovirus E1B transforming protein inhibits programmed cell death and prevents cytolysis by tumor necrosis factor alpha. *Mol. Cell. Biol.* **12,** 2570–2580.

Willey, R., Maldarelli, F., Martin, M., and Strebel, K. (1992). Human immunodeficiency virus type I Vpu protein induces a rapid degradation of CD4. *J. Virol.* **66,** 7193–7200.

Wilson, B., Blanke, S., Murphy, J., Pappenheimer, A., and Collier, R. (1990). Does diphtheria toxin have nuclease activity? *Science* **250,** 834–835.

Xue, D., and Horvitz, R. (1995). Inhibition of the *Caenorhabditis elegans* cell death protease ced-3 by a ced-3 cleavage site in baculovirus p35 protein. *Nature* **377,** 248–251.

Yamada, T., Yamaoka, S., Goto, T., Nakai, M., Tsujimoto, Y., and Hatanaka, M. (1994). The human T cell leukemia virus type I tax protein induces apoptosis which is blocked by the Bcl-2 protein. *J. Virol.* **68,** 3374–3379.

Yamamoto, C., Yoshida, S., Taniguchi, H., Qin, M., Miyamoto, H., and Mizuguchi, Y. (1993). Lipopolysaccharide and granulocyte colony stimulating factor delay neutrophil apoptosis and ingestion by guinea pig macrophages. *Infect. Immun.* **61,** 1972–1979.

Yao, Q., Ogan, P., Rowe, M., Wood, M., and Rickinson, A. (1989). Epstein–Barr virus infected B cells persist in the circulation of acyclovir treated virus carriers. *Int. J. Cancer* **43,** 67–71.

Yew, P. R., and Berk, A. J. (1992). Inhibition of p53 transactivation required for transformation by adenovirus early 1B protein. *Nature* **357,** 82–85.

Yew, P. R., Liu, X., and Berk, A. J. (1994). Adenovirus E1B oncoprotein tethers a transcriptional repression domain to p53. *Genes Dev.* **8,** 190–202.

Yoshida, H., Sumichika, H., Hamano, S., He, X., Minamishima, Y., Kimura, G., and Nomoto, K. (1995). Induction of apoptosis in T cells by infecting mice with murine cytomegalovirus. *J. Virol.* **69,** 4769–4775.

Yuan, J. (1997). Transducing signals of life and death. *Curr. Opin. Cell Biol.* **9,** 247–251.

Zauli, G., Gibellini, D., Milani, D., Mazzoni, M., Borgatti, P., La Placa, M., and Capitani, S. (1993). HIV-1 Tat protects lymphoid, epithelial, and neuronal cell lines from death by apoptosis. *Cancer Res.* **53,** 4481.

Zhang, Y., Takahashi, K., Jiang, G., Kawai, M., Fukada, M., and Yokochi, T. (1993). In vivo induction of apoptosis in mouse thymus by administration of lipopolysaccharide. *Infect. Immun.* **61,** 5044–5048.

Zhu, H., Shen, Y., and Shenk, T. (1995). Human cytomegalovirus IE1 and IE2 proteins block apoptosis. *J. Virol.* **69,** 7960–7970.

Zhuang, S., Shvarts, A., van Ormondt, H., Jochemsen, A., van der Eb, A., and Noteborn, M. (1995a). Apoptin, a protein derived from chicken anemia virus induces a p53 independent apoptosis in human osteosarcoma cells. *Cancer Res.* **55,** 486–489.

Zhuang, S., Landegent, J., Verschueren, C., Falkenburg, J., van Ormondt, H., van der Eb, A., and Noteborn, M. (1995b). Apoptin, a protein encoded by chicken anemia virus, induces cell death in various human hematologic malignant cells in vitro. *Leukemia* **9**(Suppl, 1), S118–S120.

Zhuang, S., Shvarts, A., Jochemsen, A., van Oorschot, A., van der Eb, A., and Noteborn, M. (1995c). Differential sensitivity to Ad5 E1B-21kD and Bcl-2 proteins of apoptin induced versus p53 induced apoptosis. *Carcinogenesis* **16,** 2939–2944.

Zychlinsky, A., Prevost, M. C., and Sansonetti, P. J. (1992). *Shigella flexneri* induces apoptosis in infected macrophages. *Nature* **358,** 167–168.

Zychlinsky, A., Fitting, C., Cavaillon, J. M., and Sansonetti, P. J. (1994a). Interleukin-1 is released by murine macrophages during apoptosis induced by *Shigella flexneri. J. Clin. Invest.* **94,** 1328–1332.

Zychlinsky, A., Kenny, B., Ménard, R., Prevost, M. C., Holland, I. B., and Sansonetti, P. J. (1994b). IpaB mediates macrophage apoptosis induced by *Shigella flexneri. Mol. Microbiol.* **11,** 619–627.

Zychlinsky, A., Thirumalai, K., Arondel, J., Cantey, J. R., Aliprantis, A., and Sansonetti, P. J. (1996). In vivo apoptosis in *Shigella flexneri* infections. *Infect. Immun.* **64,** 5357–5365.

Invertebrate Opioid Precursors: Evolutionary Conservation and the Significance of Enzymatic Processing

George B. Stefano* and Michel Salzet[†]
*Neuroscience Institute, State University of New York,
College at Old Westbury, Old Westbury, New York 11568-0210;
and [†]Laboratoire de Biologie Animale, Université des Sciences
et Techniques de Lille, 59655 Villeneuve d'Ascq
Cedex, France

Invertebrate tissues contain mammalian-like proenkephalin, prodynorphin, and proopiomelanocortin. Amino acid sequence determination of these opioid gene products reveals the presence of various opioid peptides exhibiting high sequence identity with their mammalian counterparts. These associated peptides are flanked by dibasic amino acid residues, indicating cleavage sites. Together with the presence of various processing enzymes, i.e., neutral endopeptidase 24.11 and angiotensin-converting enzymes, this suggests that opioid precursor processing is also similar to that described in mammals. It is noted that the levels and/or activity of invertebrate neutral endopeptidase 24.11 can be upregulated by signaling molecules shown to perform the same function in mammals, i.e., morphine. Critical to opioid precursor processing are immunocytes that contain the precursors and transport processing enzymes to sites of inflammation, in part, to cleave these peptide precursors, thus liberating immune-stimulating molecules. Furthermore, in response to lipopolysaccharides, Met-enkephalin levels peak immediately and hours after the exposure, revealing a release and induction process. It appears that the opioid precursors and their processing enzymes first evolved in "simple" animals and then have been maintained and embellished during the course of evolution guided by conformational matching.

KEY WORDS: Invertebrates, Proenkephalin, Prodynorphin, Proopiomelanocortin, Morphine, Neutral endopeptidase, Conformational matching, Opioid peptides.

I. Introduction

A. Chemical Signaling

It has become evident that intercellular communication is mediated primarily by chemical signal molecules. During the course of evolution, organisms in which this form of communication developed appear to have increased their chances of survival and thus passed this trait on to their descendents. A plausible explanation for the emergence/dominance of this mechanism of communication is based on its inherent level of sophistication, as noted by not only synaptic molecules that can enter into intercellular communication but also hormonal ones. A further advantage of this method is that it is not limited by spatial requirements. Mechanisms which employ direct contact, by their nature, require much contact space, whereas the only requirement of chemical communication is scaled-down space for receptors. This also allows for a greater diversity of the signal molecules and their corresponding receptors, and the same chemical communication mechanism would permit synaptic growth and plasticity. The end result of such chemical communication mechanisms would be a higher degree of sophistication and detailed information transfer, which allows for a greater number of behavioral characteristics (including afferent hormonal influences) to enhance an organism's chance for survival in a changing environment. If indeed this system was favored, it can be predicted that the organisms accumulating the greatest diversity of cellular communication would eventually begin to control their environment. Thus, these organisms would expand our concept of natural selection. Also, it could be predicted that the anatomy of the chemical communication mechanism during evolution would be diversified. That is, the distance between the origin of the signal molecule and its receptor does not have to be fixed. In other words, the closer the origin of a signal molecule is to its target receptor, the quicker the response. Therefore, at the extremes, it is possible to have ongoing long-term communication (hormonal) which is not essential for immediate action and immediate short-term communication (synaptic).

B. Mammalian Opioid Precursors

Many of the protein molecules active in this communication appear to be derived from larger polypeptide gene products. This is certainly true of the various opioid peptides (Undenfriend and Meienhofer, 1984). According to these authors, the endogenous opioids with the lowest molecular weights are the pentapeptides, Met-enkephalin, and Leu-enkephalin. The prefixes indicate the difference between these two enkephalins on the last residue. The first four residues, Tyr-Gly-Gly-Phe, are identical for both compounds. The higher molecular weight bioactive opioids in general contain one of the two enkephalin sequences at their N terminals. In the case of peptides E and F of proenkephalin, they each contain an additional copy of the enkephalin sequence.

For many years opioid peptides were classified into three families: the proopiomelanocortin (POMC), the proenkephalin (proenk), and the prodynorphin (prodyn) families. Today, other types of opioid peptides have been found but will not be discussed in this chapter. The older classification is based on the macromolecular precursors from which these neuropeptides are derived. These precursors all contain at least one copy of an enkephalin sequence. Interestingly, nearly all the opioid peptides in the three precursors are bracketed by paired basic amino acids on the N and C terminals (Figs. 1 and 2). Recently, with the use of cDNA techniques, the nucleotide sequences of the genes as well as the amino acid sequences for all three precursors have been determined. Although there is no direct evidence, the tremendous sequence homology among the precursors does strongly

```
G   65   CKDLLQVSKQELPQEGASSLRESGKQDESHLLSKKYGGFMKRYGGFMKKVDELYPVEPEE 124
M   24   CK-IFQYRLQKCPSLKASSLRESGKQDESHLLSKKYGGFMKRYGGFMKKVG-----EPE-  76
L   36   -------------------------------IAKKYGGFMKRRRVKKRYGGFLKKI----  59

G   125  EANGGEILTKRYGGFMKKDAEDGDALANSSDLLKELLGTGDDRDRENHHQEGGDSDEGVS 184
M   77   EILTKR-----YGGFMKKD-EAAQAAANSSDLLKELLGTGDDRDRENHHQEGGDSDEGVS 130
L   60   ---------LRYGGFLK-----------------------------RLIQPGTDLMN---  79

G   185  KRYGGFMRGLKRSPQVEDEAKELQKRYGGFMRRVGRPEWWMDYQKRYGGFLKRFAEFLPS 244
M   131  KRYGGFMRGLKRSP---KSRSSEQ---------V--------VQKRYGGFLKRFAEFLPS 170
L   80   KRYGGFMRGLRRS---------------------TRPEW------------KKFAEFLPS 106

G   245  DEEGESYSKEVPEMEKRYGGFMRF   268
M   171  EEEGESYSKEVPEMEKRYGGFMRF   194
L   107  EEEGESYSKEVPEMERRYGGFMRF   128
```

FIG. 1 Comparison of invertebrate and mammalian proenkephalin. The amino acid sequences of the various proenkephalin-derived molecules were aligned using MPSrch Version 1.5 algorithm Mpsrch (S. S. Sturrock and J. F. Collins, Biocomputing Research Unit, University of Edinburgh, UK, 1993). G, guinea pig; M, *Mytilus edulis*; L, leech. Boldlett represent identical sequences; underlined letters represent enkephalin sequences, and the last underlined sequence is enkelytin, ending with the opioid heptapeptide YGGFMRF.

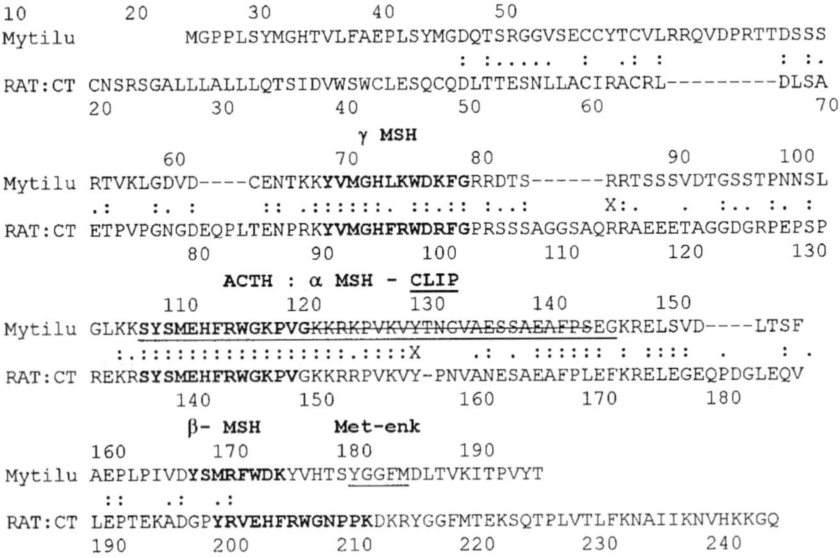

FIG. 2 Comparison of invertebrate and mammalian POMC. Colons = similarities; bold letters indicate the signal molecule, etc. The amino acid sequences of the various POMC-derived molecules were aligned using MPSrch Version 1.5 algorithm Mpsrch (S. S. Sturrock and J. F. Collins, Biocomputing Research Unit, University of Edinburgh, UK, 1993). The associated peptides are noted above the particular sequence and can be identified as indicated.

suggest a common origin. In addition, there is considerable similarity among the genes of the precursors in the arrangements of the coding.

The peptides of the POMC family are derived from a single macromolecular precursor (Fig. 1). POMC is an interesting molecule because its processed products include hormones such as adrenocorticotropin (ACTH), α-melanocyte-stimulating hormone (α-MSH), β-MSH, and the opioid peptide, β-endorphin. β-Endorphin possesses a very strong analgesic effect and contains the Met-enkephalin sequence at its N terminal. The evolutionary and physiological significance of the structural association between β-endorphin and the other opioid peptides remains unclear. However, later we will discuss a possible rationale for its presence in POMC. In many animals POMC appears to be processed to different final products in the anterior and neurointermediate lobes of the pituitary gland; it is possible that POMC may serve different functions in different tissues. In POMC, β-endorphin represents the last 31-amino acid sequence of the precursor. The 2 amino acids at its N terminal are Lys and Arg. Thus, β-endorphin is released by neuropeptide-processing enzyme with trypsin-like activity. The absence of a basic paired amino acid after the Met-enkephalin in

β-endorphin makes it an unlikely precursor for Met-enkephalin. In addition, immunocytochemistry studies have shown that β-endorphin and enkephalin do not coexist in the same regions of the brain.

The cDNA analyses of a number of intermediate-size peptides in bovine adrenal chromaffin cells led to the complete identification of the macromolecular precursor proenkephalin (Fig. 2). These studies raised the question whether the enkephalins are the only "intended" final products of the precursor. This is especially true since some of the "intermediates" do exist at a relatively high level in the tissues and have very strong opioid activities. Nevertheless, the primary structure of proenkephalin indicated that of the six copies of Met-enkephalin sequence and one copy of Leu-enkephalin sequence, only two Met-enkephalin sequences are not bracketed at the N and C terminals by paired basic amino acids. They are Met-enkephalin-Arg6-Phe7 and Met-enkephalin-Arg6-Gly7-Leu8. Thus, enkephalins can clearly be generated from proenkephalin neuropeptide-processing enzymes. In addition, time course studies with bovine chromaffin granules showed that with time the proenkephalin intermediates in the granules are processed to Met- and Leu-enkephalin as well as the heptapeptide and octapeptide and their relative proportions are also found to remain constant.

The ratio of Met-enkephalin sequence to Leu-enkephalin sequence in proenk is 6:1. On the other hand, prodyn contains three copies of Leu-enkephalin sequence but no Met-enkephalin sequence. Interestingly, the Leu-enkephalin sequences of prodynorphin are also bracketed by paired amino acid sequences. Similar to proenkephalin, prodynorphin in the tissues appears to be processed first to various dynorphins which have potent opioid activities and contain the Leu-enkephalin sequence. Again, Leu-enkephalin may only be one of the intended products of prodyn.

Thus, these large gene products can be specifically cleaved into smaller active or nonactive components. For example, 240 amino acids are incorporated into POMC, which may then be cleaved into smaller biologically active peptides whose structure can vary depending on the tissue and species. In addition to the coding regions of DNA (exons), these larger polypeptide-producing genes contain noncoding regions (introns). It has been hypothesized that introns (found only in eukaryotic genes) may function as "spacers," contain codes for enzymatic cleavage, allow for proper folding of the smaller peptide, and/or allow for different combinations of the smaller peptides to possibly make a slightly larger one. An additional function for introns can be proposed: Introns may be viewed as structures which have the ability to increase an organism's life span because they can act as "damage buffers." Eukaryotic cells contain 10 times the amount of genes as prokaryotic cells as well as have 100 times the amount of DNA. If genetic mutations occur, they will most likely affect intron regions. Organisms that have these regions would therefore stand a better chance of survival. In short, introns may be viewed as a naturally occurring antiaging mechanism.

Eukaryotic genes tend to have numerous copies of certain exons, as is the case of Met-enkephalin in proenk. Thus, this phenomenon, following enzymatic cleavage, may yield multiple copies of a signal molecule, indicating its significance. These repetitions of important exons may also be regarded as a mechanism to ensure signal fidelity in older organisms. Again, if an important signal molecule is going to be used during an organism's life span, precautions have to be incorporated into stabilizing the molecule. By repeating the sequence of a biologically important molecule many times, the chances of retaining a good copy are enhanced. If this molecule is hit by misfortune its replicates will take its place. Thus, molecular redundancy of key sequences may also be regarded as another antiaging mechanism.

Other phenomena which are equally important are the mechanism of simultaneous expression of the signal molecule and the receptor in different cell types. It is known that the same signal system can be used in different ways in the same animal and different phyla. What causes the precise expression of both complementary systems? This dual expression certainly suggests the existence of a functional interaction of these two "separate" aspects of the same signal system. It would be interesting to speculate that since both up- and downregulation of a receptor population can be regulated by the concentration of the signal molecule, the signal molecule itself can induce the presence of its receptor in a distant cell (nonspecific pinocytosis coupled to DNA disinhibition or initiation, etc., activating a dormant recognition system).

With these considerations in mind, we now review the scientific literature to determine whether opioid precursors are unique to vertebrates. It will become quite clear that they are not.

II. Invertebrate Opioid Precursors

The presence of biologically active neuropeptides in invertebrates, which are comparable to those of vertebrates, has been known for a considerable period of time (Scharrer, 1967, 1978; Frontali and Gainer, 1977; Haynes, 1980). However, detailed information on a specific class of these peptides, the endogenous opioids, is almost exclusively confined to the mammalian nervous system as noted previously. The recent interest in the diverse roles and modes of operation of these molecules, including immune actions, has sparked a search for their evolutionary history. While several reports on the occurrence of endogenous opioids in submammalian vertebrates are available (Audigier *et al.*, 1980), comparable data in invertebrates are still emerging. They consist of the demonstration in certain invertebrate ganglia of opioid peptides and their specific receptor sites (Leung and Stefano, 1987). Additionally, ACTH and β-endorphin-like amino acid sequences were detected immunologically in protozoa as part of a high-molecular-weight macromolecule (LeRoith *et al.*, 1982).

Regarding invertebrates, the presence of an opiate receptor mechanism in the central nervous system of the marine mollusc, *Mytilus edulis,* was first suggested because of a rise in ganglionic dopamine levels following intracardiac administration of exogenous Met- and Leu-enkephalin, an effect reversible by naloxone (Stefano, 1980,1982). Interestingly, this study was first submitted for scientific review in 1976. However, the author was told that opioid mechanisms do not exist in invertebrates. It was not until 1979 that the study was finally accepted for publication (Stefano and Catapane, 1979), demonstrating the strength of conventional wisdom over experimental data. The first actual demonstration of high-affinity opiate binding sites in an invertebrate ganglion was accomplished by Stefano and colleagues (Stefano, 1980) in *M. edulis.* The biochemical characteristics of this system, analyzed in detail by Kream and colleagues (1980), have been found to parallel those of mammalian systems. With respect to specific binding sites in insects, Pert and Taylor (1980) and Edley *et al.,* (1982) showed that suspensions prepared from *Drosophila* heads avidly bind [^{3}H]Leu-enkephalin and the opioid ligand [^{3}H]-diprenorphine. Specific high-affinity binding sites for a synthetic enkephalin analog, D-Ala2-Met5-enkephalinamide, were demonstrated in the cerebral ganglia and midgut of the insect *Leucophaea* (Stefano *et al.,* 1989a). The results strongly suggest the presence, in *Mytilus,* of opiate receptors that are confined to certain areas of the nervous tissue. Again, these opioid receptors were found to resemble those described in mammalian systems.

A. Proenkephalin

The presence of a mammalian-like proenkephalin peptide in invertebrates has been surmised from studies demonstrating the presence of smaller peptides that are found within this precursor (Fig. 2). In the past, Stefano and colleagues biochemically sequenced Met- and Leu-enkephalin as well as Met-enkephalin-Arg-Phe from *Mytilus* neural tissues (Leung and Stefano, 1984; Stefano and Leung, 1984). These signaling molecules were later isolated and sequenced in arthropods (Luschen *et al.,* 1991; Rothe *et al.,* 1991), annelids (Laurent and Salzet, 1996c), and the mollusc *Lymnaea stagnalis*

(Ewandinger *et al.,* 1996), thereby providing evidence for the presence of an invertebrate proenkephalin-like molecule.

It has been demonstrated that invertebrate proenkephalin is quite similar to that found in mammals (Udenfriend and Kilpatrick, 1984; Fig. 2). Recently, we have demonstrated the presence of proenkephalin in two representative invertebrates, namely, in the leech *Theromyzon tessulatum* and in the marine mussel *M. edulis* (Salzet and Stefano, 1997a; Fig. 2). This opioid precursor was found in the animals immunocytes. The structure of the leech proenkephalin material demonstrates considerable amino acid sequence similarity with amphibian proenkephalin (26.2%). *Mytilus* proenkephalin exhibits a higher sequence identity with human and guinea pig proenkephalin (39 and 50%, respectively). This proenkephalin contains Met- and Leu-enkephalin in a ratio of 3:1 for *Mytilus* and 1:2 in the leech. They also possess Met-enkephalin-Arg-Gly-Leu and Met-enkephalin-Arg-Phe that are flanked by dibasic amino acid residues, demonstrating cleavage sites. Furthermore, using both sequence comparison and a specific antiserum raised against bovine proenkephalin A (amino acids 209–237), the enkelytin peptide, FAEPLPSEEEGESYSKEVPEMEKRYGGFM, was identified in invertebrate proenkephalin (Fig. 2) and it exhibited a sequence identity of 98% with mammalian enkelytin (Goumon *et al.,* 1996).

This demonstration of proenkephalin in invertebrates supports the observations of the previous studies (Leung and Stefano, 1984; Stefano and Leung, 1984; Luschen *et al.,* 1991; Rothe *et al.,* 1991; Ewandinger *et al.,* 1996; Laurent and Salzet, 1996c) that identified proenkephalin-derived peptides as "free" signaling molecules since they are flanked by basic amino acid residues in the precursor, demonstrating, as in mammals, that they are products of enzymatic processing. In this regard, the difficulty in obtaining these pentapeptides in invertebrate tissues is due to the presence of proteolytic enzymes, i.e., neutral endopeptidase (Shipp *et al.,* 1990; Turner *et al.,* 1994; Salzet *et al.,* 1995; Laurent and Salzet, 1996c).

Of equal importance is the phenomenon of multiple copies of a repeating sequence in a precursor, i.e., Met-enkephalin (Udenfriend and Kilpatrick, 1984). This has been regarded as a simple amplification mechanism, arising from gene crossover. This phenomenon may indicate what is really different regarding proenkephalin during evolution. This is demonstrated by the leech having two the mussel three, and bovine adrenal proenkephalin four copies of Met-enkephalin. We may conclude that Met-enkephalin is singularly important in opioid processing. This is also surmised by the presence of δ_1 and δ_2 binding sites on mammalian and invertebrate tissues (Stefano *et al.,* 1989b,c,1992,1996).

The presence of the enkelytin (Goumon *et al.,* 1996) in invertebrate proenkephalin with a nearly perfect sequence match to that found in bovine chromaffin cells (98%; Stefano *et al.,* 1998a) supports the hypothesis that these molecules first evolved in simpler animals. Indeed, enkelytin, with its high antibacterial activity (Goumon *et al.,* 1996), further associates opioid peptides with immune-related activities (Stefano *et al.,* 1998a). We surmise that immune signaling or alerting may lead to enhanced proenkephalin proteolytic processing freeing both opioid peptides and enkelytin (Stefano *et al.,* 1998a). In this scenario the opioid peptides would stimulate immunocyte chemotaxis and phagocytosis as well as the secretion of classical cytokines (Stefano *et al.,* 1996). During this process the simultaneously liberated enkelytin would attack bacteria immediately, allowing time for the immune-stimulating capabilities of opioid peptides to manifest themselves. This hypothesis is further supported by the presence of specific Met-enkephalin receptors on these cells (Liu *et al.,* 1996a). Interestingly, this same scenario may occur in neural tissues (Stefano *et al.,* 1989b; Sonetti *et al.,* 1994) given the presence of glial cell types, i.e., microglia. Thus, it appears that many of the mammalian molecular and cellular survival strategies first appeared in organisms that evolved at least 500 million years ago.

B. Prodynorphin

A mammalian prodyn-derived peptide, α-neoendorphin, was purified from *T. tessulatum* central nervous system and suckers (Salzet *et al.,* 1996), suggesting the presence of a larger precursor peptide similar to prodyn of vertebrates. In mammals, processing of prodyn yields a number of bioactive peptides including Leu-enkephalin, neo-endorphins (α and β), and dynorphins (A and B) (Patey and Rossier, 1986).

Sequence alignment of the entire prodyn opioid precursor with vertebrate prodyn reveals a 28.8% sequence identity with rat and 22% with the human and pig (Table I; Civelli *et al.,* 1985). In leech prodyn, α-neo-endophin is found at position 67–76 and it exhibits a 100% sequence identity with the respective mammalian material. Dynorphin A-like material at 93–105 exhibits a 50% sequence identity and dynorphin B-like material at 106–117 exhibits a 76.6% sequence homology with the mammalian counterpart (Table I). Although the α-neo-endorphin is identical to the one found in vertebrate, the dynorphins are slightly shorter. The amount of Leu-enkephalin is similar to that found in vertebrates, i.e., 3. Moreover, the C terminus of leech prodyn is similar to that of vertebrates, whereas the N terminus is shorter. This explains the difference in mass observed between the leech (14 kDa), rat (23 kDa) (Civelli *et al.,* 1985), pig (28 kDa) (Horikawa *et al.,* 1983), and human prodyn (28 kDa) (Kakidani *et al.,* 1982), suggesting that these additions occurred later in evolution.

TABLE I
Percentage of Sequence Identity of *Mytilus* Prodynorphin-Derived Peptides with Those of Human, Rat, and Leech Materials

Mytilus	Human (%)	Rat (%)	Leech (%)
Prodynorphin	35	55	21.8
Nociceptin	25	35	ND
Orphanin FQ	50	50	
β-Neo-endorphin	100	100	100
α-Neo-endorphin	100	100	100
Dynorphin A	70.5	70.5	100
Dynorphin B	85.7	85.7	100

Note. The amino acid sequences of the various prodynorphin-derived molecules were aligned using MPSrch Version 1.5 algorithm Mpsrch (S. S. Sturrock and J. F. Collins, Biocomputing Research Unit, University of Edinburgh, UK, 1993).

As with the proenkephalin-derived peptides, the leech prodyn-derived peptides are found in positions flanked by basic amino acids, indicating cleavage sites. Furthermore, the N terminus of leech prodyn exhibits a 54.5% sequence homology with that of rat (Civelli *et al.,* 1985). Also, Leu-enk, α-neo-endorphin, dynorphin A, and dynorphin B are present at the C-terminal side of the protein.

We have recently characterized a prodyn molecule in hemocytes of the free-living bivalve mollusc *M. edulis* (Stefano *et al.,* 1998b). The ca. 16-kDa protein was purified by cutoff filtration prepurification, anti-Leu-enkephalin affinity column separation, followed by reversed-phase high-performance liquid chromatography. Its primary sequence was determined by Edman degradation, endoproteinase Glu-C digestion, and CNBr treatment. *Mytilus* prodyn contains α-neo-endorphin, dynorphin-A, and dynorphin-B at the C terminus, exhibiting 100, 70.5, and 85% sequence identity with the rat prodyn-derived counterparts, respectively (Table I). The number of Leu-enkephalins in this precursor is identical to that found in vertebrates. *Mytilus* prodyn is distinguished from that of leeches in that the N terminus is longer. Additionally, by sequence comparison, the presence of an orphanin FQ-like peptide, exhibiting 50% sequence homology with that found in mammals, was demonstrated (Stefano *et al.,* 1998b). This was the first report of the complete biochemical characterization of a prodyn in a nonparasitic invertebrate and mollusc.

C. POMC

Duvaux-Miret and colleagues (1990) demonstrated the presence of β-endorphin and of a POMC-related gene in *Schistosoma mansoni.* Dot blots of cercarial genomic DNA, hybridized with two oligonucleotide probes complementary to highly conserved POMC sequences, showed a POMC-related gene in this trematode. Northern blot analysis of adult worm RNA indicated that this gene was actively transcribed and β-endorphin, ACTH, and α-MSH were detected in all developmental stages of the parasite by radioimmunoassay. Furthermore, *S. mansoni* secretes ACTH-like and β-endorphin-like peptides into its incubation medium (Duvaux-Miret *et al.,* 1992a,b). This study constituted the first demonstration of a POMC-related gene transcribed in an invertebrate.

Salzet and colleagues (1997) sequenced a mammalian-like POMC and six of its derived peptides, including ACTH and MSH, in the immune tissues of the leech *T. tessulatum.* Of the six peptides, three showed high sequence similarity to their vertebrate counterparts, namely, Met-enkephalin, α-MSH, and ACTH (100, 84.6, and 70%, respectively), whereas γ-MSH, β-endorphin, and γ-LPH exhibited only 45, 20, and 10% sequence identity. No dibasic amino acid residues were found at the C terminus of the γ- and β-MSH peptides. In contrast, the leech α-MSH was flanked at its C terminus by the Gly-Arg-Lys amidation signal. ACTH and CLIP were also C-terminally flanked by dibasic amino acid residues. The coding region of leech POMC was also reported by RT-PCR using degenerated oligonucleotide primers (Salzet *et al.,* 1997).

In recent report we note that *M. edulis* hemocytes also contain a mammalian-like POMC (Stefano *et al.,* 1998c; Fig. 2). Of the six peptides found in this opioid precursor, Met-enkephalin, γ-MSH, α-MSH, and ACTH exhibited 100, 100, 90, and 74% sequence identity, respectively (Table II). The β-endorphin-like and γ-LPH-like molecules exhibit only 25 and 10% sequence identity. Dibasic amino acid residues are found at the C terminus of MSH and ACTH, indicating cleavage sites. The α-MSH is flanked at the C terminus by Gly-Arg-Lys, representing the amidation signal. ACTH and CLIP are also C-terminally flanked by dibasic amino acid residues. Of interest is the fact that the Met-enkephalin present in the "β-endorphin-like peptide" is not flanked by basic amino acids, as is the case in mammals, suggesting that functionally it may be present. Taken together, the results from parasites and a free-living mollusc conclusively demonstrate, without the presence of vertebrate "contamination," that

TABLE II

Percentage of Sequence Identity of *Mytilus* POMC-Derived Peptides with Those of Leech, Human, Rat, and *X. laevis*

Mytilus	Leech (%)	Human (%)	Rat (%)	*X. laevis* (%)
POMC	51.3	35.3	40.2	31.6
			6	
ACTH	91.2	94.7	93.3	92.2
γ-MSH	32	93	95	92
α-MSH	80	100	100	100
CLIP	90	95	95	88
γ-LPH	38	5	5	8
β-Endorphin	25	18	20	21
Met-enkephalin	100	100	100	100

Note. The amino acid sequences of the various POMC-derived molecules were aligned using MPSrch Version 1.5 algorithm Mpsrch (S. S. Sturrock and J. F. Collins, Biocomputing Research Unit, University of Edinburgh, UK, 1993).

POMC and many of the derived bioactive peptides, i.e., α-MSH, are present in invertebrates. Furthermore, in regard to their function, i.e., immune regulatory actions, they appear to be conserved as well (Stefano *et al.,* 1996; Salzet *et al.,* 1997).

III. Opioid Processing

A. General Considerations

What is the significance of opioid precursors being found in invertebrate hemocytes? We surmise that these large precursor proteins are sequestered in the cell and only processed into their smaller active peptides, such as Met-enkephalin, when required. The functions of these molecules in invertebrates can be deduced from the processing of the precursor molecules in various tissues.

Since precursor processing involves enzymes, the presence of specific enzymes becomes important. An examination of the literature reveals the presence of many types of enzymes in both vertebrates and invertebrates, some of which are important in processing neuropeptides, e.g., neutral endopeptidase (NEP) and angiotensin-converting enzyme (ACE) (Shipp *et al.,* 1990; Turner *et al.,* 1994; Laurent and Salzet, 1995, 1996a,b; Laurent *et al.,* 1998).

The peptide precursors, as stated earlier, represent large gene products which contain active proteins; these are often flanked by dibasic amino acids, indicating cleavage sites. The precursor proteins also contain multiple copies of particular biologically active peptides. For example, in both mammals and invertebrates, proenkephalin contains multiple copies of Met-enkephalin. Based on these two observations, we can conclude that immune tissues can rapidly produce high levels of these peptides without the necessity for gene expression to produce the final protein *de novo* (Fig. 3). Since immune cells hold these large proteins in their "inactive" state, it seems reasonable to speculate that they are important for initial immune responses, i.e., proinflammatory events. Since many of the processing enzymes are extracellular, it also seems likely that the cell secretes the precursor molecules. For example, neutral endopeptidase 24.11 (NEP, CD10, CALLA, and enkephalinase) is found on the surface of granulocytes and invertebrate immunocytes (Shipp *et al.,* 1990). In invertebrates, opioid precursor molecules are also found in the hemolymph (Salzet *et al.,* 1997).

The enzyme NEP appears to be quite important. For example, it may be responsible not only for cleaving the precursor proenkephalin or POMC/

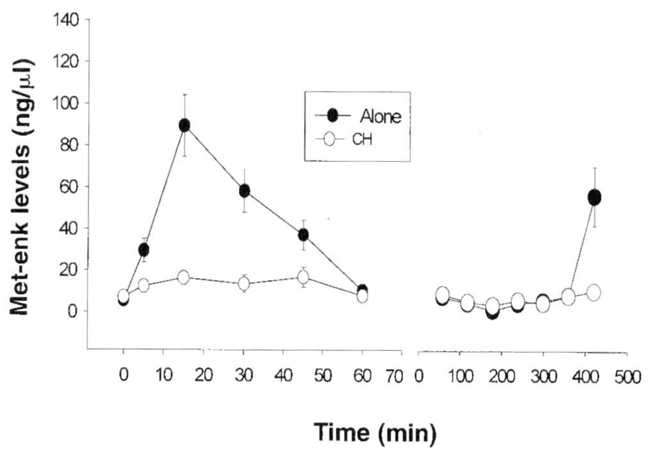

FIG. 3 Demonstration of opioid dynamics in response to an antigenic challenge. LPS (1 μM/ml; designated alone) stimulates at least two waves of hemolymph Met-enkephalin level increases. We surmise the first Met-enkephalin peak comes from a preexisting pool, i.e., proenk, since it occurs immediately, whereas the second, occurring hours later, is induced as surmised by its inhibition by cycloheximide (10 μM/10 μl injected into the hemolymph every 90 min), a protein synthesis inhibitor. Each experiment was repeated four times and Met-enkephalin levels were determined by RIA as noted elsewhere (reproduced with permission from Stefano *et al.,* 1995).

ACTH but also for the active processed peptides, i.e., Met-enkephalin and MSH, respectively (Stefano and Scharrer, 1991; Duvaux-Miret *et al.,* 1992ab; Smith *et al.,* 1992; Stefano and Smith, 1996; (Fig. 4). This is a marvelously sophisticated processing mechanism because the same enzyme breaks down the precursor to generate active peptides and then inactivates the same active molecules producing inactive products. This represents a multidimensional process that requires less DNA "message" since the same enzyme performs these tasks. Furthermore, in some cases the actual inactive products may act as competitive inhibitors to further limit the activity of the prime enzyme, adding another degree of microenvironmental control (Fig. 4). This has been noted in our laboratory by NEP processing of Met-enkephalin-Arg-Phe (Stefano and Scharrer, 1991).

Having the processing enzymes extracellularly also means that the concentration of enzyme in the microenvironment can be controlled by local microenvironmental conditions, i.e., the number of granulocytes carrying NEP. The immunocytes can be viewed as enzyme transporters, used to perform and/or enhance processing events by changing the local enzyme

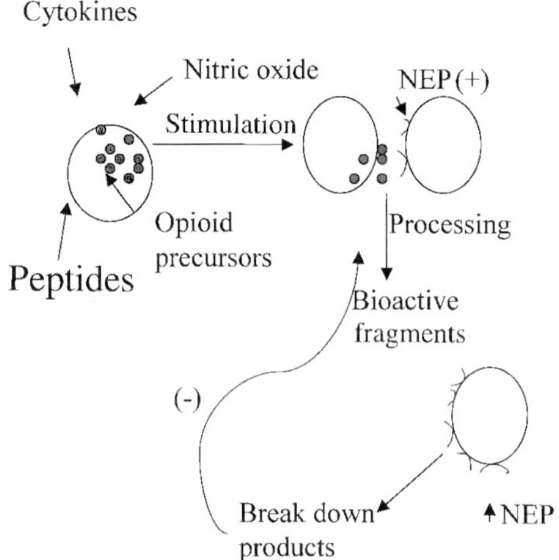

FIG. 4 Illustration for opioid peptide involvement in an immune response. Invertebrate and vertebrate immunocytes contain opioid precursors. Upon an initial stimulus, i.e., antigenic challenge, trauma, etc., the precursors are released. The stimuli may come from cytokines, nitric oxide (NO), and other signaling molecules, i.e., peptides. This initiates the release of the opioid precursors which are then processed by enzymes, i.e., NEP. This releases the smaller bioactive peptides, i.e., Met-enkephalin. During this process NEP is upregulated, reducing the time required for this processing. In time, these enzymatically driven processes break down the smaller bioactive peptides, thus reducing the level of stimulation. In some cases, these breakdown products may also interfere with the enzyme's activity, reducing it even more.

concentration. Keeping this in mind, when an antigenic challenge emerges, the "damage" may directly or indirectly—via cytokines—signal immunocytes in the area to start their activation process. For example, an immunocyte will secrete proenk that is processed to Met-enkephalin. Met-enkephalin or Met-enkephalin-Arg-Phe can diffuse away from this source since large numbers of immunocytes are not present during the initial event, and in so doing they can create a concentration gradient for additional immunocyte recruitment since it is chemotaxic (Stefano *et al.,* 1996). With more cells arriving, the enzyme concentration increases. With this, additional opioid signaling molecules are broken down, and when the cells enter this area, they process even more material. They initially produce more Met-enkephalin. It may be equally true that the initial damage event/agent may release proenk directly without cytokine involvement.

In a full proinflammatory situation, enkephalin is constantly being processed from the precursor because of more incoming NEP. However, the precursor may be used up and this process may have to wait for the *de novo* synthesis of this material, constituting a second wave of precursor presence (Fig. 3). This situation has the potential to be self-limiting since if there are enough enzyme-bearing immunocytes, this area will appear to stop producing signals since the same enzyme is rapidly cleaving the active peptide into inactive products. Thus, counterintuitively, beyond a certain point, continued recruitment of immunocytes may serve to decrease the proinflammatory process, making it a self-limiting event. However, if some cells are destroyed as they enter the region, then the inflammatory process would continue because the enzymes would also be destroyed. In this situation the inflammatory process would continue to escalate. Thus, "winning" and/or "losing" a proinflammatory process can be explained, in part, by the presence of processing enzymes, e.g., immunocytes. If the trauma is widespread/diffuse, enough cells cannot be brought in to stop the process.

The complexity of this process increases with the understanding that POMC liberates ACTH and MSH—two immunocyte inhibitory molecules (Stefano and Smith, 1996). We surmise that there is a balance of stimulatory and inhibitory molecules. Indeed, this balance may have a strong time factor associated with it. As noted previously, in an inflammatory situation enkephalin appears rapidly; ACTH processing is slow and the levels of α-MSH, a potent immunocyte downregulating signal, increase gradually (Stefano *et al.,* 1996; Stefano and Smith, 1996), ensuring that the beneficial initial immune response can occur. This is even augmented by POMC processing since upon its initial processing the relatively enzyme-resistant chemotactic factor β-endorphin is liberated, ensuring that distant signaling can take place. Thus, we surmise there is a balance between precursor processing and active endproduct degradation. This balance can be shifted by the total number of immunocytes that are present in the local area. In so doing, the "chemotactic trail" may be lost as can the immunocyte-stimulating properties of the active processed neuropeptides.

This scenario offers a hypothesis as to why immune cells carry and process opioid proteins. This scenario unfolds with the use of aprotinin, a serine protease inhibitor. Using this compound can and does diminish the diffuse inflammatory response associated with surgery (Soeparwata *et al.,* 1996; Chopin *et al.,* 1997), demonstrating the significance of processing enzymes. In patients undergoing major heart surgery, we found that just before surgery, plasma ACTH levels dropped below the level of detection (Fricchione *et al.,* 1996), indicating the activation of the processing enzymes (Stefano and Smith, 1996). In this regard, it is widely known that various immune and neural-type signaling molecules can upregulate enzymes such as NEP (Shipp *et al.,* 1990, 1991; Stefano *et al.,* 1996). This response can be considered biphasic. A mechanism for enhanced neuropeptide precursor processing is followed by enhanced processed peptide degradation due to a further increase in enzyme levels, i.e., immunocyte recruitment.

Based on this, the outcome of an immune proinflammatory response depends on the precise time-dependent buildup of enzyme levels followed by their dissipation and resulting lack of immunocyte recruitment due to diminished peptide presence that also stops further immunocyte stimulation, i.e., cytokine secretion. Clearly, with this scenario, cascading immune responses can be better understood. In this regard, it is important to realize that invertebrate immune/defense systems have been utilizing these processes for over 500 million years (Salzet *et al.,* 1997; Salzet and Stefano, 1997a,b; Stefano *et al.,* 1996, 1998abc). Furthermore, we can surmise that there is an overall logic to the process. As can be observed in Fig. 5, the enkephalins can exert an immediate overall immunostimulatory action, whereas MSH exerts an immunoinhibitory action. To avoid conflict the processing to generate MSH is slow, thus allowing for immunostimulation to occur first.

B. Specific Regulatory Processes

There is a growing body of evidence demonstrating that morphine influences ACTH processing in vertebrates and invertebrates (Stefano *et al.,* 1996). This is especially important since it is a naturally occurring signal molecule found in human plasma and invertebrate hemolymph (Stefano *et al.,* 1993, 1995; Liu *et al.,* 1996b; Brix-Christensen *et al.,* 1997). In this regard, exogenously applied morphine can increase the release of ACTH in rat hypothalamus (Nikolarakis *et al.,* 1989; Stefano *et al.,* 1996). Recent work from our laboratory demonstrates that nitric oxide (NO) controls neurohormonal release from median eminence neuroendocrine nerve terminals in

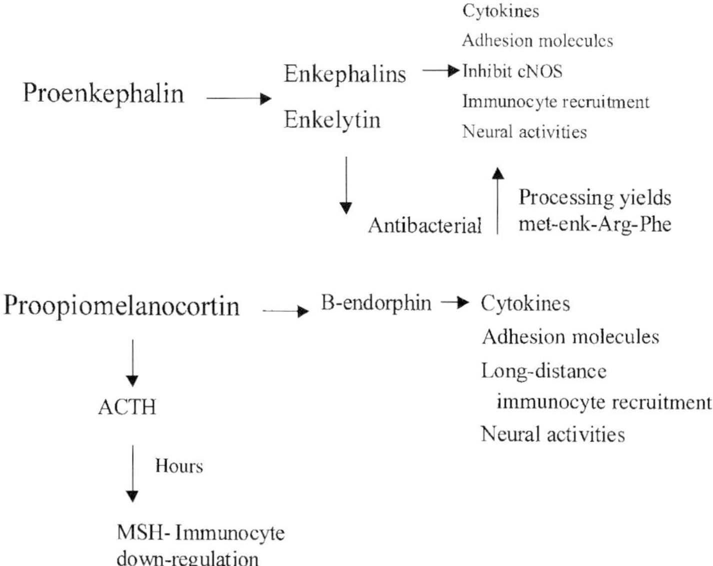

FIG. 5 Immediate processing of proenkephalin and proopiomelanocortin results in several products exhibiting diverse but focused functions. (Top) Immune stimulation: The enkephalins stimulate immunocyte chemotaxis and activation as well as cytokine secretion and production, i.e., interleukin-6 (Zhong *et al.*, 1998). It increases endothelial adherence of immunocytes and blocks NO production (Stefano *et al.*, 1998d). Enkelytin represents an immediate antibacterial strategy prior to a large-scale immunocyte activation (Stefano *et al.*, 1998a). In addition, enkephalins have critical neuroregulatory actions, i.e., inhibiting neural transmitter release and pain modulation (Stefano *et al.*, 1996; Liu *et al.*, 1996c). (Bottom) Immune downregulation: ACTH is converted to α-MSH, which downregulates invertebrate and vertebrate immunocytes (Stefano and Smith, 1996; Salzet *et al.*, 1997). This process occurs over hours, thus allowing time for immunoactivation. The β-endorphin liberation is rapid, and since the molecule is resistant to proteolytic attack, it can travel greater distances from the site of processing to recruit distant immunocytes. Thus, the sequential and timely nature of the processing allows for an appropriate immune response.

the rat (Prevot *et al.*, 1998). The stimulation of this release from median eminence fragments, including vascular tissues, occurs by μ_3 receptor activation by morphine (Stefano *et al.*, 1993, 1996). Furthermore, morphine by the NO-dependent process influences neurohormonal release from ME nerve terminals within 10 min, releasing corticotropin, which can then account for the action of morphine noted earlier.

In another mechanism associated with ACTH, morphine, in a dose-dependent manner, by way of NO increases leech processing of POMC as noted by higher hemolymph levels of α-MSH and ACTH (Salzet *et al.*, 1997). In *Mytilus* we also demonstrated that morphine stimulates the processing of ACTH (amino acids 1–39) to MSH (amino acids 1–13) by NEP as determined by phosphoramidon inhibition. The ability of morphine to enhance enzyme levels has also been noted in other studies using mammalian and human tissues (Malfroy *et al.,* 1978; Stefano *et al.,* 1996). The mechanism for this morphine action, based on these reports, is by increasing the processing of the precursor, stimulating the release of the precursor, or both.

The significance and specificity of opiate molecules in these studies are enhanced by the observation that lipopolysaccharide stimulation results in ACTH (amino acids 1–24) in the hemolymph, indicating that other enzymatic processes can occur by way of different signaling molecules. Furthermore, ACTH (amino acids 1–24) processing occurred by an enzymatic process independent of NEP, i.e., renin-type enzyme (Stefano *et al.,* 1998c). Taken together, as in mammals, differential processing of ACTH occurs in invertebrates. Additionally, invertebrate immunocytes are capable of displaying different responses to ACTH fragments, including those of *Mytilus* (Genedani *et al.,* 1993), further supporting the differential processing pattern and its potential significance as a meaningful event.

In conclusion, POMC and its derived peptides appear to have originated in "simple" animals. The same sequential order of the derived peptides in POMC and their flanking by dibasic amino acids is of noteworthy significance in this regard. POMC can be found in hemocytes, demonstrating a role for these peptides in immune regulation. Furthermore, POMC processing and release can be stimulated by the same signaling molecules that perform this activity in vertebrates. Immunocytes from diverse animals, including invertebrates, can respond to POMC-derived peptides in a highly specific manner. Taken together, we surmise that POMC and its associated peptides had their origin much earlier than previously thought. Therefore, aside from its historical origin, it may be more correct to speak of mammalian POMC as invertebrate-like.

IV. Stabilization of a Signal System within Evolution

Why should these signaling molecules and their apparent systems/mechanisms be retained relatively intact during the course of evolution? In order to answer this question, we must briefly review some basic principles of intercellular signaling. It should also be noted that the same principles may apply to intracellular signal systems. Major requirements of a compound to be established as a signaling molecule, be it as a neurotransmitter or a hormone, are (i) the presence of the molecule in a particular cell; (ii) its release from that cell upon appropriate stimulation; (iii) high-affinity, stereospecific binding to a receptor on the target cell; (iv) a specific physiological effect of the molecule on the effector cell; and (v) a specific inactivation mechanism.

In peptidergic signal systems, these characteristics are directly gene determined. The enzymes that synthesize and process such signal molecules must be present in their cells of origin and the information to produce these enzymes resides in the DNA of the cell. This is also true of the stereoselective receptor molecules found on the target cell as well as all stereoselective components of a given intercellular communication system. The entire sequence, from synthesis of the signal molecule to its inactivation, is based on sequential stereospecific events, including, in another cell, receptor recognition. Therefore, the components of the system had to evolve simultaneously in order for the system to be operational. Compatible structural conformations had to be found in the synthesizing enzymes, the signal molecule, the receptor molecule, and the inactivation enzymes. Such conformational "matching" of molecules within each signal system is difficult and time-consuming to achieve. In addition, to be operational within an organism, all components had to be expressed simultaneously. Thus, evolutionary changes had to occur on the corresponding genes of the components within a particular signal system if large-scale changes were to take place. The conformational complexity and rigidity of the match among the sequential components for a given signal system would thus seem to exert a determining influence during evolution to maintain the conformational integrity of the signal system given the degree of difficulty in obtaining it originally (Stefano, 1980, 1986; Makman and Stefano, 1984). Thus, "ancient" communication systems, e.g., opioids, would tend to remain relatively intact in increasingly complex animal phyla, especially the structure or conformation of the bioactive portions of the molecules themselves.

This principle of conservation does not preclude events that may lead to an old signal system being used in a new functional capacity. In summary, the determining force during evolution which appears to maintain signal systems may well be the number of highly precise stereospecific conformational matching events associated with intercellular communication mechanisms.

This same consideration may also be applied to intracellular communication systems being maintained throughout evolution. Certainly, cAMP appears to occur in all organisms. In addition, the primary structure of calmodulin, a multifunctional intracellular messenger, is believed to be retained since mammalian antibodies appear to react with extracts from coelenterates (Cheung, 1980).

The list of "mammalian-type" signaling molecules in "simpler" organisms is steadily growing. In prokaryotes, chorionic gonadotropin-like material has been detected (Acevedo *et al.,* 1978). In protozoans, not only have similar mammalian-type neuropeptides been detected but also opioids have been shown to alter feeding behavior, an effect inhibited by naloxone, the opiate receptor blocker (Josefsson and Johansson, 1979). Taken together, the evidence indicates that signal effector and receptor communication mechanisms may be present in unicellular organisms. This in turn suggests that the origin of signal systems may have occurred during prokaryotic development. Indeed, many of these systems may have started as intracellular communication mechanisms. Thus, the term neuropeptides may be totally erroneous, even in vertebrates, since many serve and are found in locations other than neural tissues, i.e., immune.

The accumulating evidence also strongly indicates that this is not a "static" system in invertebrates, but rather one that is highly variable in regard to quantitative changes, even within the same species and in the same tissue. Aging variations have been shown to occur in regard to opioid levels and high-affinity opioid binding site densities (Leung and Stefano, 1987) as also noted in mammals (Codd and Byrne, 1980). Seasonal variation in opioid-binding densities occurs in *M. edulis* (Stefano and Leung, 1986). These variations manifest themselves in changing ratios of the three detected opioids, including periods of time when they are below the level of detection. Seasonal opioid variations also occur in leeches (Flanagan and Zipser, 1986) and in the anterior byssus retractor muscle response in morphine (Bianchi and Wang, 1986). Clearly, the opioid systems in invertebrates are complex and show striking similarities with those of mammals, thus indicating a common evolutionary trend.

The question is how change or diversification can be introduced into this conservative concept of signal system evolvement. It would seem as though change has occurred in various signal system "families." Even though substance P can exert physiological effects in invertebrates, attempts to isolate it biochemically have proven unsuccessful (Kream *et al.,* 1986). However, closely related molecules, the tachykinens, have been found in lower life forms (physaelamine in Amphibians and eledoisin in Octopus). Given the rather conservative nature of signal system evolvement noted in this review, how can one account for the diversified signal molecules found in higher organisms? It is likely that the answer lies in the concept of conformational matching. The most significant aspect of conformational recognition within the stereospecific components of a signal system is by definition its functional shape. Substitution of amino acids would or could be tolerated only if the characteristic conformation of the signal system component (enzymes, receptor, messenger molecule, etc.) retains its functional shape, within very narrow limits, at each recognition event. This is true of deltorphin and Met-enkephalin in both vertebrates and invertebrates since they have different amino acid sequences (Stefano *et al.,* 1992). There is also an orphanin-FQ-like peptide exhibiting 50% sequence homology with the mammalian material in *Mytilus* prodyn, suggesting that its occurrence in animals can also be traced back to invertebrates (Stefano *et al.*, 1998b). Thus, change may be retained and incorporated into the signal system as long as the molecule maintains its proper conformation. We speculate that the pliable nature of a signal molecule, especially peptides, can be modified at a recognition step by the normal microenvironmental factors (pH, ionic composition, etc.). This implies that there exists a degree of conformational "give." It would be possible to initiate diversification of particular signal systems within this narrow give range. Given the small amount of "play" allowed, change would still occur at a very slow pace. The immunoreactive histochemical studies have certainly demonstrated that strong structural similarities exist with regard to peptides of vertebrate origin (in which they were first discovered) being found in invertebrates. Thus, signal system families are centralized around an essential conformation. In invertebrates we see the evolvement of β-endorphin as noted earlier since the Met-enkephalin molecule is not surrounded by cleavage sites, indicating that it may be active.

V. Conclusion

In summary, the data on invertebrate opioids indicate that posttranslational variation exists in addition to posttranslational modification processes. There is diversity in the number of opioid products originating from the precursors. The opioid system is dynamic and varies with age and seasons. Variations in the processing of proopiomelanocortin are known to occur in different higher vertebrate neural tissues (Smyth, 1986). This demonstrates flexibility and tissue specificity in processing mechanisms, as also noted in invertebrates (Martin *et al.*, 1986; Duvaux-Miret *et al.*, 1992ab; Salzet *et al.*, 1997).

Since many opioid characteristics are common to both invertebrates and vertebrates, their ancient status appears to be established. Again, we must question why this signal system would remain intact, especially the shorter molecules. We believe the answer, in part, is the need for conformational matching of the stereospecific nature of these systems as mentioned earlier. Each signal molecule contains a critical area responsible for the initiation of biological activity, and the conformation of this critical area is the most important aspect of the molecule. For example, only the first 24 of 39 amino acids of ACTH are required for its full activity and this important region is stable among various mammals and invertebrates as well. However, at the C-terminal end (amino acids 25–39) variation of amino acid composition occurs and seems tolerable. If this serves as an example, we should expect to find more short-chain sequences of biologically important molecules or parts of these molecules to be identical in most phyla, whereas variation may be found in the larger signaling molecules especially in regions distant from the active "center." Somatostatin also appears to be highly conserved in that 35 of the 42 nucleotides coding for it are conserved in rat and angler fish (Goodman *et al.,* 1982). These organisms are believed to have diverged in evolution about 400 million years ago. The same appears to be true for gastrin/CCK, as noted recently (Roth *et al.,* 1982; Hansen *et al.,* 1987). Indeed, these modifications may be considered to be the evolutionary changes or "advances" occurring during evolution. In brief, over long periods of time, additions to essential "key" conformations may occur, increasing the efficiency of a particular signal system. Thus, as in the case of the opioid substances, it may be more proper to speak of a "family" of compounds. The establishment of such a family lies in the rigidity of the conformational components and events occurring simultaneously in the involvement of vital biochemical reactions. Therefore, conformational matching becomes a determining force and not a selective force in establishing a basic signal system that can be enriched as time goes on.

Acknowledgments

This work was supported in part by the following grants: NIMH COR 17138, NIDA 09010, and the Research Foundation and Central Administration of the State University of New York (GBS) and NIH Fogarty INT 00045 (MS and GBS).

References

Acevedo, H. F., Sliifkin, M., Pouchet, G. R., and Pardo, M. (1978). Immunohistochemical localization of a choriogonadotropin-like protein in bacteria isolated from cancer patients. *Cancer* **41,** 1217–1229.

Audigier, Y., Deprat, A. M., and Cros, J. (1980). Comparative study of opiate and enkephalin receptors on lower vertebrates. *Comp. Biochem. Physiol.* **67,** 191–194.

Bianchi, C. P., and Wang, Z. (1986). Morphine enhancement of the cholinergic response of anterior byssus retractor muscle of *Mytilus edulis. In* "Handbook of Opioid and Related Neuropeptide Mechanisms" (G. B. Stefano, ed.), Vol. 2, pp. 59–64. CRC Press, Boca Raton, FL.

Brix-Christensen, V., Tonnesen, E., Sanchez, R. G., Bilfinger, T. V., and Stefano, G. B. (1997). Endogenous morphine levels increase following cardiac surgery as part of the antiinflammatory response? *Int. J. Cardiol.* **62,** 191–197.

Cheung, W. Y. (1980). Calmodulin plays a pivotal role in cellular regulation. *Science* **207,** 19–27.

Chopin, V., Stefano, G. B., and Salzet, M. (1999). Tessulin: A new trypsin–cathepsine G inhibitor from the leech Theromyzon tessulatum. *J. Enz. Inhibition,* in press.

Civelli, O., Douglass, J., Goldstein, A., and Herbert, E. (1985). *Proc. Natl. Acad. Sci. USA* **82,** 4291–4295.

Codd, E. E., and Byrne, W. L. (1980). Seasonal variation in the apparent number of H-naloxone binding sites. *In* "Endogenous and Exogenous Opiate Agonists" (E. L. Way, ed.), pp. 67–71. Pergamon, New York.

Duvaux-Miret, O., Dissous, C., Guatron, J. P., Pattou, E., Kordon, C., and Capron, A. (1990). The helminth Schistosoma mansoni expresses a peptide similar to human beta-endorphin and possesses a POMC-related gene. *New Biol.* **2,** 93–99.

Duvaux-Miret, O., Stefano, G. B., Smith, E. M., Dissous, C., and Capron, A. (1992a). Immunosuppression in the definitive and intermediate hosts of the human parasite *Schistosoma mansoni* by release of immunoactive neuropeptides. *Proc. Natl. Acad. Sci. USA* **89,** 778–781.

Duvaux-Miret, O., Stefano, G. B., Smith, E. M., Mallozzi, L., and Capron, A. (1992b). Proopiomelanocortin-derived peptides as tools of immune evasion for the human trematode *Schistosoma mansoni. Acta Biol. Hungari* **43,** 281–286.

Edley, S. M., Hall, L., Herkenham, M., and Pert, C. B. (1982). Evolution of striated opiate receptors. *Brain Res.* **249,** 184–188.

Ewandinger, N. M., Ridgway, R. L., Syed, N. I., Lukowiak, K., and Bulloch, A. G. M. (1996). Identification and localization of a (Met5)-enkephalin-like peptide in the mollusc, Lymnaea stagnalis. *Brain Res.* **737,** 1–15.

Flanagan, T., and Zipser, B. (1986). Opioid–peptide and substance P immunoreactivity in cyotlogical preparations and tissue homogenates of the leech. *In* "Handbook of Comparative Opioid and Related Neuropeptide Mechanisms" (G. B. Stefano, ed.), pp. 165–180. CRC Press, Boca Raton, FL.

Fricchione, G. L., Bilfinger, T. V., Jandorf, L., Smith, E. M., and Stefano, G. B. (1996). Surgical anticipatory stress manifests itself in immunocyte desensitization: Evidence for autoimmunoregulatory involvement. *Int. J. Cardiol.* **53,** S65–S74.

Frontali, N., and Gainer, H. (1977). Peptides in invertebrate nervous system. *In* "Peptides in Neurobiology" (H. Gainer, ed.), pp. 259–271. Plenum, New York.

Genedani, S., Bernardi, M., Ottaviani, E., Franceschi, C., Leung, M. K., and Stefano, G. B. (1993). Differential modulation of invertebrate hemocyte motility by CRF, ACTH, and its fragments. *Peptides* **15,** 203–206.

Goodman, R. H., Jacobs, J. W., Dee, P. C., and Habener, J. R. (1982). Somatostatin-28 encoded in a cloned cDNA obtained from a rat medulary thyroid carcinoma. *J. Biol. Chem.* **257,** 1156–1159.

Goumon, Y., Strub, J. M., Moniatte, M., Nullans, G., Poteur, L., Hubert, P., Van Dorsselaer, A., Aunis, D., and Metz-Boutigue, M. H. (1996). The C-terminal biophosphorylated proenkephalin-A-(209–237)-peptide from adrenal medullary chromaffin granules possesses antibacterial activity. *Eur. J. Biochem.* **235,** 516–525.

Hansen, G. N., Hansen, B. L., and Scharrer, B. (1987). Gastrin/CCK-like immunoreactivity in the corpus cardiacum–corpus allatum complex of the cockroach *Leucophaea maderae. Cell Tissue Res.* **248,** 595–598.

Haynes, L. W. (1980). Peptide neruoregulation in invertebrate. *Prog. Neurobiol.* **15,** 205–223.

Horikawa, S., Takai, T., Toyosata, M., Takahashi, H., Noda, M., Kakidani, H., Kubo, T., Hirose, T., Inayama, S., Hayashida, H., and Miyata, T. (1983). *Nature* **306,** 611–614.

Josefsson, J. O., and Johansson, P. (1979). Naloxone reversible effect of opioid on pinocytosis in Amoeba proteus. *Nature (London)* **78,** 283–292.

Kakidani, H., Furutani, Y., Takahashi, H., Noda, M., Morimoto, Y., Hirose, T., Asai, M., Inayama, S., Nakanishi, S., and Numa, S. (1982). *Nature* **298,** 245–249.

Kream, R. M., Zukin, R. S., and Stefano, G. B. (1980). Demonstration of two classes of opiate binding sites in the nervous tissue of the marine mollusc Mytilus edulis. Positive homotropic cooperativity of lower affinity binding sites. *J. Biol. Chem.* **255,** 9218–9224.

Kream, R. M., Leung, M. K., and Stefano, G. B. (1986). Is there authentic substance p in invertebrates? *In* "Comparative Opioid and Related Neuropeptide Mechanisms" (G. B. Stefano, ed.), pp. 65–72. CRC Press, Boca Raton, FL.

Laurent, V., and Salzet, M. (1995). Isolation of a neuropeptide-degrading endopeptidase from the leech *Theromyzon tessulatum*. *Eur. J. Biochem.* **233,** 186–191.

Laurent, V., and Salzet, M. (1996a). Identification and properties of an angiotensin-converting enzyme in the leech *Theromyzon tessulatum*. *Peptides* **17,** 737–745.

Laurent, V., and Salzet, M. (1996b). Metabolism of angiotenisns in head membranes of the leech *Theromyzon tessulatum* by peptidases. *FEBS Lett.* **384,** 123–127.

Laurent, V., and Salzet, M. (1996c). Metabolism of enkephalins in head membranes of the leech *Theromyzon tessulatum* by peptidases. *Regul. Peptides* **65,** 123–131.

Laurent, V., Stefano, G. B., and Salzet, M. (1998). The leech angiotensin-converting enzyme. *Ann. N.Y. Acad. Sci.* **839,** 500–502.

LeRoith, D., Liotta, A. S., Roth, J., Shilozch, J., Lewis, M. E., Pert, C. B., and Krieger, D. T. (1982). ACTH and β-endorphin-like materials are native to unicellular organisms. *Proc. Natl. Acad. Sci. USA* **79,** 2086–2090.

Leung, M. K., and Stefano, G. B. (1984). Isolation and identification of enkephalin in pedal ganglia of Mytilus edulis (mollusca). *Proc. Natl. Acad. Sci. USA* **81,** 955–958.

Leung, M. K., and Stefano, G. B. (1987). Comparative neurobiology of opioids in invertebrates with special attention to senescent alterations. *Prog. Neurobiol.* **28,** 131–159.

Liu, Y., Casares, F., and Stefano, G. B. (1996a). $\delta 2$ opioid receptor mediates immunocyte activation. *Chinese J. Neuroimmunol. Neurol.* **3,** 69–72.

Liu, Y., Bilfinger, T. V., and Stefano, G. B. (1996b). A rapid and sensitive quantitation method of endogenous morphine in human plasma. *Life Sci.* **60,** 237–243.

Liu, Y., Shenouda, D., Bilfinger, T. V., Stefano, M. L., Magazine, H. I., and Stefano, G. B. (1996c). Morphine stimulates nitric oxide release from invertebrate microglia. *Brain Res.* **722,** 125–131.

Luschen, W., Buck, F., Willig, A., and Jaros, P. P. (1991). Isolation, sequence analysis, and physiological properties of enkephalins in the nervous tissue of the shore crab, Carcinus maenas L. *Proc. Natl. Acad. Sci. USA* **88,** 8671–8675.

Makman, M. H., and Stefano, G. B. (1984). Marine mussels and cephalopods as models for study of neuronal aging. *In* "Invertebrate Models in Aging Research" (D. H. Mitchell and T. E. Johnson, eds.), pp. 165–190. CRC Press, Boca Raton, FL.

Malfroy, B., Swerts, J. P., Guyon, A., Roques, B. P., and Schwartz, J. C. (1978). High affinity enkephalin degrading peptidase in brain is increased after morphine. *Nature* **276,** 523–526.

Martin, R., Haas, C., and Voight, K. H. (1986). Opioid and related neuropeptides in molluscan neurons. *In* "Handbook of Comparative Opioid and Related Neuropeptide Mechanisms" (G. B. Stefano, ed.), pp. 49–64. CRC Press, Boca Raton, FL.

Nikolarakis, K. E., Pfeiffer, A., Stalla, G., and Herz, A. (1989). Facilitation of ACTH secretion by morphine is mediated by activation of CRF releasing neurons and sympathetic neuronal pathways. *Brain Res.* **498,** 385–388.

Patey, G., and Rossier, J. (1986). Decouverte, anatomie et biosynthese des differentes familles de peptides opioides endogenes. *Ann. Endocrinol.* **47,** 71–87.

Pert, C. B., and Taylor, D. (1980). Type 1 and type 2 opiate receptors: A subclassification scheme based upon GTP's differential effects on binding. *In* "Endogenous and Exogenous Opiate Agonists and Antagonists" (E. L. Way, ed.), pp. 87–94. Pergamon, New York.

Prevot, V., Rialas, C., Croix, D., Salzet, M., Dupouy, J.-P., Puolain, P., Beauvillain, J. C., and Stefano, G. B. (1998). Morphine and anandamide coupling to nitric oxide stimulated GnRH and CRF release from rat median eminence: Neurovascular regulation. *Brain Res.*, **790,** 236–244.

Roth, J., LeRoith, D., Shiloach, J., Rosenziveig, J. L., Lesniak, M. A., and Havrankova, J. (1982). The evolutionary origins of hormones, neurotransmitters and other extracellular chemical messengers. *N. Engl. J. Med.* **306,** 523–529.

Rothe, H., Luschen, W., Asken, A., Willig, A., and Jaros, P. P. (1991). Purified crustacean enkephalins inhibits release of hyperglycemic hormone in the crab, Carcinus maenas L. *Comp. Biochem. Physiol. C.* **99,** 57–62.

Salzet, M., and Stefano, G. B. (1997a). Invertebrate proenkephalin: Delta opioid binding sites in leech ganglia and immunocytes. *Brain Res.* **768,** 232.

Salzet, M., and Stefano, G. B. (1997b). Prodynorphin in invertebrates. *Mol. Brain Res.* **52,** 46–52.

Salzet, M., Bulet, P., Verger-Bocquet, M., and Malecha, J. (1995). Isolation and structural characterization of enkephalin-related peptides in the brain of the Rhynchobdellid leech Theromyzon tessulatum. *FEBS Lett.* **357,** 187–191.

Salzet, M., Verger-Bocquet, M., Bulet, P., Beauvillain, J. C., and Malecha, J. (1996). Purification, sequence analysis and cellular localization of a prodynorphin-derived peptide related to the α neo-endorphin in the rhynchobdellid leech *Theromyzon tessulatum. J. Biol. Chem.* **271,** 13191–13196.

Salzet, M., Cocquerelle, C., Verger-Bocquet, M., Pryor, S. C., Rialas, C. M., Laurent, V., and Stefano, G. B. (1997). Leech immunocytes contain proopiomelanocortin: Nitric oxide mediates hemolymph POMC processing. *J. Immunol.* **159,** 5400–5411.

Scharrer, B. (1967). The neurosecretory neuron in neuroendocrine regulatory mechanisms. *Am. Zool.* **7,** 161–168.

Scharrer, B. (1978). Peptidergic neuron: Facts and trends. *Gen. Comp. Endocrinol.* **34,** 50–62.

Shipp, M. A., Stefano, G. B., D'Adamio, L., Switzer, S. N., Howard, F. D., Sinisterra, J. I., Scharrer, B., and Reinhertz, E. L. (1990). Downregulation of enkephalin-mediated inflammatory response by CD10/neutral endopeptidase 24.11. *Nature* **347,** 394–396.

Shipp, M. A., Stefano, G. B., Switzer, S. N., Griffin, J. D., and Reinherz, E. (1991). CD10 (CALLA)/neutral endopeptidase 24.11 modulates inflammatory peptide-induced changes in neutrophil morphology, migration, and adhesion proteins and is itself regulated by neutrophil activation. *Blood* **78,** 1834–1841.

Smith, E. M., Hughes, T. K., Hashemi, F., and Stefano, G. B. (1992). Immunosuppressive effects of ACTH and MSH and their possible significance in human immunodeficiency virus infection. *Proc. Natl. Acad. Sci. USA* **89,** 782–786.

Smyth, D. G. (1986). Flexibility in the processing of beta-endorphin. *In* "Handbook of Comparative Opioid and Related Neuropeptide Mechanisms" (G. B. Stefano, ed.), pp. 37–40. CRC Press, Boca Raton, FL.

Soeparwata, R., Hartman, A., Frerichmann, U., Stefano, G. B., Scheld, H. H., and Bilfinger, T. V. (1996). Aprotinin diminishes inflammatory processes. *Int. J. Cardiol.* **53,** S55–S64.

Sonetti, D., Ottaviani, E., Bianchi, F., Rodriguez, M., Stefano, M. L., Scharrer, B., and Stefano, G. B. (1994). Microglia in invertebrate ganglia. *Proc. Natl. Acad. Sci. USA* **91,** 9180–9184.

Stefano, G. B. (1980). Opiates and neuroactive pentapeptides: Binding characteristics and interactions with dopamine stimulated adenylate cyclase in the pedal ganglia of Mytilus edulis. *In* "Neurotransmitters in Invertebrates" (K. S. Rozsa, ed.), Vol. 22, pp. 423–453. Pergamon, London.

Stefano, G. B. (1982). Comparative aspects of opioid-dopamine interaction. *Cell. Mol. Neurobiol.* **2,** 167–178.

Stefano, G. B. (1986). Conformational matching: A determining force in maintaining signal molecules. *In* "Comparative Opioid and Related Neuropeptide Mechanisms" (G. B. Stefano, ed.), Vol. 2, pp. 271–277. CRC Press, Boca Raton, FL.

Stefano, G. B., and Catapane, E. J. (1979). Enkephalin increases dopamine levels in the CNS of a marine mollusc. *Life Sci.* **24,** 1617–1622.

Stefano, G. B., and Leung, M. K. (1984). Presence of met-enkephalin-Arg-Phe in molluscan neural tissues. *Brain Res.* **298,** 362–365.

Stefano, G. B., and Leung, M. K. (1986). Opioid aging and seasonal variations in invertebrate ganglia: Evidence for an opioid compensatory mechanism. *In "Comparative Opioid and Related Neuropeptide Mechanisms"* (G. B. Stefano, ed.), Vol. 2, pp. 233–242. CRC Press, Boca Raton, FL.

Stefano, G. B., and Scharrer, B. (1991). A possible immunoregulatory function for [Met]-enkephalin-Arg6-Phe7 involving human and invertebrate granulocytes. *J. Neuroimmunol.* **31,** 97.

Stefano, G. B., and Smith, E. M. (1996). Adrenocorticotropin, a central trigger in immune responsiveness: Tonal inhibition of immune activation. *Med. Hypotheses* **46,** 471–478.

Stefano, G. B., Scharrer, B., and Leung, M. K. (1989a). Neurobiology of opioids in Leucophaea maderae. *In "Cockroaches as Models for Neurobiology: Applications in Biomedical Research,"* pp. 85–102. CRC Press, Boca Raton, FL.

Stefano, G. B., Cadet, P., and Scharrer, B. (1989b). Stimulatory effects of opioid neuropeptides on locomotory activity and conformational changes in invertebrate and human immunocytes: Evidence for a subtype of delta receptor. *Proc. Natl. Acad. Sci. USA* **86,** 6307–6311.

Stefano, G. B., Leung, M. K., Zhao, X., and Scharrer, B. (1989c). Evidence for the involvement of opioid neuropeptides in the adherence and migration of immunocompetent invertebrate hemocytes. *Proc. Natl. Acad. Sci. USA* **86,** 626–630.

Stefano, G. B., Melchiorri, P., Negri, L., Hughes, T. K., and Scharrer, B. (1992). (D-Ala2)-Deltorphin I binding and pharmacological evidence for a special subtype of delta opioid receptor on human and invertebrate immune cells. *Proc. Natl. Acad. Sci. USA* **89,** 9316–9320.

Stefano, G. B., Digenis, A., Spector, S., Leung, M. K., Bilfinger, T. V., Makman, M. H., Scharrer, B., and Abumrad, N. N. (1993). Opiate-like substances in an invertebrate, a novel opiate receptor on invertebrate and human immunocytes, and a role in immunosuppression. *Proc. Natl. Acad. Sci. USA* **90,** 11099–11103.

Stefano, G. B., Leung, M. K., Bilfinger, T. V., and Scharrer, B. (1995). Effect of prolonged exposure to morphine on responsiveness of human and invertebrate immunocytes to stimulatory molecules. *J. Neuroimmunol.* **63,** 175–181.

Stefano, G. B., Scharrer, B., Smith, E. M., Hughes, T. K., Magazine, H. I., Bilfinger, T. V., Hartman, A., Fricchione, G. L., Liu, Y., and Makman, M. H. (1996). Opioid and opiate immunoregulatory processes. *Crit. Rev. Immunol.* **16,** 109–144.

Stefano, G. B., Salzet, B., and Fricchione, G. L. (1998a). Enkelytin and opioid peptide association in invertebrates and vertebrates: Immune activation and pain. *Immunol. Today* **19,** 265–268.

Stefano, G. B., Salzet-Raveillon, B., and Salzet, M. (1998b). *Mytilus edulis* hemocytes contain prodynorphin. *Immunol. Lett.* **63,** 33–39.

Stefano, G. B., Salzet-Raveillon, B., and Salzet, M. (1998c). *Mytilus edulis* hemocytes contain pro-opiomelanocortin: LPS and morphine stimulate differential processing. *Mol. Brain Res.,* in press.

Stefano, G. B., Salzet, M., Hughes, T. K., and Bilfinger, T. V. (1998d). δ_2 opioid receptor subtype on human vascular endothelium uncouples morphine stimulated nitric oxide release. *Int. J. Cardiol.* **64,** S61–66.

Turner, A., Leung, M. K., and Stefano, G. B. (1994). Peptidases of significance in neuroimmunoregulation. *In* "Neuropeptides in Neuroimmunology" (B. Scharrer, E. M. Smith, and G. B. Stefano, eds.), pp. 152–169. Springer-Verlag, New York.

Udenfriend, S., and Kilpatrick, D. L. (1984). Proenkephalin and the products of its processing: Chemistry and biology. *In "The Peptides"* (S. Udenfriend and J. Meienhofer, eds.), Vol. 6, pp. 26–69. Academic Press, San Diego.

Udenfriend, S., and Meienhofer, J. (1984). Opioid peptides: Biology, chemistry and genetics. *In* "The Peptides," Vol. 6. Academic Press, San Diego.

Zhong, F., Li, X. Y., Yang, S., Stefano, G. B., Fimiani, C., and Bilfinger, T. V. (1998). Methionine-enkephalin stimulates interleukin-6 mRNA expression: Human plasma levels in coronary artery bypass grafting. *Int. J. Cardiol.* **64,** S53–60.

INDEX

A

B

C

D

E

F

G

H

I

K

L

M

N

O

P

R

S

T

V

Y